Water Conservation, Recycling and Reuse: Issues and Challenges

Rajeev Pratap Singh • Alan S. Kolok
Shannon L. Bartelt-Hunt
Editors

Water Conservation, Recycling and Reuse: Issues and Challenges

Springer

Editors
Rajeev Pratap Singh
Institute of Environment and Sustainable Development
Banaras Hindu University
Varanasi, Uttar Pradesh, India

Alan S. Kolok
Idaho Water Resources Research Institute
University of Idaho
Moscow, ID, USA

Shannon L. Bartelt-Hunt
Department of Civil Engineering
University of Nebraska–Lincoln
Lincoln, NE, USA

ISBN 978-981-13-3178-7 ISBN 978-981-13-3179-4 (eBook)
https://doi.org/10.1007/978-981-13-3179-4

Library of Congress Control Number: 2018968145

This Springer imprint is published by the registered company Springer Nature Singapore Pte Ltd.
The registered company address is: 152 Beach Road, #21-01/04 Gateway East, Singapore 189721, Singapore

This book is dedicated to our parents and other family members.

Foreword

Water contamination is as old as civilization, with metals, industrial effluent, and human sewage taking the lead in stamping our historical footprint on the planet's lakes, streams, and rivers. Over time, we have become a bit wiser as global citizens, successfully reducing our legacy of contamination from these traditional sources. Like peeling an onion, however, each layer of contamination we remove unmasks contaminants below.

These emergent contaminants share a common story line: they are present in municipal and industrial wastewater discharge. As we attempt to change the water resource paradigm from divided sectors of water supply and wastewater discharge to a One Water world, research into the removal, neutralization, and bioremediation of traditional and emergent contaminants is critical to wastewater' reuse as part of the world's water supply.

As stewards of our planet's finite freshwater resources, it is our responsibility to see that our water is safe and clean to use now and for future generations. All are related.

Associate Director
Idaho Water Resources Research
Institute, Moscow, ID, USA

Mark Solomon PhD

Preface

Increasing global population combined with diminishing freshwater supplies exert growing pressures on available water resources. Significant non-potable uses of water include crop irrigation and energy production. As agricultural production and energy production intensify to meet the growing global needs, the pressures on freshwater supplies are accelerated. An alternative to increasing withdrawals of freshwater supplies to meet future agricultural and energy demands is the use of nonconventional water sources including wastewater reuse. Effective long-term use of alternative waters requires an understanding of temporal, spatial, and climate-induced trends in their quantity and quality as well as any regulatory or economic barriers to their adoption. Concerns associated with the use of alternative water sources are based on questions about the safety and quality of food crops growing using recycled wastewater, the safety of agricultural workers, and the potential for contaminant accumulation and impacts on soil productivity. Additionally, regulations controlling the use of alternative water sources vary significantly from across the globe. These economic, regulatory, and technical concerns must be addressed with research efforts to identify sustainable strategies that link collection, storage, and treatment of recycled or reused water. This book explores various issues related to water recycling and reuse in the context of agricultural production and energy applications.

Chapter 1 entitled "Occurrence and Health Impacts of Emerging Contaminants in Municipal Wastewater Reuse" focuses on the occurrence of pharmaceuticals and personal care products (PPCPs) and perfluorinated compounds (PFCs) in wastewater and describes treatment technologies for removal of these compounds prior to wastewater reuse.

Chapter 2 entitled "Promises and Challenges of Growing Microalgae in Wastewater" describes the potential for growing algae as a feedstock for biofuel production using wastewater, which can provide a valuable source of nutrients. The microalgae grown in wastewater can also serve as a remediation strategy for water-borne pollutants.

Chapter 3 is entitled “Risk of Metal Contamination in Agriculture Crops by Reuse of Wastewater: An Ecological and Human Health Risk Perspective.” This chapter focuses on the potential for accumulation of metals in the topsoil of plants and subsequent uptake into food crops. This chapter also explores human and ecological health risks from this wastewater reuse paradigm.

Chapter 4 entitled “Biological Wastewater Treatment for Prevention of River Water Pollution and Reuse: Perspective and Challenges” outlines the potential for wastewater remediation strategies that involve oxidation ponds, plants, and constructed wetlands, which can potentially reduce the costs and infrastructure associated with wastewater treatment.

Chapter 5 entitled “Climate Change and Sustainable Management of the River System with Special Reference to the Brahmaputra River” describes the results of a sampling campaign along the Brahmaputra River and the influence of climate on mineral weathering and metal release within the river system.

Chapter 6 entitled “Effects of Climate Change on Reuse of Wastewater for Aquaculture Practices” evaluates the impact of climate change, including higher water temperatures and more extreme precipitation events on wastewater quality and the potential for reuse of wastewater through aquaculture.

Chapter 7 entitled “Bioprocesses for Wastewater Reuse: Closed Loop System for Energy Options” describes various biologic processes for wastewater treatment linked with generation of biomass, which can be used to produce bioenergy and/or other biobased products.

Chapter 8 entitled “Subsurface Processes Controlling Reuse Potential of Treated Wastewater Under Climate Change Conditions” evaluates biogeochemical processes controlling solute fate in the subsurface as well as a case study of groundwater contamination from land application of wastewater.

Chapter 9 entitled “Removal of Organic Pollutants from Industrial Wastewaters Treated by Membrane Techniques” describes how textile industrial wastewaters can be treated by a variety of liquid membrane techniques.

Chapter 10 entitled “Assessing the Impacts of Temperature, Precipitation and Land Use Chanange on Open Water Bodies of Middle Ghaghara River Basin” describes how variation in temperature and precipitation can influence the available area of surface water body area within a wetland system.

Chapter 11 entitled “Climate Change, Water and Wastewater Treatment: Interrelationship and Consequences” focuses on how climate will influence available water resources and wastewater treatment technologies.

Chapter 12 entitled “Treatment of Wastewater Using Vermifiltration Technology” focuses on a biological method of wastewater treatment using vermifiltration. This technology may be appropriate for wastewater treatment in developing countries due to its ease of use.

Chapter 13 entitled “Reuse of Wastewater in Agriculture” describes various aspects of the reuse wastewater in agricultural production for irrigation, including water quality criteria, water demand, irrigation scheduling and planning.

Chapter 14 entitled “Application of the Ecological Network Analysis (ENA) Approach in Water Resource Management Research: Strengths, Weaknesses and

Future Research Directions" outlines water resource management strategies using the ecological network analysis approach. This chapter includes a case study of the Heihe River Basin in China.

Varanasi, India
Moscow, ID, USA
Lincoln, NE, USA

Rajeev Pratap Singh
Alan S. Kolok
Shannon L. Bartelt-Hunt

Acknowledgment

The editors extend their sincere gratitude to all contributing authors for raising their views in the form of chapters on every current important issue of water conservation in the present book. We would like to acknowledge the valuable contributions of all reviewers who played a pivotal role in improving the quality and presentation of manuscripts. Our thanks go to all of the members of the Editorial Advisory Board for their guidance and support in completion of this project.

The editors are extremely grateful to Prof. Mark Solomon, associate director of Idaho Water Resources Research Institute for writing the foreword that helped in providing glimpse of the book. Lastly, our special thanks go to God for providing us strength and fortitude and also for giving this opportunity. Dr. Rajeev Pratap Singh is also extremely thankful to Daugherty Water for Food Institute (DWFI), UNL; Institute of Environment and Sustainable Development, Banaras Hindu University; and Indo-US Science and Technology Forum (IUSSTF) as without being "Water Advance and Innovation (WARI) fellow" this collaborative outcome was not possible.

Banaras Hindu University, Varanasi, India — Rajeev Pratap Singh

University of Idaho, Moscow, ID, USA — Alan S. Kolok

University of Nebraska-Lincoln, Lincoln, NE, USA — Shannon L. Bartelt-Hunt

Contents

About the Editors

Dr. Rajeev Pratap Singh is an Assistant Professor at the Institute of Environment and Sustainable Development (IESD), Banaras Hindu University, India. He completed his Ph.D. at the same university. His main research interests include solid waste management, bio-composting, and green technologies. He has received several international awards, including the "Green Talent" award from the Federal Ministry of Education and Research (BMBF), Germany; Prosper.Net Scopus Young Scientist award, DST Young Scientist Award, etc. Dr. Singh is a member of reviewer and editorial team of several important scientific journals and has co-authored 4 books and more than 40 highly cited research and review articles on solid waste management. Dr. Singh also received a Water Advanced Research and Innovation (WARI) Fellowship from DST, Govt. of India, IUSSTF, University of Nebraska-Lincoln (UNL), and the Robert Daugherty Water for Food Institute (DWFI).

Dr. Alan S. Kolok is the Director of Idaho Water Resources Research Institute, University of Idaho, Moscow, USA. He received his Ph.D. in 1991 from the Department of Environmental, Population, and Organismic Biology, University of Colorado, Boulder; his M.S. in 1981 from the Department of Fisheries and Oceanic Science, University of Washington Seattle; and his B.S. in 1978 from the Department of Zoology, Miami University, Oxford, Ohio. His research interests include land use and its impact on water quality and environmental health, gene expression in environmental sentinel organisms as indicators of exposure to toxic compounds, the role of sediment in the fate and transport of emerging contaminants, cancer incidence and its geographical relationship to watersheds, and citizen scientist and its role in data acquisition.

Dr. Shannon L. Bartelt-Hunt is a Professor of Civil Engineering at the University of Nebraska, Omaha. She is also the Associate Dean of Graduate Studies. She completed her Ph.D. in Civil Engineering (Environmental) at the University of

Virginia in 2004. Her areas of research and professional interest include the physicochemical fate of contaminants in soil and water, contaminant fate and transport in landfills, and water reuse in agricultural systems. She has been awarded several prestigious awards, such as the Student Impact Award: Outstanding Advisor (Engineers Without Borders), 2017; Grand Prize for University Research, American Academy of Environmental Engineers and Scientists, 2015; R. Vernon McBroom Engineering Leadership Fellow, 2014; Association of Environmental Engineering and Science Professors (AEESP), Distinguished Service Award, 2014; College of Engineering Faculty Research and Creative Activity Award, 2013; National Science Foundation CAREER award, 2012; McGraw-Hill/AEESP Award for Outstanding Teaching in Environmental Science and Engineering, 2011.

Chapter 1
Occurrence and Health Impacts of Emerging Contaminants in Municipal Wastewater Reuse

Wei Chen, Shu-Yuan Pan, Zihao Wang, and Xiaoping Zhang

Abstract Municipal wastewater reuse offers the potential to significantly increase the total available water resources. Recently, the occurrence of emerging contaminants (ECs), like pharmaceuticals and personal care products (PPCPs) and perfluorinated compounds (PFCs), in water resources is of continued concern for the public health and safety. However, the existing conventional wastewater treatment plants (WWTPs) were not originally conceived to eliminate these unidentified contaminants, which have not been monitored routinely because of the absence of stringent-specific regulation. This chapter focuses on the occurrence of these ECs and feasible opportunities for guidelines in municipal wastewater reclamation and reuse. An environmental risk assessment posed by a common means of the risk quotient shows that 27 pharmaceuticals pose high or medium risk. The concept of source control and source separation could reduce the manufacture and produce a wastewater with an optimal composition for further centralized treatment. Additional and integrated technologies for synergic treatment units are found necessary to provide high-quality recycled water and sustainable water resources.

W. Chen (✉) · Z. Wang
School of Environment and Energy, South China University of Technology, Guangzhou Higher Education Mega Centre, Guangzhou, People's Republic of China
e-mail: wei_ch810@163.com

S.-Y. Pan
Energy Technologies Area, Lawrence Berkeley National Laboratory, Berkeley, CA, USA

X. Zhang
School of Environment and Energy, South China University of Technology, Guangzhou Higher Education Mega Centre, Guangzhou, People's Republic of China

The Key Lab of Pollution Control and Ecosystem Restoration in Industry Clusters, Ministry of Education, South China University of Technology, Guangzhou Higher Education Mega Centre, Guangzhou, People's Republic of China

R. P. Singh et al. (eds.), *Water Conservation, Recycling and Reuse: Issues and Challenges*, https://doi.org/10.1007/978-981-13-3179-4_1

1.1 Significance and Importance

Water resource scarcity could be the major challenge of civilization, because of the unbalanced distribution of resources, environmental pollutants, climate change, population growth, rapid urbanization, and so on. In this sense, the concept of water resources sustainability has created increasing interest in water reuse and recycling. Water resources sustainability was defined in the report of the Brundtland Commission (World Commission on Environment and Development) as (Holdgate 1987):

> Water resources sustainability is the ability to use water in sufficient quantities and quality from the local to the global scale to meet the present needs of humans and environmental ecosystems, while not impairing the needs of future generations to do the same.

It involves the aspects of freshwater allocation, urban water use cycle, wastewater reuse, and risk management. Among various concepts, municipal wastewater reuse is quite promising in densely inhabited district, offering irrigation, landscape, recreation, humidification, cooling, and surface or groundwater replenishment, due to mechanical treatment for suspended solids and conventional organic pollutants in the mature chemical and biological process of municipal wastewater treatment plants (WWTPs) (Valipour 2014).

1.1.1 Occurrence of Emerging Contaminants

Emerging contaminants as new products or chemicals without regulatory status in sewage and aquatic environments could cause known or suspected adverse ecological and/or human health effects (high toxicity, carcinogenic and mutagenic effects) (Ratola et al. 2012). Two important chemical constituents, including pharmaceuticals and personal care products (PPCPs) and perfluorinated compounds (PFCs), have attracted global concern due to their ubiquitous occurrence in the multimedia, biota, and humans. WWTPs as "initial biological treatments" were inefficient in the removal of PPCPs and PFCs (Stamatis and Konstantinou 2013; Chen et al. 2017a). PPCPs have been entering into the environment for over 25 years from individual human and social activities (e.g., waste excretion, bathing, residues of manufacturing, agribusiness, and hospitals). About 95% of the total amount of PPCP consumption are:

Analgesics: 5-aminosalicylic acid, acetaminophen, acetylsalicylic acid, codeine, dextropropoxyphene, diclofenac, dipyrone, fenoprofen, flurbiprofen, hydrocodone, ibuprofen, indomethacin, ketoprofen, mefenamic acid, naproxen, propyphenazone, salicylic acid, tramadol

Antibiotics: amoxicillin, azithromycin, cefaclor, cefalexin, cefotaxime, chlortetracycline, ciprofloxacin, clarithromycin, doxycycline, erythromycin, lincomycin, norfloxacin, ofloxacin, roxithromycin, sulfamethoxazole, trimethoprim
Antidiabetics: glibenclamide
Antihypertensives: diltiazem, hydrochlorothiazide
Barbiturates: phenobarbital
Beta-blocker: atenolol, metoprolol, propranolol, sotalol
Lipid regulators: bezafibrate, gemfibrozil
Psychiatric drugs: carbamazepine, fluoxetine
Receptor antagonists: ranitidine, valsartan
Hormones: estradiol, estriol, estrone
Antineoplastics: ifosfamide, tamoxifen
Antiseptic: triclosan

It is difficult to assess typical concentration levels of PPCPs as monitoring data of PPCPs varies with the time, the type of waste water, and geographical areas. Table 1.1 presents the molecular structure and the variability ranges for the concentrations of common compounds grouped with their therapeutic class in municipal WWTP influent based on the previous scientific literature data, which reflects common pharmaceuticals (analgesics/anti-inflammatories, antibiotics, antihypertensives, beta-blockers, lipid regulators, psychiatric drugs, receptor antagonists, and contrast agent) could be detected in wastewater that receives each municipal WWTP. The differences among these results could also be related with the types of products in different geographic markets or the possibility in some countries with limited supervision to purchase medicines without medical prescriptions.

PFCs that are a class of anthropogenic compounds with repellency of water and oil have been commonly used in industrial and household products for more than 60 years (Fujii et al. 2007). There is also a long-term public concern due to their being persistent and accumulative and biotoxicity (Giesy and Kannan 2001). For the most environmental concerning and frequently detected predominant compounds, the concentration levels determined for perfluorooctane sulfonate (PFOS) and perfluorooctanoic acid (PFOA) in municipal influent and secondary treated wastewater are presented in Table 1.2. They are ubiquitously detected in influent wastewater and secondary treated wastewater due to the high bond energy of C–F long chains and their high-water solubility made by hydrophilic functional groups (such as sulfonate and carboxyl).

Besides, the presence of PPCPs and PFCs as environmental pollutants was reported in the Antarctic environment and Antarctic biota with concentrations similar to those detected in urban living areas throughout the world (Corsolini 2009). The concerning issue with PPCPs and PFCs is not their acute toxic effects but their chronic toxicity due to their ubiquitously distributed and large quantities in both consumer and industrial settings.

Table 1.1 Ranges of concentrations and molecular structure in the influent for the common PPCPs

Type of PPCPs	Compound	Influent (μg/L)	Molecular structure	References
Analgesics/anti-inflammatories	Acetaminophen	18–71	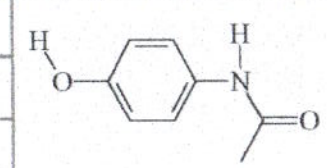	Verlicchi et al. (2012)
		1.57–37.5		Rosal et al. (2010)
		7.1–11.4		Radjenovic et al. (2009)
		29–246		Gomez et al. (2007)
		104		Rosal et al. (2010)
	Acetylsalicylic acid	1.32–5.44	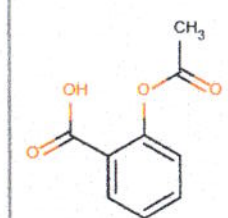	Kasprzyk-Hordern et al. (2009)
	Codeine	2.8–11	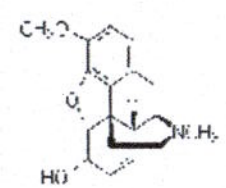	Gomez et al. (2007)
		2.49–12.6		Kasprzyk-Hordern et al. (2009)
		0.15–2.09		Rosal et al. (2010)
	Ibuprofen	2.6–5.7	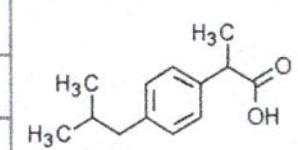	Carballa et al. (2004)
		34–168		Gomez et al. (2007)
		9.8–19.8		Lindqvist et al. (2005)
		14.6–31.3		Radjenovic et al. (2009)
	Indomethacin	0.95	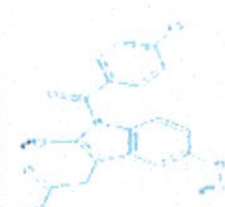	Stumpf et al. (1999)
		0.66–1		Radjenovic et al. (2009)
	Ketoprofen	1.3–3	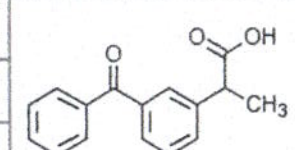	Lindqvist et al. (2005)
		0.15–0.41		Tauxe-Wuersch et al. (2005)
		0.41–0.52		Thomas and Foster (2005)

	Mefenamic acid	0.8–1.2	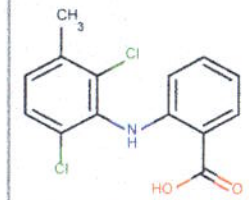	Radjenovic et al. (2009)
		0.75–2.9		Tauxe-Wuersch et al. (2005)
	Naproxen	1.79–4.6	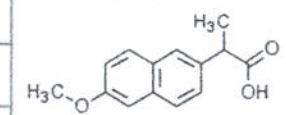	Carballa et al. (2004)
		0.62–3.5		Kasprzyk-Hordern et al. (2009)
		3.5–4.5		Rodriguez et al. (2003)
	Propyphenazone	0.04–0.09	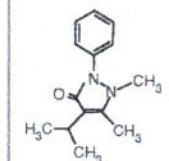	Radjenovic et al. (2009)
	Salicylic acid	5.6–32.08	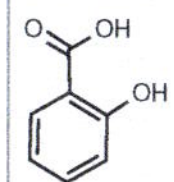	Kasprzyk-Hordern et al. (2009)
Antibiotics	Amoxicillin	0.19–0.28	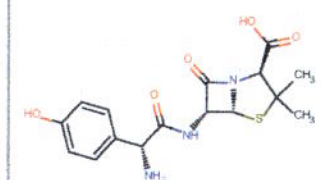	Watkinson et al. (2007)
	Azithromycin	0.16–1.34	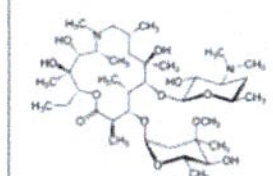	Ghosh et al. (2009)
	Cefaclor	0.5–0.98	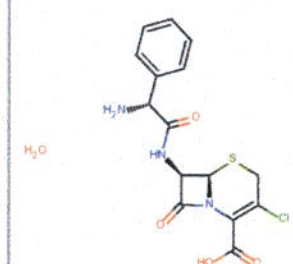	Watkinson et al. (2007)

(continued)

Table 1.1 (continued)

Type of PPCPs	Compound	Influent (μg/L)	Molecular structure	References
	Cefalexin	0.67–2.9	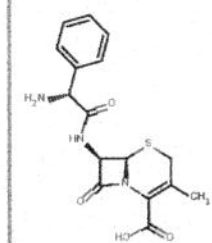	Gulkowska et al. (2008)
	Chloramphenicol	0.15–0.45	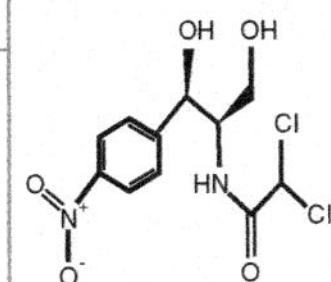	Kasprzyk-Hordern et al. (2009)
		1.73–2.43		Peng et al. (2006)
	Clarithromycin	0.33–0.6	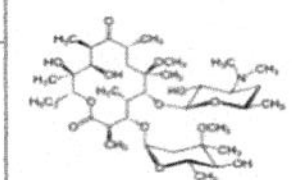	Gobel et al. (2005)
		1.13–4.82		Ghosh et al. (2009)
	Erythromycin	0.47–0.74	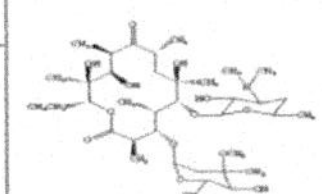	Gulkowska et al. (2008)
		0.48–1.2		Karthikeyan and Meyer (2006)

	Lincomycin	0.06–0.08	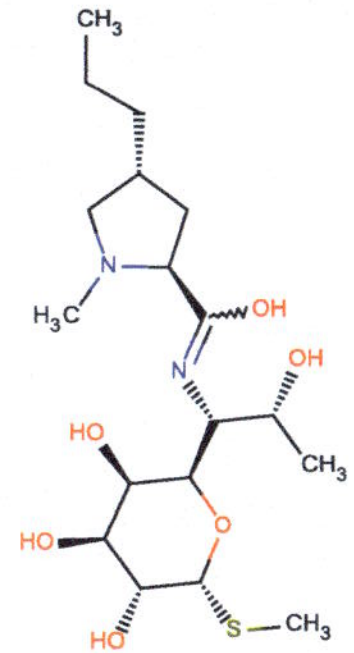	Watkinson et al. (2007)
	Metronidazole	0.34–0.96	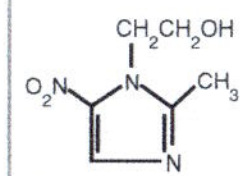	Kasprzyk-Hordern et al. (2009)
	Ofloxacin	0.52–5.56	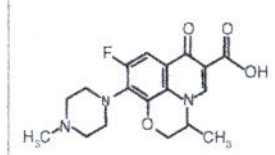	Peng et al. (2006)
		0.89–31.7		Radjenovic et al. (2009)
	Sulfamethoxazole	0.23–0.57	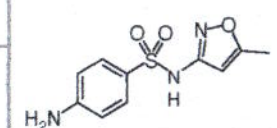	Gobel et al. (2005)
		0.17–1.25		Karthikeyan and Meyer (2006)
	Trimethoprim	0.21–0.44	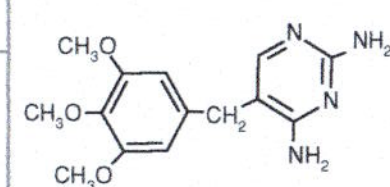	Gobel et al. (2005)
		0.15–0.43		Radjenovic et al. (2009)

(continued)

Table 1.1 (continued)

Type of PPCPs	Compound	Influent (μg/L)	Molecular structure	References
Antihypertensives	Diltiazem	0.41–5.26	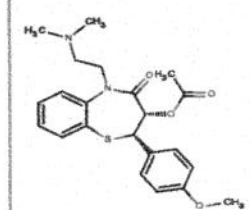	Kasprzyk-Hordern et al. (2009)
Beta-blockers	Propranolol	0.11–1.9	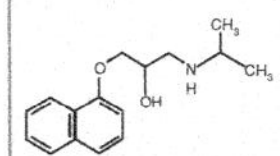	Kasprzyk-Hordern et al. (2009)
Lipid regulators	Bezafibrate	1.9–29.8	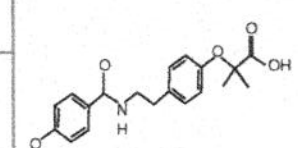	Radjenovic et al. (2009)
		1.55–7.6		Clara et al. (2005)
	Clofibric acid	0.17–0.37	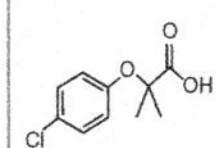	Tauxe-Wuersch et al. (2005)
	Gemfibrozil	0.6–1.1	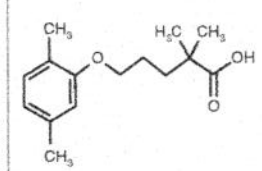	Verlicchi et al. (2012)
		0.41–17.1		Rosal et al. (2010)
	Pravastatin	0.46–1.5	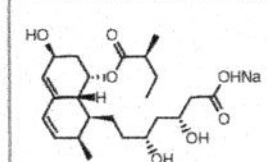	Radjenovic et al. (2009)

Psychiatric drugs	Carbamazepine	0.32–1.2	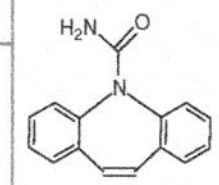	Clara et al. (2005)
		0.1–3.11		Kasprzyk-Hordern et al. (2009)
	Fluoxetine	0.12–2.3	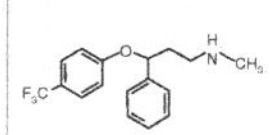	Radjenovic et al. (2009)
Receptor antagonists	Cimetidine	0.68–6.5	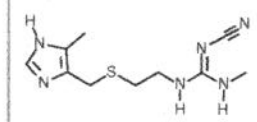	Kasprzyk-Hordern et al. (2009)
	Ranitidine	0.072–0.54	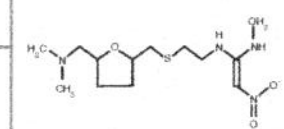	Radjenovic et al. (2009)
		2–11.15		Kasprzyk-Hordern et al. (2009)
Contrast agent	Iopromide	0.03–3.84	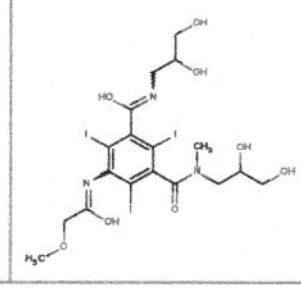	Clara et al. (2005)

Table 1.2 Concentration of common PFCs in the influent and secondary effluent of WWTPs

Location	Influent wastewater (ng/L)		Secondary treated wastewater (ng/L)		References
	PFOS	PFOA	PFOS	PFOA	
USA	1.4–400.0	1.7–49.0	1.1–130.0	2.5–97.0	Schultz et al. (2006)
Kentucky, USA	7.0–16.0	22.0–184.0	8.0–28.0	122.0–183.0	Loganathan et al. (2007)
Georgia, USA	2.5–7.9	2.0–50.0	1.8–13.0	6.7–102.0	Loganathan et al. (2007)
Germany	1.0–85.0	1.8–40.0	12.0–140.0	8.7–9.3	Becker et al. (2008)
Denmark	1.5–10.1	2.0–23.5	1.5–18.1	2.0–24.4	Bossi et al. (2008)
Switzerland	18.0–449.0	4.9–35.0	16.0–303.0	8.9–35.0	Huset et al. (2008)
Greece	2.4–26.3	10.2–20.7	5.2–21.0	12.7–34.0	Arvaniti et al. (2012)
Spain	78.1	22.4	91.0	14.9	Campo et al. (2014)
Japan	14.0–336.0	14.0–41.0	42.0–635.0	10.0–68.0	Murakami et al. (2009)
Singapore	7.9–374.5	14.1–638.2	7.3–461.7	15.8–1057.1	Yu et al. (2009a)
Korea	n.d.–68.1	4.3–615.0	n.d.–5.7	6.4–591.0	Guo et al. (2010)
Taiwan	175–216.7	17.6–23.6	162.7–5663.3	19.3–480.3	Lin et al. (2010)
China	0.03–12	0.05–54.0	0.03–7.3	0.09–26.2	Pan et al. (2011)
Thailand	381.3	6.6	552.8	16.9	Kunacheva et al. (2011)
Australia	/	/	23.0–38.6	15.0–27.0	Thompson et al. (2011)

1.1.2 Municipal Wastewater Reuse

Global water shortage is placing an unprecedented pressure due to the inferior quality and insufficient quantity of water resource. Many regions, including the Middle East, Northern Africa, parts of China and India, and the southwest quarter of the USA, are experiencing high levels of water stress (with annual per capita water supplies below 1700 m^3). Projections predict that by 2025, 2/3 of the world's population will be living under the conditions of moderate to high water stress (Wu et al. 2014). Nowadays, the common municipal WWTPs consist of physical preliminary, primary treatment and secondary biochemical system with the treated sewage being discharged into natural water or reused in many aspects. The application of municipal sewage (treated to remove pathogens, organic matter, and nutrients) represents an additional opportunity in augmenting freshwater supply for irrigation, landscape, recreation, humidification, cooling, and surface or groundwater replenishment, especially in semiarid and arid regions. Unlike many organic contaminants, the existing WWTPs are ineffective at removing emerging contaminants. The potential risks of the reuse of treated municipal wastewater associated with emerging contaminants, especially PPCPs and PFCs, are new issues garnering public attention.

Surface or groundwater replenishment: Reclamation of water after treatment in modern WWTPs will likely be an important yet currently unrealized part of sustainable water resource management. Most treated municipal wastewater discharged into the receiving aquatic environment shows a relatively reliable water source, but it can convey emerging contaminants to cities and drinking water intakes located downstream. WWTP discharges are also the main source of PPCPs and PFCs in surface water environment. The surface ocean was assumed to be the main reservoir for PFCs. According to the difference in bioconcentration in natural aquatic background or different profiles, fish and other seafood seem to be the food group in which more PFCs are detected and where the detectable concentrations of typical compounds are higher. The levels of PFCs in the analyzed foodstuffs decreased in the order seafood > animal liver > freshwater/marine fish > egg > meat > butter in the samples of five European countries (Belgium, Czech Republic, Italy, and Norway) (D'Hollander et al. 2010). This means that individuals consuming great amounts of seafood are assuming certain risks in certain countries, which are not currently quantified.

Irrigation: Up to 49–90% of consumptive water is used in agriculture irrigation; municipal wastewater reuse appears to be a valuable water resource to supplement agriculture irrigation (Wu et al. 2014). The most challenging aspect of this application is the safety attention of produce due to the residual contaminants. PPCPs and PFCs persist in both aqueous effluent and treated biosolids. Irrigation land application of secondary treated wastewater and biosolids through root uptake on crops can facilitate the PPCPs and PFCs into the terrestrial food web, especially for fresh produce that may be consumed raw (e.g., vegetables, fresh fruits) (Blaine et al. 2014; Krippner et al. 2015).

Humidification: Reclaimed municipal wastewater began to be used for sprinkling roads to mitigate heat island and reduce road dust in urban areas. The results of measured temperatures of air and road with/without sprinkling show that sprinkling reclaimed wastewater lowered the road surface temperature by eight degrees during the daytime in Japan (Ogoshi et al. 2001). Inhalation is one of the most important routes of transmission of inflammatory agents, and it is noteworthy that the evaporation process can affect human health directly through inhalation during this application (Xue et al. 2016).

1.1.3 Exposure Routes of PPCPs and PFCs

As emerging contaminants, there are multiple potential input pathways for PPCPs and PFCs into the environment as shown schematically in Fig. 1.1.

Industrial wastewater has been regarded as a main point source for PPCPs and PFCs as well as their precursors entering into the receiving water bodies (Arvaniti and Stasinakis 2015). Besides, other sources of PPCPs and PFCs intake by humans

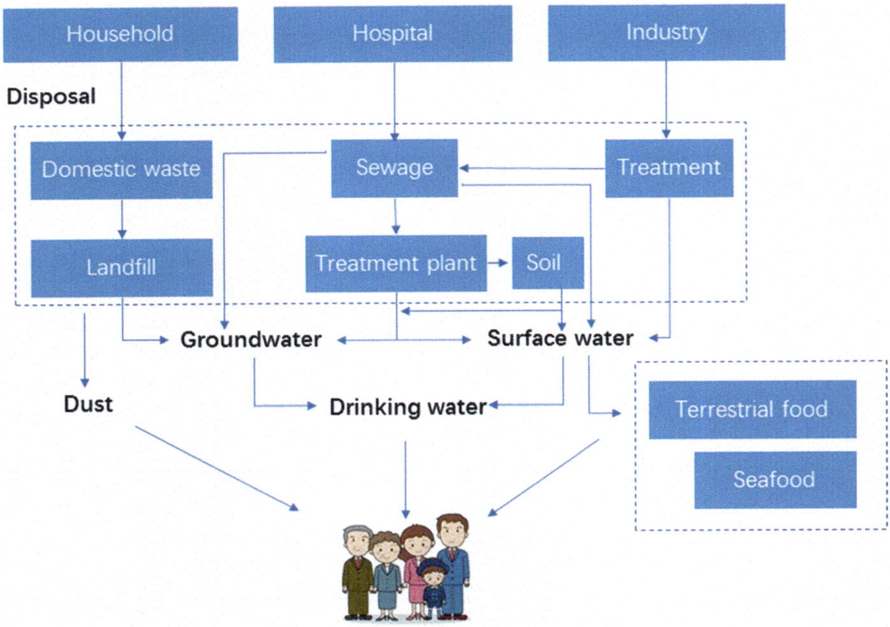

Fig. 1.1 Exposure routes of emerging PPCPs and PFCs for humans

could be through consumption of contaminated food and inhalation of air or fine particles contaminated with volatile PPCPs and PFCs precursor that would be degraded to become PFOS, PFOA, or other intermediates. Unlike the PFCs, parts of the PPCPs are usually disposed as solid waste or in wastewater effluent by improper disposal from private households and hospitals. Most hospital sewers are directly connected to the municipal drainage system, and no additional treatment is performed. Sorption of lipophilic PPCPs and hydrophobic PFCs onto sewage sludge may be a relevant removal pathway from the wastewater stream. Partial PPCPs can be assumed to be destroyed by chemical oxidation in sludge incineration process, while PFCs could be retained due to their highly polar and strong carbon–fluorine bonds. When sewage sludge is used as a fertilizer, PPCPs and PFCs may return to the surrounding aquatic environment through the surface runoff (De Sanctis et al. 2017).

WWTPs are generally not equipped to deal with complex PPCPs and PFCs, and irrigation of treated wastewater on arable land can also lead to a pollution of groundwater. As a result, effective environmental risk assessment and tertiary treatment technologies of upgrading WWTPs should be implemented in order to ensure the safe usage of municipal wastewater.

1.2 Environmental Risk Assessment

Risk assessment is a key component of many environmental regulatory decisions, which provides an evaluation of the probability and impact of detrimental effects attributable to the potential toxicity of contaminants (Lammerding and Fazil 2000). Figure 1.2 shows the general principles of the environmental risk assessment, which are accepted extensively, including five steps (hazard identification, exposure assessment, effects assessment, risk characterization, and risk management).

It is noted in the effect assessment that the risk for humans from environmental contaminants is not discussed in this step but rather the risk for organisms in the environment caused by anthropogenic use of these compounds. Besides, the possible long-term, low-level effects of PPCPs and PFCs should not be overlooked, which can be helpful to test the standard scheme of environmental risk assessment with "familiar contaminants" and to supplement potentially hazardous characterization.

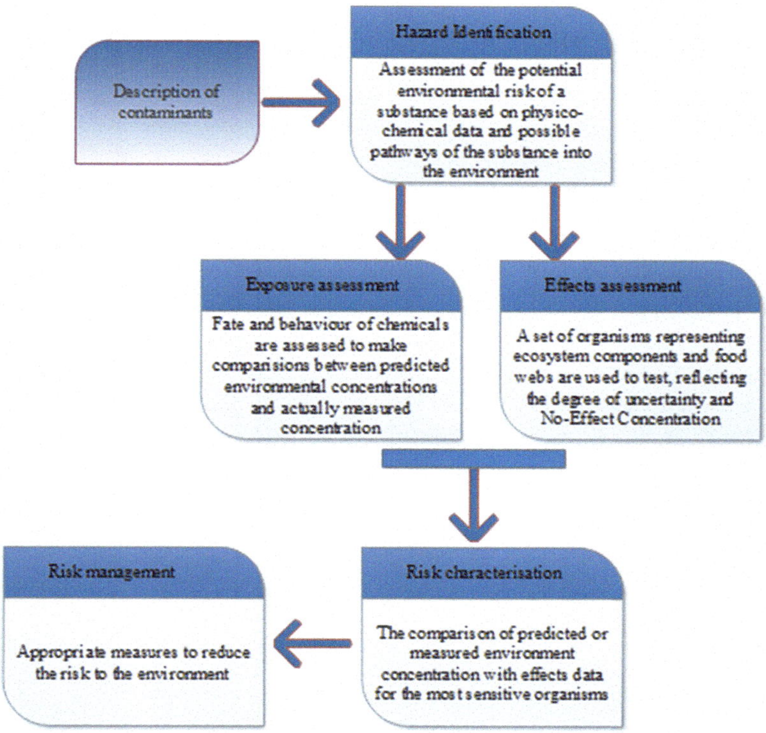

Fig. 1.2 Steps of emerging contaminants' environmental risk assessment

1.2.1 Predicted Environmental Concentrations (PEC) of PPCPs

A program for the risk assessment of PPCPs has been outlined by the European Medicines Agency (EMA). There are several models for PEC prediction in different draft versions of the EMA, which can provide options for prioritizing species on which to focus the concentrated research efforts (Liebig et al. 2005). The EMA model for initial exposure assessment presented in 2001 is based on the predicted, prescribed, or used volumes of pharmaceuticals in Eq. (1.1).

$$\mathrm{PEC_{SW}}[\mathrm{g/L}] = \frac{A \times (100 - R)}{365 \times P \times V \times D \times 100} \tag{1.1}$$

where A (kg) is the predicted use/prescription amount per year, R (%) is the removal rate, P is the number of inhabitants of the specific geographic area, V (m^3) is the volume of wastewater per capita and day, and D is the factor for dilution of wastewater by surface water flow.

The second EMA model for initial PEC assessment is based on the maximum daily dose of the active ingredient of a pharmaceutical ($\mathrm{DOSE_{ai}}$) in Eq. (1.2). A factor for the market penetration F_{pen} represents the proportion of the population being treated daily with a specific drug.

$$\mathrm{PEC_{SW}}[\mathrm{mg/L}] = \frac{\mathrm{DOSE_{ai}} \times F_{\mathrm{pen}}}{V \times D \times 100} \tag{1.2}$$

where the defined daily dose (DDD) referred to the World Health Organization Collaborating Centre for Drug Statistics Methodology was defined as $\mathrm{DOSE_{ai}}$ (mg) and the recommended default value of 1% was defined as F_{pen}. For V and D, the same default values as in Eq. (1.1) were used; hence the $\mathrm{DOSE_{ai}}$ (DDD) was the only variable in this formula which results in a direct correlation of the $\mathrm{PEC_{sw}}$ to the DDD.

1.2.2 Environmental Risk Assessment of PPCPs in Secondary Biological Effluent

The environmental risk assessment with the presence of PPCPs in natural water should be considered after predicting the environmental concentration. Although the safety threshold values have been defined for a limited number of PPCPs, more research need to be further strengthened for many compounds themselves and their mixture toxicity, because of the previous studies only in single compound-single organism toxicity. Table 1.3 presents the predicted no-effect concentrations (PNECs) and the average concentrations of predominant PPCPs for fish in the

Table 1.3 PNECs and average concentrations of predominant PPCPs for fish in the secondary effluent (Sanderson et al. 2003; Stuerlauridsen et al. 2000; Jones et al. 2002; Lee et al. 2008; Ferrari et al. 2004; Hallingsorensen 2000; Kim et al. 2007)

Compounds	Toxicity (mg/004C)	Average secondary effluent concentration (μg/L)	PNEC (μg/L)
Acetaminophen	1	0.89	1
Acetylsalicylic acid	796	0.36	61
Codeine	238	1.7	16
Dextropropoxyphene	13	0.10	1
Ibuprofen	5	3.6	1.65
Indomethacin	3.9	0.47	3.9
Ketoprofen	32	0.21	15.6
Mefenamic acid	0.43	0.63	0.43
Naproxen	34	1.0	2.62
Phenazone	3	0.16	1.1
Propyphenazone	0.8	0.04	0.8
Salicylic acid	1.28	0.17	1.28
Amoxicillin	0.1	0.01	0.0037
Azithromycin	0.15	0.16	0.15
Cefaclor	11,524	0.01	687.42
Cefalexin	2.5	0.13	2.5
Cefotaxime	0.04	0.02	0.04
Chloramphenicol	1.6	0.05	1.6
Ciprofloxacin	246,000	0.86	938
Clarithromycin	20	0.29	0.07
Clindamycin	0.5	0.01	0.5
Doxycycline	0.3	0.04	0.3
Erythromycin	61	0.73	0.02
Lincomycin	1391	0.06	82
Metronidazole	2.5	0.25	2.5
Ofloxacin	10	0.45	0.016
Oxytetracycline	16,600	0.01	0.207
Roxithromycin	50	0.50	4
Sulfadimethoxine	3.5	0.09	3.5
Sulfamethoxazole	890	0.28	0.027
Trimethoprim	795	0.36	2.6
Diltiazem	23	0.12	1.9
Metoprolol	116	0.32	8
Propranolol	29.5	0.17	0.244
Bezafibrate	5.3	0.90	5.3
Clofibric acid	53	0.21	40.2
Fenofibric acid	7.6	11	7.6
Gemfibrozil	0.9	0.93	0.9
Pravastatin	1.8	0.02	1.8
Carbamazepine	101	1.04	13.8

(continued)

Table 1.3 (continued)

Compounds	Toxicity (mg/004C)	Average secondary effluent concentration (μg/L)	PNEC (μg/L)
Diazepam	28	9.1	2
Fluoxetine	11	0.24	0.05
Cimetidine	571	3.5	35
Ranitidine	1076	0.51	63
Ifosfamide	140	0.97	11
Iopromide	86,500	2.5	370,000

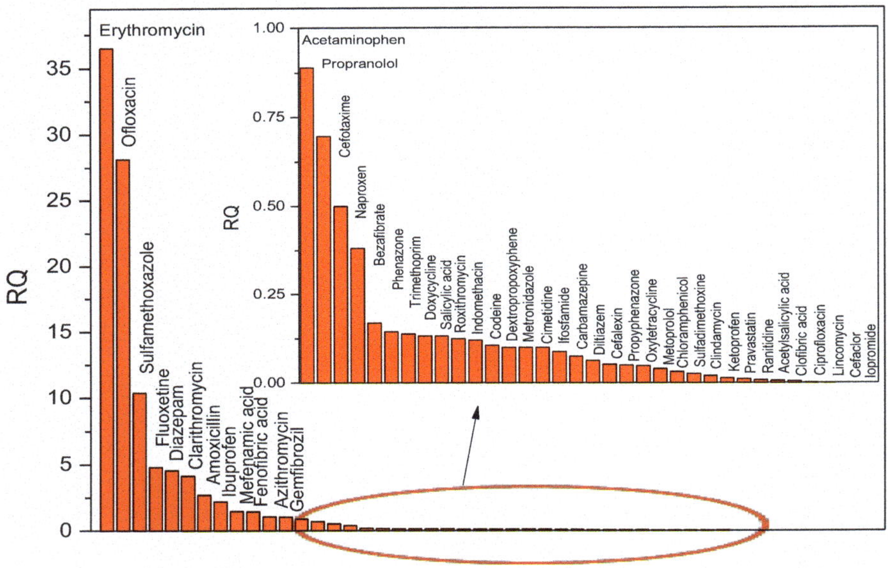

Fig. 1.3 Risk quotient (RQ) of the predominant PPCPs

secondary effluent. A common means of the risk quotient (RQ), the ratio between the average PPCPs concentration measured and its corresponding PNECs, can be used to evaluate the environmental risk by PPCPs: RQ < 0.1, low risk to aquatic organisms; $0.1 \leq RQ \leq 1$, medium risk; and RQ > 1, high risk (Caldwell et al. 2012).

The RQ of the predominant PPCPs in descending order is reported in Fig. 1.3. Several PPCPs have been found in high levels (27 compounds pose high or medium risk), thereby evidencing the risk that environmental concentrations of PPCPs can be higher than their PNECs, especially in effluent-dominant rivers whose dilution capacity and self-purifying processes are insufficient to temper the risk to aquatic life.

1.3 PPCPs and PFCs Removal from Wastewater

The initial purpose of treatment plants of municipal wastewater was conceived to reduce the organic load into aquatic environment. WWTPs are ineffective to removal PPCPs and PFCs due to their recalcitrant characteristic. Centralized WWTPs are considered as important micropollutant point sources of PPCPs and PFCs for the receiving aquatic environment, which provide the potential of implementing centralized removal processes within urban drainage systems for facilitating the municipal wastewater reuse (Wang and Wang 2016; Arvaniti and Stasinakis 2015).

1.3.1 Removal of PPCPs

Removal technologies of PPCPs can be mainly composed of physical, biological, and chemical methods. It is noteworthy that the different categories may be involved in the same removal process, such as sorption, biodegradation, stripping, and photodegradation, which would be included in the aerobiotic/anaerobic activated sludge processes. Besides, advanced oxidation processes (AOPs) (or combination with other technologies) have also been widely tested against recalcitrant contaminants.

1.3.1.1 Removal of PPCPs by Physical Sorption

Carbonaceous materials (e.g., activated carbon, graphene and graphene oxide, and carbon nanotubes) as suitable adsorbents have been widely applied for removing trace (pseudo)-persistent PPCPs. These carbonaceous adsorbents have high sorption capacity for the typical PPCPs, which mainly depend on the hydrophobicity and charge of adsorbates. Many factors such as organic matter existing, contact time, solution pH, and structure of carbonaceous adsorbents have remarkable effects on the removal efficiency (Mailler et al. 2015; Meinel et al. 2014; Kyzas et al. 2015; Liu et al. 2014; Jung et al. 2015; Wan and Wang 2016). Notwithstanding, many problems should be solved for the large-scale applications.

Due to the competing sorption and steric effect, the deterioration of these porous adsorbents may occur with the increase of operation time in the complex wastewater systems.

More efforts should be focused on producing the high-surface-area graphene and nanotubes with relative low cost.

The toxic effect (nanotoxicity) of the coexistence of graphene/nanotubes and PPCPs should be investigated and prevented.

1.3.1.2 Removal of PPCPs by Biological Degradation and Chemical Advanced Oxidation

Microbial degradation is the most prevalent process to remove organic contaminants in the environment. Activated sludge treatment has been regarded as the effective technology to degrade PPCPs easily, which contains combined effects of volatilization, sorption, and biodegradation. Compared with the contribution of biodegradation, the volatilization with aeration and sorption by activated sludge play only a small part in the whole process. However, the concentrations of PPCPs are low; biodegradation would not be always effective for degrading them, which is caused by the low abundance (or lack) of degraders and the contaminant load (Gobel et al. 2005; Suarez et al. 2010; Li et al. 2015). Some measures, such as biological acclimation and bioaugmentation, can conquer the above limitations.

Based on the previous studies on the PPCPs concentration and contamination profiles, different compounds of PPCPs have been frequently tested in the secondary effluent of WWTPs, indicating that the PPCPs would not be removed completely in the conventional WWTPs. Chemical advanced oxidation technologies, such as ozonation (estrone (>95%), sulfamethoxazole (>99%), bezafibrate (>89%), trimethoprim (100%)) (Lin et al. 2009; Garoma et al. 2010; Tootchi et al. 2013; Kuang et al. 2013), UV treatment (estrone (>60%), sulfamethoxazole (100%), bezafibrate (>90%), trimethoprim (>99%)) (Kim et al. 2009; Ma et al. 2015; Wols et al. 2015), Fenton or Fenton-like oxidation (estrone (>98.4%), sulfamethoxazole (100%), bezafibrate (100%), trimethoprim (100%)) (Dirany et al. 2010; Feng et al. 2005; Ternes et al. 2002), and ionizing irradiation (Chu et al. 2015), were effective for the PPCPs. However, the toxic and resistant intermediates appeared during the operating process are another issue in order to complete mineralization. Therefore, the combination of advanced chemical processes and biological technologies as tertiary treatment can provide satisfactory performance.

1.3.2 Removal of PFCs

As two most typical PFCs, both PFOS and PFOA have been included in the forbidden/limited list or the relevant regulations. Some regions and organizations still have expressly restricted the use of PFCs due to the lag in regulation and lack of effective substitute. Unlike PPCPs, biological degradation is effective in removing PFCs attributed to the high-energy carbon–fluorine (C–F) bonds (110 kcal/mol) (Rahman et al. 2014). Therefore, effective remediation technologies should be developed continually.

1.3.2.1 Sorption and Sorption Mechanisms

Sorption is considered as an effective and economical method to remove PFCs from wastewater. Besides conventional adsorbents, many synthetic adsorbents have been applied to remove PFCs. Table 1.4 summarizes the sorption of PFOS and PFOA onto different adsorbents reported in the literature.

Nonpolar hydrophobic interaction between the adsorbents and PFCs existed during the sorption processes is the main affinity, because the C–F chains in PFCs molecule exhibit hydrophobic properties. Self-assembly of hemi-micelle and micelle has been proved to be present in the sorption process, which can block the inner pores of porous adsorbents. In addition, there is a negatively charged shell around a positively charged core of the PFCs molecules due to the high electronegativity of fluorine atoms, and weak interactions may be generated between negative dipole and charged adsorbent surface. Since most PFCs are anionic species by their low pK_a, the sorption processes, such as electrostatic attraction and ion exchange column, can be involved in the treatment. The tails of PFCs molecule as paired groups can replace the hydroxyl groups on aluminum oxides by ligand exchange, as described in Eq. (1.3) (Wang et al. 2012):

$$\mathrm{Al-OH+CF_3(CF_2)_{\mathit{n}}COO^- \rightarrow Al-OOC(CF_2)_{\mathit{n}}CF_3+OH^-} \tag{1.3}$$

1.3.2.2 Filtration Processes

Membrane filtration is a widely used technology in wastewater and drinking water treatment due to its high efficiency in the reduction of wide range of contaminants.

Commercially available reverse osmosis membranes are effective to remove PFOS in a wide range of inlet concentrations from 0.5 to 1600 ppm. Rejection is not affected by membrane zeta potential, but membranes flux would decrease with the increase of PFOS concentration. Nanofiltration membrane such as NF270 has a higher rejection (>95%) to PFCs than microfiltration and ultrafiltration membranes, as well as the higher water flux than reverse osmosis (Zhao et al. 2016). Increased calcium chloride concentrations also can improve PFCs rejection due to increasing permeation of the smaller anions, as well as pore blockage and calcium-bridging between PFCs and calcium ions (Zhao et al. 2013).Sorption, coagulation, and other technologies as a pretreatment can be chosen to enhance the PFCs permeate flux rejection and prevent severe membrane fouling. Multistage filtration membrane arrays can be conceived to further increase the removal efficiency.

1.3.2.3 Advanced Treatment Technologies

Advanced oxidation technologies such as Fenton and Fenton-like oxidation, UV/H_2O_2, sonolysis, as well as photolysis have been widely applied in recalcitrant

Table 1.4 Sorption capacities of PFOS and PFOA on different adsorbents

Adsorbents	Adsorbates	C_0 (mg/g)	pH	q_m (mg/g)	References
Powdered activated carbon	PFOS	20–300	5–7	374–550	Yu et al. (2009b), Liang et al. (2011), Punyapalakul et al. (2012) and Rattanaoudom and Visvanathan (2012)
	PFOA	20–300	5–7	175–524	
Granular activated carbon	PFOS	15–250	4.4–7.2	160–229	Yu et al. (2009b), Ochoaherrera and Sierraalvarez (2008) and Carter and Farrell (2010)
	PFOA	15–250	5–7.2	112–161	
Zeolites	PFOS	15–300	3–5	8–126	Punyapalakul et al. (2012) and Ochoaherrera and Sierraalvarez (2008)
	PFOA	15–300	3	34–37	
Anion-exchange	PFOS	20–400	3–5	210–2575	Yu et al. (2009b); Deng et al. (2010)
	PFOA	20–250	5	1206	
Hydrotalcite	PFOS	1–1000	/	998	Rattanaoudom et al. (2012)
	PFOA	1–1000	/	1033	
Quaternized cotton	PFOS	95–459	5–9	1647	Deng et al. (2012)
	PFOA	78–380	5–9	1280	
Aminated rice husk	PFOS	0–250	5	1322	Deng et al. (2013)
	PFOA	0–207	5	1028	
Quaternized Polyacrylonitrile fiber	PFOS	20–300	2–10	1496	Chen et al. (2017b)
	PFOA	20–300	2–10	1086	
Chitosan bead	PFOS	46–371	3	2745	Zhang et al. (2011)
Mesoporous carbon nitride	PFOS	280	3.25	388	Yan et al. (2013)
Maize straw-origin ash	PFOS	1–500	7	811	Chen et al. (2011)
Molecular imprinted polymer	PFOS	50	5	550	Yu et al. (2008)

contaminants. The ideal treatment for the recalcitrant PFCs should be the cleavage of long C–F chains to F^- ions, which can easily combine with Ca^{2+} to form environmental harmless CaF_2. Hydroxyl radicals (HO·) are generated in most advanced oxidation processes in situ, which can attack the PFCs to form short C–F chain or carbon center radicals through direct electron transfer. Table 1.5 shows the conditions and removal efficiencies under different advanced treatment processes.

Based on the high removal efficiency, it seems that some of the aforementioned processes can achieve better removal efficiency for PFCs. It should be noted that most of these researches have been performed under high temperature, high pressure, or high radiation source. These harsh laboratory reaction conditions indicate

Table 1.5 PFCs removal using different advanced treatment processes

	Conditions	Removal (%)	Reference
UV/H_2O_2(PFOA)	[PFOA] = 1.35 mM $[H_2O_2]$ = 1.0 M λ = 240–460 nm (24 h)	35.6%	Hori et al. (2004)
UV-($S_2O_8^{2-}$) (PFOA)	[PFOA] = 1.35 mM[$S_2O_8^{2-}$] = 50.0 mM λ = 240–460 nm (4 h)	100%	Hori et al. (2005)
UV-photocatalyst ($H_3PW_{12}O_{40}$•$6H_2O$)(PFOA)	[PFOA] = 1.35 mM [$H_3PW_{12}O_{40}$•$6H_2O$] = 6.68 mM λ = 240–460 nm (24 h)	100%	Hori et al. (2004)
Zerovalent iron (PFOS)	[PFOS] = 372 mM[Fe (0)] = 9.60 mM T = 350°°C, 20 MPa (6 h)	100%	Hori et al. (2006)
UV-alkaline isopropanol (IPA) (PFOS)	[PFOS] = 40.0 μM [IPA] = 68.0 mM λ = 254 nm	76% (1 d) 92% (10 d)	Yamamoto et al. (2007)
Sonolysis (PFOS)	[PFOS] = 0.19 mM Ultrasonic generator (200 kHz) (60 min)	60%	Cheng et al. (2008)

that the degradation of PFCs would be unrealistically costly and be incomplete under conditions commonly found in the field or practical PFCs-containing wastewater in large-scale application.

1.4 Final Considerations and Future Perspectives

The aforementioned advanced treatment technologies are mainly directed toward the removal of priority PPCPs or PFCs, as well as partial other types of contaminants. Since both inorganic and organic contaminants would occur in wastewaters, combined treatment processes or advanced technologies should be used for municipal wastewater. Therefore, future efforts should concentrate on the integrated technologies for synergic treatments as tertiary treatment processes. Centralized treatment approach for emerging contaminants can separate promptly and has high efficiency by controlling the removal process and treatment scale. However, the end control treatment requires vast occupancy, high cost of construction, and operation. Emergency situations, such as storm overflow, leakage of the treatment system, and influent loads exceeding the treatment capacity may lead to more serious environmental risks. Therefore, more strategies and legislation for management of PPCPs/PFCs need to be further developed and strengthened for source control and source separation.

The concept of source control is to reduce the manufacture and use through strict legislations and regulations, optimizing PPCPs/PFCs use. PPCPs are usually disposed as solid waste or in wastewater effluent by improper disposal from private households and hospitals; many measures such as supporting studies on quantifying the effectivity, reducing the over-the-counter sale, and limiting the authorization of new pharmaceuticals to products with improved efficacy or application range may be conducive to reduce probability of disposing of rest drug. For PFCs, PFOS with its salts and perfluorooctane sulfonyl fluoride (PFOSF) have been added to the list of "persistent organic pollutants" in the Stockholm Convention. They are still allowed to be limitedly used in some areas including electroplating, polytetrafluoroethylene manufacturing, and optoelectronic industries. More environmentally friendly substitutes should be developed in the next step.

From the pollution sources point of view, the ideal of source separation is to produce a wastewater with an optimal composition for further centralized treatment, preventing PPCPs/PFCs-containing wastewater from being discharged into WWTPs. Nowadays, "cleaner production" programs which contribute to an ecologically, economically, and societal sustainable future are adapted in many industries (Lora et al. 2000). These preventive applications can realize recycling through "green chemistry." Specific treatment technologies toward the priority PPCPs or PFCs can be easily selected to apply in centralized treatment process after source separation. With the development of the advanced analytical instrument and standard analytical procedures, more emerging micro-contaminants would be detected. The environmental legislations, standards, and related removal technologies need to be updated, preventing emerging contaminants transferring into the environment.

References

Arvaniti OS, Stasinakis AS (2015) Review on the occurrence, fate and removal of perfluorinated compounds during wastewater treatment. Sci Total Environ 524:81–92. https://doi.org/10.1016/j.scitotenv.2015.04.023

Arvaniti OS, Ventouri EI, Stasinakis AS, Thomaidis NS (2012) Occurrence of different classes of perfluorinated compounds in Greek wastewater treatment plants and determination of their solid-water distribution coefficients. J Hazard Mater 239:24–31. https://doi.org/10.1016/j.jhazmat.2012.02.015

Becker AM, Gerstmann S, Frank H (2008) Perfluorooctane surfactants in waste waters, the major source of river pollution. Chemosphere 72(1):115–121. https://doi.org/10.1016/j.chemosphere.2008.01.009

Blaine AC, Rich CD, Sedlacko EM, Hyland KC, Stushnoff C, Dickenson ER, Higgins CP (2014) Perfluoroalkyl acid uptake in lettuce (Lactuca sativa) and strawberry (Fragaria ananassa) irrigated with reclaimed water. Environ Sci Technol 48(24):14361–14368. https://doi.org/10.1021/es504150h

Bossi R, Strand J, Sortkjaer O, Larsen MM (2008) Perfluoroalkyl compounds in Danish wastewater treatment plants and aquatic environments. Environ Int 34(4):443–450

Caldwell DJ, Mastrocco F, Anderson PD, Lange R, Sumpter JP (2012) Predicted-no-effect concentrations for the steroid estrogens estrone, 17β-estradiol, estriol, and 17α-ethinylestradiol. Environ Toxicol Chem 31(6):1396–1406

Campo J, Masia A, Pico Y, Farre M, Barcelo D (2014) Distribution and fate of perfluoroalkyl substances in Mediterranean Spanish sewage treatment plants. Sci Total Environ 472:912–922. https://doi.org/10.1016/j.scitotenv.2013.11.056

Carballa M, Omil F, Lema JM, Llompart M, Garcia-Jares C, Rodriguez I, Gomez M, Ternes T (2004) Behavior of pharmaceuticals, cosmetics and hormones in a sewage treatment plant. Water Res 38(12):2918–2926. https://doi.org/10.1016/j.watres.2004.03.029

Carter KE, Farrell J (2010) Removal of perfluorooctane and perfluorobutane sulfonate from water via carbon adsorption and ion exchange. Sep Sci Technol 45(6):762–767

Chen X, Xia X, Wang X, Qiao J, Chen H (2011) A comparative study on sorption of perfluorooctane sulfonate (PFOS) by chars, ash and carbon nanotubes. Chemosphere 83 (10):1313–1319. https://doi.org/10.1016/j.chemosphere.2011.04.018

Chen H, Wang X, Zhang C, Sun R, Han J, Han G, Yang W, He X (2017a) Occurrence and inputs of perfluoroalkyl substances (PFASs) from rivers and drain outlets to the Bohai Sea China. Environ Pollut 221:234–243. https://doi.org/10.1016/j.envpol.2016.11.070

Chen W, Zhang X, Zhang Y, Mamadiev M (2017b) Facile and efficient synthesis of polyacrylonitrile-based functional fibers and its sorption properties of perfluorooctane sulfonate and perfluorooctanoate. J Mol Liq 241:1013–1022. https://doi.org/10.1016/j.molliq.2017.06.090

Cheng J, Vecitis CD, Park H, Mader BT, Hoffmann MR (2008) Sonochemical degradation of perfluorooctane sulfonate (PFOS) and Perfluorooctanoate (PFOA) in landfill groundwater: environmental matrix effects. Environ Sci Technol 42(21):8057–8063

Chu L, Wang J, Liu Y (2015) Degradation of sulfamethazine in sewage sludge mixture by gamma irradiation. Radiat Phys Chem 108:102–105

Clara M, Strenn B, Gans O, Martinez E, Kreuzinger N, Kroiss H (2005) Removal of selected pharmaceuticals, fragrances and endocrine disrupting compounds in a membrane bioreactor and conventional wastewater treatment plants. Water Res 39(19):4797–4807. https://doi.org/10.1016/j.watres.2005.09.015

Corsolini S (2009) Industrial contaminants in Antarctic biota. J Chromatogr A 1216(3):598–612. https://doi.org/10.1016/j.chroma.2008.08.012

De Sanctis M, Del Moro G, Chimienti S, Ritelli P, Levantesi C, Di Iaconi C (2017) Removal of pollutants and pathogens by a simplified treatment scheme for municipal wastewater reuse in agriculture. Sci Total Environ 580:17–25. https://doi.org/10.1016/j.scitotenv.2016.12.002

Deng S, Yu Q, Huang J, Yu G (2010) Removal of perfluorooctane sulfonate from wastewater by anion exchange resins: effects of resin properties and solution chemistry. Water Res 44 (18):5188–5195

Deng S, Zheng YQ, Xu FJ, Wang B, Huang J, Yu G (2012) Highly efficient sorption of perfluorooctane sulfonate and perfluorooctanoate on a quaternized cotton prepared by atom transfer radical polymerization. Chem Eng J 193-194:154–160. https://doi.org/10.1016/j.cej.2012.04.005

Deng S, Niu L, Bei Y, Wang B, Huang J, Yu G (2013) Adsorption of perfluorinated compounds on aminated rice husk prepared by atom transfer radical polymerization. Chemosphere 91 (2):124–130. https://doi.org/10.1016/j.chemosphere.2012.11.015

D'Hollander W, de Voogt P, De Coen W, Bervoets L (2010) Perfluorinated substances in human food and other sources of human exposure. Rev Environ Contam Toxicol 208:179–215. https://doi.org/10.1007/978-1-4419-6880-7_4

Dirany A, Aaron SE, Oturan N, Sires I, Oturan MA, Aaron J (2010) Study of the toxicity of sulfamethoxazole and its degradation products in water by a bioluminescence method during application of the electro-Fenton treatment. Anal Bioanal Chem 400(2):353–360

Feng X, Ding S, Tu J, Wu F, Deng N (2005) Degradation of estrone in aqueous solution by photo-fenton system. Sci Total Environ 345(1):229–237

Ferrari B, Mons R, Vollat B, Fraysse B, Paxēaus N, Giudice RL, Pollio A, Garric J (2004) Environmental risk assessment of six human pharmaceuticals: are the current environmental

risk assessment procedures sufficient for the protection of the aquatic environment? Environ Toxicol Chem 23(5):1344–1354
Fujii S, Tanaka S, Hong Lien NP, Qiu Y, Polprasert C (2007) New POPs in the water environment: distribution, bioaccumulation and treatment of perfluorinated compounds – a review paper. J Water Supply Res Technol AQUA 56(5):313. https://doi.org/10.2166/aqua.2007.005
Garoma T, Umamaheshwar SK, Mumper A (2010) Removal of sulfadiazine, sulfamethizole, sulfamethoxazole, and sulfathiazole from aqueous solution by ozonation. Chemosphere 79 (8):814–820
Ghosh GC, Okuda T, Yamashita N, Tanaka H (2009) Occurrence and elimination of antibiotics at four sewage treatment plants in Japan and their effects on bacterial ammonia oxidation. Water Sci Technol 59(4):779–786. https://doi.org/10.2166/wst.2009.067
Giesy JP, Kannan K (2001) Global distribution of perfluorooctane sulfonate in wildlife. Environ Sci Technol 35(7):1339–1342. https://doi.org/10.1021/es001834k
Gobel A, Thomsen A, Mcardell CS, Joss A, Giger W (2005) Occurrence and sorption behavior of sulfonamides, macrolides, and trimethoprim in activated sludge treatment. Environ Sci Technol 39(11):3981–3989
Gomez MJ, Bueno MJM, Lacorte S, Fernandez-Alba AR, Aguera A (2007) Pilot survey monitoring pharmaceuticals and related compounds in a sewage treatment plant located on the Mediterranean coast. Chemosphere 66(6):993–1002. https://doi.org/10.1016/j.chemosphere.2006.07.051
Gulkowska A, Leung HW, So MK, Taniyasu S, Yamashita N, Yeunq LWY, Richardson BJ, Lei AP, Giesy JP, Lam PKS (2008) Removal of antibiotics from wastewater by sewage treatment facilities in Hong Kong and Shenzhen, China. Water Res 42(1–2):395–403. https://doi.org/10.1016/j.watres.2007.07.031
Guo R, Sim WJ, Lee ES, Lee JH, Oh JE (2010) Evaluation of the fate of perfluoroalkyl compounds in wastewater treatment plants. Water Res 44(11):3476–3486. https://doi.org/10.1016/j.watres.2010.03.028
Hallingsorensen B (2000) Algal toxicity of antibacterial agents used in intensive farming. Chemosphere 40(7):731–739
Holdgate MW (1987) Our common future: the report of the world commission on environment and development. Oxford University Press. Environ Conserv 14(03):282–282
Hori H, Hayakawa E, Koike K, Einaga H, Ibusuki T (2004) Decomposition of nonafluoropentanoic acid by heteropolyacid photocatalyst H3PW12O40 in aqueous solution. J Mol Catal A Chem 211(1–2):35–41. https://doi.org/10.1016/j.molcata.2003.09.029
Hori H, Yamamoto A, Hayakawa E, Taniyasu S, Yamashita N, Kutsuna S (2005) Efficient decomposition of environmentally persistent perfluorocarboxylic acids by use of persulfate as a photochemical oxidant. Environ Sci Technol 39(7):2383–2388. https://doi.org/10.1021/es0484754
Hori H, Nagaoka Y, Yamamoto A, Sano T, Yamashita N, Taniyasu S, Kutsuna S, Osaka I, Arakawa R (2006) Efficient decomposition of environmentally persistent perfluorooctanesulfonate and related fluorochemicals using zerovalent iron in subcritical water. Environ Sci Technol 40 (3):1049–1054. https://doi.org/10.1021/es0517419
Huset CA, Chiaia AC, Barofsky DF, Jonkers N, Kohler HPE, Ort C, Giger W, Field JA (2008) Occurrence and mass flows of fluorochemicals in the Glatt Valley watershed, Switzerland. Environ Sci Technol 42(17):6369–6377. https://doi.org/10.1021/es703062f
Jones OAH, Voulvoulis N, Lester JN (2002) Aquatic environmental assessment of the top 25 English prescription pharmaceuticals. Water Res 36(20):5013–5022
Jung C, Son A, Her N, Zoh K, Cho J, Yoon Y (2015) Removal of endocrine disrupting compounds, pharmaceuticals, and personal care products in water using carbon nanotubes: a review. J Ind Eng Chem 27:1–11
Karthikeyan KG, Meyer MT (2006) Occurrence of antibiotics in wastewater treatment facilities in Wisconsin, USA. Sci Total Environ 361(1–3):196–207. https://doi.org/10.1016/j.scitotenv.2005.06.030

Kasprzyk-Hordern B, Dinsdale RM, Guwy AJ (2009) The removal of pharmaceuticals, personal care products, endocrine disruptors and illicit drugs during wastewater treatment and its impact on the quality of receiving waters. Water Res 43(2):363–380. https://doi.org/10.1016/j.watres.2008.10.047

Kim SD, Cho J, Kim IS, Vanderford BJ, Snyder SA (2007) Occurrence and removal of pharmaceuticals and endocrine disruptors in South Korean surface, drinking, and waste waters. Water Res 41(5):1013–1021

Kim I, Yamashita N, Tanaka H (2009) Performance of UV and UV/H2O2 processes for the removal of pharmaceuticals detected in secondary effluent of a sewage treatment plant in Japan. J Hazard Mater 166(2):1134–1140

Krippner J, Falk S, Brunn H, Georgii S, Schubert S, Stahl T (2015) Accumulation potentials of perfluoroalkyl carboxylic acids (PFCAs) and perfluoroalkyl sulfonic acids (PFSAs) in maize (Zea mays). J Agric Food Chem 63(14):3646–3653. https://doi.org/10.1021/acs.jafc.5b00012

Kuang J, Huang J, Wang B, Cao Q, Deng S, Yu G (2013) Ozonation of trimethoprim in aqueous solution: Identification of reaction products and their toxicity. Water Res 47(8):2863–2872

Kunacheva C, Tanaka S, Fujii S, Boontanon SK, Musirat C, Wongwattana T, Shivakoti BR (2011) Mass flows of perfluorinated compounds (PFCs) in central wastewater treatment plants of industrial zones in Thailand. Chemosphere 83(6):737–744. https://doi.org/10.1016/j.chemosphere.2011.02.059

Kyzas GZ, Koltsakidou A, Nanaki SG, Bikiaris DN, Lambropoulou DA (2015) Removal of beta-blockers from aqueous media by adsorption onto graphene oxide. Sci Total Environ 537:411–420

Lammerding AM, Fazil A (2000) Hazard identification and exposure assessment for microbial food safety risk assessment. Int J Food Microbiol 58(3):147–157

Lee Y, Lee S, Lee D, Kim Y (2008) Risk assessment of human antibiotics in Korean aquatic environment. Environ Toxicol Pharmacol 26(2):216–221

Li W, Shi Y, Gao L, Liu J, Cai Y (2015) Occurrence, fate and risk assessment of parabens and their chlorinated derivatives in an advanced wastewater treatment plant. J Hazard Mater 300:29–38

Liang X, Gondal MA, Chang X, Yamani ZH, Li N, Lu H, Ji G (2011) Facile preparation of magnetic separable powdered-activated-carbon/Ni adsorbent and its application in removal of perfluorooctane sulfonate (PFOS) from aqueous solution. J Environ Sci Health A Tox Hazard Subst Environ Eng 46(13):1482–1490

Liebig M, Moltmann J, Knacker T (2005) Evaluation of measured and predicted environmental concentrations of selected human pharmaceuticals and personal care products (10 pp). Environ Sci Pollut Res Int 13(2):110–119. https://doi.org/10.1065/espr2005.08.276

Lin Y, Peng Z, Zhang X (2009) Ozonation of estrone, estradiol, diethylstilbestrol in waters. Desalination 249(1):235–240

Lin AYC, Panchangam SC, Ciou PS (2010) High levels of perfluorochemicals in Taiwan's wastewater treatment plants and downstream rivers pose great risk to local aquatic ecosystems. Chemosphere 80(10):1167–1174. https://doi.org/10.1016/j.chemosphere.2010.06.018

Lindqvist N, Tuhkanen T, Kronberg L (2005) Occurrence of acidic pharmaceuticals in raw and treated sewages and in receiving waters. Water Res 39(11):2219–2228. https://doi.org/10.1016/j.watres.2005.04.003

Liu F, Zhao J, Wang S, Du P, Xing B (2014) Effects of solution chemistry on adsorption of selected pharmaceuticals and personal care products (PPCPs) by graphenes and carbon nanotubes. Environ Sci Technol 48(22):13197–13206

Loganathan BG, Sajwan KS, Sinclair E, Kumar KS, Kannan K (2007) Perfluoroalkyl sulfonates and perfluorocarboxylates in two wastewater treatment facilities in Kentucky and Georgia. Water Res 41(20):4611–4620. https://doi.org/10.1016/j.watres.2007.06.045

Lora EES, Arrieta FP, Carpio RC, Nogueira LAH (2000) Clean production: efficiency and environment. Int. Sugar J. 102(1219):343–351

Ma Y, Wang L, Liu L, Zhang X (2015) Biodegradation of tylosin residue in pharmaceutical solid waste by a novel Citrobacter amalonaticus strain. Environ Prog 34(1):99–104

Mailler R, Gasperi J, Coquet Y, Deshayes S, Zedek S, Crenolive C, Cartiser N, Eudes V, Bressy A, Caupos E (2015) Study of a large scale powdered activated carbon pilot: removals of a wide range of emerging and priority micropollutants from wastewater treatment plant effluents. Water Res 72:315–330

Meinel F, Ruhl AS, Sperlich A, Zietzschmann F, Jekel M (2014) Pilot-scale investigation of micropollutant removal with granular and powdered activated carbon. Water Air Soil Pollut 226(1):2260

Murakami M, Shinohara H, Takada H (2009) Evaluation of wastewater and street runoff as sources of perfluorinated surfactants (PFSs). Chemosphere 74(4):487–493. https://doi.org/10.1016/j.chemosphere.2008.10.018

Ochoaherrera V, Sierraalvarez R (2008) Removal of perfluorinated surfactants by sorption onto granular activated carbon, zeolite and sludge. Chemosphere 72(10):1588–1593

Ogoshi M, Suzuki Y, Asano T (2001) Water reuse in Japan. Water Sci Technol 43(10):17–23

Pan YY, Shi YL, Wang JM, Cai YQ (2011) Evaluation of perfluorinated compounds in seven wastewater treatment plants in Beijing urban areas. Sci China Chem 54(3):552–558. https://doi.org/10.1007/s11426-010-4093-x

Peng XZ, Wang ZD, Kuang WX, Tan JH, Li K (2006) A preliminary study on the occurrence and behavior of sulfonamides, ofloxacin and chloramphenicol antimicrobials in wastewaters of two sewage treatment plants in Guangzhou, China. Sci Total Environ 371(1–3):314–322. https://doi.org/10.1016/j.scitotenv.2006.07.001

Punyapalakul P, Suksomboon K, Prarat P, Khaodhiar S (2012) Effects of surface functional groups and porous structures on adsorption and recovery of perfluorinated compounds by inorganic porous silicas. Sep Sci Technol 48(5):775–788

Radjenovic J, Petrovic M, Barcelo D (2009) Fate and distribution of pharmaceuticals in wastewater and sewage sludge of the conventional activated sludge (CAS) and advanced membrane bioreactor (MBR) treatment. Water Res 43(3):831–841. https://doi.org/10.1016/j.watres.2008.11.043

Rahman MF, Peldszus S, Anderson WB (2014) Behaviour and fate of perfluoroalkyl and polyfluoroalkyl substances (PFASs) in drinking water treatment: a review. Water Res 50:318–340. https://doi.org/10.1016/j.watres.2013.10.045

Ratola N, Cincinelli A, Alves A, Katsoyiannis A (2012) Occurrence of organic microcontaminants in the wastewater treatment process. A mini review. J Hazard Mater 239-240:1–18. https://doi.org/10.1016/j.jhazmat.2012.05.040

Rattanaoudom R, Visvanathan C (2012) Removal of PFOA by hybrid membrane filtration using PAC and hydrotalcite. Desalin Water Treat 32:262–270

Rattanaoudom R, Visvanathan C, Boontanon SK (2012) Removal of concentrated PFOS and PFOA in synthetic industrial wastewater by powder activated carbon and hydrotalcite. J Water Sustainability 2(4):245–248

Rodriguez I, Quintana JB, Carpinteiro J, Carro AM, Lorenzo RA, Cela R (2003) Determination of acidic drugs in sewage water by gas chromatography-mass spectrometry as tert.-butyldimethylsilyl derivatives. J Chromatogr A 985(1–2):265–274. https://doi.org/10.1016/S0021-9673(02)01528-5

Rosal R, Rodriguez A, Perdigon-Melon JA, Petre A, Garcia-Calvo E, Gomez MJ, Aguera A, Fernandez-Alba AR (2010) Occurrence of emerging pollutants in urban wastewater and their removal through biological treatment followed by ozonation. Water Res 44(2):578–588. https://doi.org/10.1016/j.watres.2009.07.004

Sanderson H, Johnson DJ, Wilson CJ, Brain RA, Solomon KR (2003) Probabilistic hazard assessment of environmentally occurring pharmaceuticals toxicity to fish, daphnids and algae by ECOSAR screening. Toxicol Lett 144(3):383–395

Schultz MM, Barofsky DF, Field JA (2006) Quantitative determination of fluorinated alkyl substances by large-volume-injection liquid chromatography tandem mass spectrometry – Characterization of municipal wastewaters. Environ Sci Technol 40(1):289–295. https://doi.org/10.1021/es051381p

Stamatis NK, Konstantinou IK (2013) Occurrence and removal of emerging pharmaceutical, personal care compounds and caffeine tracer in municipal sewage treatment plant in Western Greece. J Environ Sci Health B 48(9):800–813. https://doi.org/10.1080/03601234.2013.781359

Stuerlauridsen F, Birkved M, Hansen LP, Lutzhoft HCH, Hallingsorensen B (2000) Environmental risk assessment of human pharmaceuticals in Denmark after normal therapeutic use. Chemosphere 40(7):783–793

Stumpf M, Ternes TA, Wilken R, Rodrigues S, Baumann W (1999) Polar drug residues in sewage and natural waters in the state of Rio de Janeiro Brazil. Sci Total Environ 225(1):135–141

Suarez S, Lema JM, Omil F (2010) Removal of pharmaceutical and personal care products (PPCPs) under nitrifying and denitrifying conditions. Water Res 44(10):3214–3224

Tauxe-Wuersch A, De Alencastro LF, Grandjean D, Tarradellas J (2005) Occurrence of several acidic drugs in sewage treatment plants in Switzerland and risk assessment. Water Res 39 (9):1761–1772. https://doi.org/10.1016/j.watres.2005.03.003

Ternes TA, Meisenheimer M, Mcdowell D, Sacher F, Brauch H, Haistgulde B, Preuss G, Wilme U, Zuleiseibert N (2002) Removal of pharmaceuticals during drinking water treatment. Environ Sci Technol 36(17):3855–3863

Thomas PM, Foster GD (2005) Tracking acidic pharmaceuticals, caffeine, and triclosan through the wastewater treatment process. Environ Toxicol Chem 24(1):25–30. https://doi.org/10.1897/04-144r.1

Thompson J, Eaglesham G, Reungoat J, Poussade Y, Bartkow M, Lawrence M, Mueller JF (2011) Removal of PFOS, PFOA and other perfluoroalkyl acids at water reclamation plants in South East Queensland Australia. Chemosphere 82(1):9–17. https://doi.org/10.1016/j.chemosphere.2010.10.040

Tootchi L, Seth R, Tabe S, Yang P (2013) Transformation products of pharmaceutically active compounds during drinking water ozonation. Water Sci Technol Water Supply 13 (6):1576–1582

Valipour M (2014) Future of agricultural water management in Africa. Arch Agron Soil Sci 61 (7):907–927

Verlicchi P, Al Aukidy M, Zambello E (2012) Occurrence of pharmaceutical compounds in urban wastewater: Removal, mass load and environmental risk after a secondary treatment-A review. Sci Total Environ 429:123–155. https://doi.org/10.1016/j.scitotenv.2012.04.028

Wan Z, Wang J (2016) Ce-Fe-reduced graphene oxide nanocomposite as an efficient catalyst for sulfamethazine degradation in aqueous solution. Environ Sci Pollut R 23(18):18542–18551

Wang J, Wang S (2016) Removal of pharmaceuticals and personal care products (PPCPs) from wastewater: a review. J Environ Manag 182:620–640. https://doi.org/10.1016/j.jenvman.2016.07.049

Wang F, Liu C, Shih K (2012) Adsorption behavior of perfluorooctanesulfonate (PFOS) and perfluorooctanoate (PFOA) on boehmite. Chemosphere 89(8):1009–1014. https://doi.org/10.1016/j.chemosphere.2012.06.071

Watkinson AJ, Murby EJ, Costanzo SD (2007) Removal of antibiotics in conventional and advanced wastewater treatment: implications for environmental discharge and wastewater recycling. Water Res 41(18):4164–4176. https://doi.org/10.1016/j.watres.2007.04.005

Wols BA, Harmsen DJH, Beerendonk EF, Hofmancaris CHM (2015) Predicting pharmaceutical degradation by UV (MP)/H2O2 processes: a kinetic model. Chem Eng J 263:336–345

Wu X, Conkle JL, Ernst F, Gan J (2014) Treated wastewater irrigation: uptake of pharmaceutical and personal care products by common vegetables under field conditions. Environ Sci Technol 48(19):11286–11293. https://doi.org/10.1021/es502868k

Xue J, Zhang J, Xu B, Xie J, Wu W, Lu Y (2016) Endotoxins: the critical risk factor in reclaimed water via inhalation exposure. Environ Sci Technol 50(21):11957–11964. https://doi.org/10.1021/acs.est.6b02395

Yamamoto T, Noma Y, Sakai S, Shibata Y (2007) Photodegradation of perfluorooctane sulfonate by UV irradiation in water and alkaline 2-propanol. Environ Sci Technol 41(16):5660–5665

Yan T, Chen H, Wang X, Jiang F (2013) Adsorption of perfluorooctane sulfonate (PFOS) on mesoporous carbon nitride. RSC Adv 3(44):22480. https://doi.org/10.1039/c3ra43312a
Yu Q, Deng SB, Yu G (2008) Selective removal of perfluorooctane sulfonate from aqueous solution using chitosan-based molecularly imprinted polymer adsorbents. Water Res 42(12):3089–3097. https://doi.org/10.1016/j.watres.2008.02.024
Yu J, Hu JY, Tanaka S, Fujii S (2009a) Perfluorooctane sulfonate (PFOS) and perfluorooctanoic acid (PFOA) in sewage treatment plants. Water Res 43(9):2399–2408. https://doi.org/10.1016/j.watres.2009.03.009
Yu Q, Zhang R, Deng S, Huang J, Yu G (2009b) Sorption of perfluorooctane sulfonate and perfluorooctanoate on activated carbons and resin: Kinetic and isotherm study. Water Res 43 (4):1150–1158. https://doi.org/10.1016/j.watres.2008.12.001
Zhang Q, Deng S, Yu G, Huang J (2011) Removal of perfluorooctane sulfonate from aqueous solution by crosslinked chitosan beads: sorption kinetics and uptake mechanism. Bioresour Technol 102(3):2265–2271. https://doi.org/10.1016/j.biortech.2010.10.040
Zhao C, Zhang J, He G, Wang T, Hou D, Luan Z (2013) Perfluorooctane sulfonate removal by nanofiltration membrane the role of calcium ions. Chem Eng J 233:224–232
Zhao C, Tang CY, Li P, Adrian P, Hu G (2016) Perfluorooctane sulfonate removal by nanofiltration membrane-the effect and interaction of magnesium ion/humic acid. J Membr Sci 503:31–41. https://doi.org/10.1016/j.memsci.2015.12.049

Chapter 2
Promises and Challenges of Growing Microalgae in Wastewater

Olumayowa Osundeko, Preeti Ansolia, Sanjay Kumar Gupta, Pushan Bag, and Amit K. Bajhaiya

Abstract Microalgae have been theoretically described as a sustainable feedstock for biofuel production. However, there are still some concerns and obstacles that need to be overcome in order to translate the theoretical promise into commercial and economic success. These obstacles include a high requirement for nutrients and sustainable water source and the identification of affordable cultivation conditions. It has been suggested that growing microalgae in wastewater can potentially offset some of these obstacles. Microalgae can perform a dual role for remediation of nutrient pollutants and biomass production when grown in wastewater. However, there are huge challenges to overcome before this route can be exploited in an economically and environmentally sustainable manner. In the present chapter, the potentials and challenges of growing microalgae in wastewater and its future implications are discussed in detail.

Keywords Bioremediation · Biofuel · Biomass · Microalgae · Wastewater

O. Osundeko
Faculty of Life Sciences, University of Manchester, Manchester, UK

P. Ansolia
Department of Pharmacy, Radharaman Institute of Technology & Science (RITS), Bhopal, India

S. K. Gupta
Environmental Engineering, Department of Civil Engineering, Indian Institute of Technology, Delhi, India

P. Bag
Department of Plant Physiology, Umeå Plant Science Centre, Umeå University, Umea, Sweden

A. K. Bajhaiya (✉)
Department of Chemistry, Chemical Biological Centre (KBC), Umeå University, Umeå, Sweden
e-mail: amit.bajhaiya@umu.se

R. P. Singh et al. (eds.), *Water Conservation, Recycling and Reuse: Issues and Challenges*, https://doi.org/10.1007/978-981-13-3179-4_2

2.1 Introduction

There is an increasing global demand for sustainable biofuel production to mitigate greenhouse gas emissions and reduce reliance on fossil fuels. Many sources of biofuel feedstock like wheat, corn, sugarcane, sugar beets, or molasses have been criticized because of their competition with food crops for agriculture land, water, and nutrients (Doornbosch and Steenblik 2007). The use of some non-food crops like buffalo grass (*Buchloe dactyloides*), *Jatropha curcas*, and switch grass for biofuel production has also been examined, but these may still not fully solve the issue of not using agricultural resources (Sujatha et al. 2008; Robins 2010). To obviate this competition of biofuel production and agriculture resources, researchers have considered using algae biomass for biofuel production.

Algae are a large and diverse group of photosynthetic organisms ranging from unicellular to multicellular forms. Algae are broadly divided into two groups: macroalgae and microalgae. The macroalgae include the large, multicellular seaweeds generally divided into three groups: green (Chlorophyta), red (Rhodophyta), and brown kelps (Phaeophyta – related to Chromista) (Fitzgerald et al. 2011). On the other hand, microalgae are small (~ 1 to 50 μm) and unicellular and normally grow in marine or fresh water bodies but are also found on soil or vegetation (Aresta et al. 2005). Microalgae are the most primitive life-form on earth including prokaryotic cyanobacteria and eukaryotic algae. Their simple cellular structure makes them efficient converters of solar energy, to grow photoautotrophically and produce approximately half of the global atmospheric oxygen while removing carbon dioxide. Biologists have categorized microalgae into different classes, mainly based on pigmentation, life cycle, and basic cellular structures. The four important forms of microalgae are blue-green algae (Cyanophyceae), green algae (Chlorophyceae), diatoms (Bacillariophyceae), and golden algae (Chrysophyceae) (Bajhaiya et al. 2012). These tiny biological factories harness nature's energy using photosynthesis and can double their biomass in every few hours during their exponential growth period (Chisti 2007; Bajhaiya et al. 2010; Louime et al. 2012). Faster growth rate and valuable biomass production make microalgae a promising agricultural prospect.

Photosynthetic carbon can be converted into large amounts of lipids in many microalgal species; for example, some microalgae can generate up to 50% oil by weight; the estimated yield per unit area of oil is between 5000 and 20,000 gallons per acre, per year, which is 7 to 31 times greater than the next best crop, palm oil (Demirbas 2009). Microalgae are known to be very adaptive and can be grown in salt water, freshwater, or even on contaminated industrial effluents (known as wastewater) without any extra requirements of nutrients or agricultural land (Rawat et al. 2016). On top of these advantages, microalgae can grow much better when fed with greenhouse gas (carbon dioxide) and waste nutrients like sewage. So, microalgae could produce biofuel while cleaning up the wastewater (Gupta et al. 2016).

Here we will discuss the potentials and challenges of growing microalgae in wastewater and the potential applications of microalgae grown in wastewater, including pollutant remediation, and the use of algal biomass for various industrial applications, including as a feedstock for biofuel production.

2.2 Wastewater Treatment Processes

Wastewater is a consequence of human activities, which include domestic, agricultural, and industrial waste effluents. Most wastewater emanating from domestic source contains nutrients, suspended solids, heavy metals, pathogens, oxygen-demanding material, and dissolved inorganic and refractory organic materials. Industrial wastewater could contain some xenobiotic materials in addition to those in domestic wastewater, and some require statutory pretreatment before discharge into wastewater treatment streams (Ternes 1998; Badawy et al. 2009). Treating wastewater before discharge to river or land is a regulatory obligation throughout the world due to toxic and hazardous effects on environment and human health. The treatment of wastewater usually involves a preliminary, primary, and secondary treatment stages. The treatment systems in wastewater industries are often capital intensive and result in negative economic return (Melo and Câmara 1994; Chan et al. 2009). The preliminary stage requires removing of large and coarse solids. The primary treatment is designed to remove solids containing organic nitrogen, phosphorus, and heavy metals as well as reduce BOD (biological oxygen demand) by sedimentation and floatation processes (Sonune and Ghate 2004; Abdel-Raouf et al. 2012). The effluents from this stage are characterized by high nitrogen, metal, BOD, COD (chemical oxygen demand), and organic substances. The secondary treatments require removal of nutrients and reduction of BOD and COD. Most common methods are by biological means. This involves the use of microbes, which perform aerobic and anaerobic biodegradation processes. The aerobic process relies mainly on microorganisms using the dissolved oxygen for conversion of biodegradable substances to biomass and CO_2. Whereas in anaerobic processes, microorganisms convert organic substances in the wastewater into methane, CO_2, and water via three basic steps described as hydrolysis and acidogenesis including acetogenesis and methanogenesis in the absence of oxygen (Chan et al. 2009). While biological treatments through activated sludge or biofilm systems are still the popular methods in secondary treatment of wastewater, there are challenges in terms of disposal of sludge in relation to regulation and economic and environmental hurdles (Wei et al. 2003). For example, in studies carried out in four European countries (France, Germany, Italy, and the United Kingdom) which account for 84% of the sludge disposed off in the continent, sludge disposed on a per capita basis was in the range 35–119 g dry solids per head per day (Davis and Hall 1997). The majority of the sludge ends up in landfill or is incinerated or is used in agriculture. The growing interest in the use of biosolids might help in the sustainable disposal of the sludge and possible revenue generation.

The secondary effluents are usually passed through a clarifier and then disposed into freshwater or marine waters. In some cases, they are reused. The current water scarcity in many parts of the world is a growing concern, and recycling wastewater has become an attractive option in sustainable water supply. For example, the Middle East and North African region, which is home to about 5% of the world's population, contains less than 1% of the world annual freshwater resource (Abdel-Raouf et al. 2012). To prevent adverse environmental impacts, it is important that the final effluent meets the required statutory regulation before discharge. Removing nutrients from final effluent to a safe limit for reuse or discharge has been one of the expensive tasks in wastewater treatment. The most final effluent has been characterized with potentially high concentrations of nitrogen and phosphorous and other inorganic and organic compounds which when disposed into river can cause eutrophication or environmental damage (Al-jasser 2011; Chen et al. 2012; Abdel-Raouf et al. 2012). Wastewater treatment plants in many developed countries have incorporated tertiary treatment. This is advanced biological and chemical treatment with the aim to remove nitrogen, phosphorus, and metals. The chemical treatment process works by filtration, ozonation, reverse osmosis, and activated carbon absorption but is often limited by its high implementation cost and secondary pollution contribution (Oswald 1988; Wilde and Benemann 1993; Hoffmann 1998; Sonune and Ghate 2004; Mohamed et al. 2011; Abdel-Raouf et al. 2012). Biological methods are potentially more attractive for tertiary treatment because of the low cost of implementation and value-added by-products.

Microalgae have been well known for decades for having a potential of removing contaminants from wastewater effluents (Oswald et al. 1957; McGriff Jr and McKinney 1972; Gupta et al. 2017a). Integrating microalgae into wastewater treatment can potentially offer a sustainable, cheap, and efficient process without contributing polluting substances (Hoffmann 1998, Gupta et al. 2016; Gupta et al. 2015; Ansari et al. 2017a, b). Despite these advantages, there are many challenges to overcome, which include the cost and efficiency of microalgae harvesting (Gupta et al. 2017b). Immobilizing or co-immobilizing microalgae on a matrix and the use of non-suspending microalgae species are some of the methods that can potentially offset these challenges (Hoffmann 1998; De-Bashan and Bashan 2010).

2.2.1 Microalgae Growth in Wastewater: A Sustainable Approach

The use of microalgae as a sustainable means of remediating wastewater and producing valuable sub-products has received considerable research attentions for decades (Oswald et al. 1959; Oswald 1988; Ruiz-Marin et al. 2010; Rawat et al. 2011; Park et al. 2011; Lohrey and Kochergin 2012). Microalgae growth in wastewater may offer a source of biomass for biofuel and other potential applications such as biomass production for animal feeds in addition to its remediation potential.

2.2.1.1 Remediation Opportunity

Besides having contaminant (nitrogenous and phosphorus)-removing capability, microalgae usage offers opportunity of eco-friendly and cost-effective recycling of chemical in the tertiary treatment stage (Ip et al. 1982; González et al. 1997; An et al. 2003a, b; Aslan and Kapdan 2006).

Early research has focused on the use of microalgae to remove contaminants from secondary effluent as a form of tertiary treatment before discharge into rivers, to prevent eutrophication (Tam and Wong 1989). However, a recent report has shown that microalgae can also effectively remove contaminants during secondary treatment stage (Wang et al. 2010). In this study, it was reported that *Chlorella* sp. grew well in wastewater centrate with a specific growth rate of 0.948 day^{-1} and removed 78.3%, 85.6%, and 83.0% of NH_4^-, phosphorous, and COD, respectively, within a few days. An integrated approach where two or more organisms are used simultaneously is considered as an alternative process of improving nutrient/contaminant removal compared to a single species pond system. For example, an integrated system involving *Chlorella vulgaris* and water hyacinth *Eichhornia crassipes* resulted in 23% more nitrogen removal compared to the high rate algal pond dominated by *Chlorella vulgaris* alone (Bich et al. 1999).

Microalgae can grow in wide varieties of wastewater such as municipal, industrial, artificial, and agricultural types (Bajhaiya et al. 2010). *Chlorella vulgaris* and *Scenedesmus dimorphus* were shown to remove about 95% of ammonium and 55% of phosphorus in secondary effluent of agro-industrial wastewater which emanated from dairy and pig farming (González et al. 1997). Similarly, *Chlamydomonas* species have been reported to remove 100% of NH_4^+ and NO_3^- and 33% of PO_4^{3-} when grown in raw industrial wastewater containing 38.4 mg/L NH_4^+, 3.1 mg/L NO_3^-, and 44.7 mg/L PO_4^{3-} (Wu et al. 2012). A recent study demonstrated excellent treatment of raw domestic wastewater by *Chlorella* and *Scenedesmus* sp. and production of biomass for biofuels (Gupta et al. 2016). In another study, Scenedesmus sp. showed a potential to treat flocculated wastewater as well (Gupta et al. 2017c).

Other benefits of using microalgae for remediation include their disinfection capability by increasing the pH of wastewater because of their photosynthetic activity (Abdel-Raouf et al. 2012). This helps in reducing BOD and coliform bacteria in wastewater. Furthermore, microalgae can potentially reduce the energy input associated with oxygen supply in the conventional activated sludge system and emission of CO_2 due to autotrophic metabolism (Riaño et al. 2011). Substantial removal of coliforms and fecal coliforms from wastewater by *Chlorella* and *Scenedesmus* sp. has also been demonstrated (Gupta et al. 2016). In a recent review, phycoremediation potential of various microalgae species for various types of emerging contaminants and different aspects of wastewater treatment by green microalgae have been demonstrated (Gupta et al. 2015; Rawat et al. 2016; Ansari et al. 2017a, b). A study by has shown Chabukdhara and his colleagues in (2017) has shown comprehensive application of various microalgal species for the heavy metal removal and use of production of algal biomass for biofuels. Therefore, a microalgal-based process helps wastewater industry to meet their emission targets. Growing microalgae in wastewater for remediation also offers additional advantage in terms of biomass production for biofuel (Gupta et al. 2016; Balasubramani et al. 2016). This dual approach has been well discussed as possible ways of enhancing the sustainability and economy of microalgae biofuel strategy (Pittman et al. 2011; Olguin 2012). Wastewaters contain essential nutrients for microalgae growth such as nitrogen, phosphorus, trace metals, and carbon. Although the concentrations of these nutrients vary in wastewaters depending on factors such as climate, size of the community, density of development, economics, and water conservation (Metcalf and Eddy 1991).

2.2.1.2 Biofuel Feedstock Opportunity

The biofuel potential from algae has been theoretically projected to be capable of meeting global demand for transport fuel (Chisti 2007, 2008; Amin 2009; Khan et al. 2009; Brennan and Owende 2010; Demirbas 2010; Mata et al. 2010; Bajhaiya et al. 2010; Amaro et al. 2011; Ahmad et al. 2011; Wu et al. 2012). However, this projection is still far from reality based on its economic and sustainability evaluations (Kovacevic and Wesseler 2010). The high requirements for fertilizer and freshwater, high cost of production, and negative net energy have been major setbacks (Olguin 2012; Acién et al. 2012). Also, phosphorous, an important nutrient for microalgae growth, is at risk of depletion in the world (U.S. DOE 2010). A dual purpose of growing microalgae in wastewater has been proposed to minimize these concerns and improve the economy and sustainability of algae biofuel strategy (Pittman et al. 2011). Wastewater is also a potential source of nitrogen, phosphorus, and carbon which are the major nutrients required for microalgae growth (Oswald et al. 1957; Ruiz-Marin et al. 2010).

A net energy life cycle analysis of growing *Haematococcus pluvialis* and *Nannochloropsis* sp. in synthetic medium for biofuel has shown large energy deficit even with very optimistic assumptions regarding the performance of processing units; they also suggested that some energy can be saved when microalgae are grown in wastewater by eliminating the use of fertilizer. Similarly, Lundquist et al. (2010) comprehensively analyzed different scenarios of algae-based wastewater treatment coupled with biofuel production and concluded that wastewater usage can produce a cost-competitive biofuel compared to the usage of other resources. In one of their studies, wastewater treatment credit can potentially reduce unit cost of oil or electricity from microalgae by 20–25% in addition to eliminating the cost of nutrient and water.

With fast increasing global population, water usages per year have been estimated to reach 2440 billion m^3 by year 2025 (Shiklomanov 1999). About 70% of this water is expected to be used for agriculture and 22% for industries and 8% for domestic use. Using the projections made in Chinnasamy et al. (2010), if 50% of this water usage end up in sewerage and used for growing algae, it could generate approximately 609 million tons of biomass and 91 million tons of oil with the assumption that the algae lipid productivity is 15% of its biomass.

Apart from lipid extraction from microalgal biomass for biodiesel production, biochemical conversion of algae biomass can yield bioethanol, biogas, and biohydrogen (Brennan and Owende 2010). Energy recovery from microalgae can be further optimized by anaerobic digestion to produce methane. Sialve et al. (2009) described some factors that could impede the optimization of microalgae digestion, including the biochemical composition of the microalgae cell wall and protein content. High protein content in microalgae can lead to excessive release of ammonia that can be potentially toxic. Other factors include the presence of sodium salt, which can affect digester performance. Most of these concerns can be omitted by pretreating microalgae before digestion.

2.2.1.3 Non-fuel Applications of Wastewater-Grown Algae

Microalgae growth in wastewater has other applications in addition to remediation and biofuel. Alginate from microalgae has long been used in pharmaceutical products and as a food additive (Chapman and Chapman 1980). The microalgae biomass has potential usage as animal feed and raw materials for chemical production such as methane.

2.2.1.3.1 Production of Biomass for Animal Feed

Some species of microalgae such as *Chlorella* sp., *Dunaliella* sp., and *Spirulina* have been known to contain long-chain fatty acids and protein that have nutritional benefit for animals (Spolaore et al. 2006; Kumar et al. 2010). Algae can produce vitamins and minerals required for healthy growth, better weight control, and healthy skin of pets such as cats, dogs, and ornamental birds (Certik and Shimizu 1999; Spolaore et al. 2006). Ansari et al. (2015) evaluated the feasibility of using lipid-extracted algae (LEA) as a source for protein and reduced sugar. The findings showed that even after extraction of oil from the microalgal biomass, it can be further used for the feed supplements of aquaculture or animals. In a research study, examining biomass and lipid profiling of *Chlorella vulgaris* grown in industrial and agricultural wastewater as an aquaculture feed (Bertoldi et al. 2006) has found that the use of hydroponic wastewater as an alternative culture medium can help in generating quality lipid, fatty acid, and carotenoids from *Chlorella vulgaris*.

2.2.1.3.2 Medicinal Importance

Extracting products of high medicinal value from algae is a fast-growing market today. For example, *Chlorella* is currently being produced by more than 70 companies around the world and having a world market value worth more than US$ 38 billion over a decade ago (Yamaguchi 1997) and is expected to have grown since. Some algae have been found to contain secondary metabolites and antioxidants such as carotenoids, astaxanthin, that can be a remedy of some health issues (Spolaore et al. 2006). *Chlorella* contains β-1,3-glucan, an active immunostimulant, which is a free radical scavenger and helps in reducing blood lipids. *Spirulina* extract as a food supplement has been found to have a medicinal importance in protecting liver against mercuric chloride-induced toxicity in Swiss albino mice (Kumar et al. 2005). Chlorophyll in microalgae has been found to have pharmaceutically importance as well. It was reported to have some chelating characteristics suitable for use in liver recovery and ulcer treatment (Singh and Gu 2010). However, algae grown for this sort of application are usually grown in a bioreactor with high sterility (Belarbi et al. 2000).

2.2.1.3.3 Fertilizer Application

The use of algae for fertilizer has long been a practice in many parts of the world (Booth 1966). Algae can improve soil fertility by enhancing the water retention capacity of soil and macronutrient composition. For example, a study on the effects of algal fiber on germination and growth of barley found that it enhanced the soil physical and chemical qualities (Riley 2002). In the same report, algae fiber waste from an alginate extraction plant in Norway was studied in respect to its potential for soil improvement and as a nutrient source for potato crop. Algae fiber was shown to have effects on soil aeration and water retention capability like perlite (amorphous glassy globule found in volcanic eruption) used in plant growth medium. Plant-available water was reported to increase by 3.6% by volume when 10% by volume of algae fiber was incorporated, as against 1.2% by volume with the same volume of pure perlite. Furthermore, potato yield also increased with the use of algae fiber.

2.2.2 Integrating Microalgae Cultivation in Wastewater Works

One of the more efficient ways to fast-track large-scale production of microalgae biofuel is by incorporating it into the existing wastewater works with little adaptation (Fig. 2.1). This can potentially reduce the cost of infrastructure (Lundquist et al. 2010). The choice of microalgae to be grown is very important, and finding species that can grow efficiently for high biofuel productivity in wastewater is a key to the success of this strategy (Singh and Gu 2010). While studying the effect of varying nutrient loadings on the growth and lipid accumulation adoptability, Shriwastav et al. (2014) reported that green microalgae such as of *Chlorella* are stress resistant and has potential to accumulate high amount of lipids. Therefore,

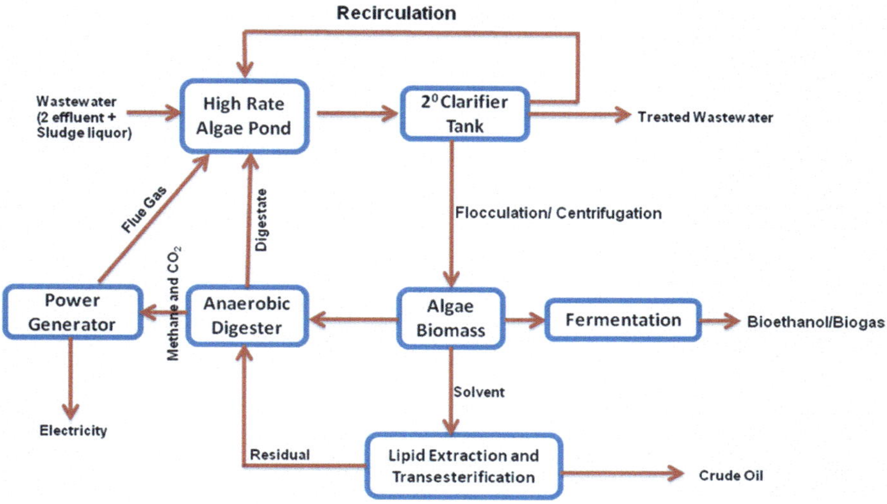

Fig. 2.1 A flow diagram showing how microalgae cultivation can be incorporated in wastewater works for dual role of remediation and biofuel production

for dual applications, i.e., wastewater treatment and biofuel production, selection of stress resistant and high lipid-producing species is very important. Some microalgae have high oil content but are slow growing. For example, *Botryococcus braunii* is well known to have a high percentage of lipid content but grows slower compared to *Chlorella* sp. (Orpez et al. 2009; Kong et al. 2010). Algae that grow faster even with little lipid content will be ideal because it will reduce the production time and energy demands.

Microalgae cultivation can either be done by an open or closed system at the secondary or tertiary wastewater treatment stage. This can produce a cheap way of remediating wastewater. Algae biomass harvesting by centrifugation can potentially increase the energy input and the cost of production and still require more research and developments. Other efficient harvesting methods include flocculation or biofilm induction.

Biofuel products from microalgae include biodiesel and bioethanol which can be obtained by transesterification and fermentation, respectively. Microalgae biomass can be integrated into wastewater digestion sludge for methane and CO_2 production (Kumar et al. 2010). The CO_2 can be recycled into the algae pond providing a source of carbon. High productivities of microalgae grown in wastewater in terms of biomass and lipids reported in many studies need to be translated to the large industrial scale for cost-effective and sustainable biofuel production. However, there are some challenges to overcome before this can be realistically achieved. While studying the comprehensive biochemical methane potential using algal biomass, Ansari et al. (2017b) reported that pretreatment could be an important step for the better methane production from microalgae.

2.2.3 Challenges Associated with Growing Microalgae in Wastewater

The uses of microalgae in the wastewater industry, either as a remediating agent or for biofuel feedstock, are still relatively limited. Some of the setbacks include the presence of growth-inhibiting factors and difficult harvesting processes. Some of these inhibitors include biotic and abiotic factors. Biotic factor can be in the form of viruses, bacteria, zooplankton, grazers, phytoplankton, and fungi (Kagami et al. 2007; Park et al. 2011), which will impede or significantly inhibit the growth of microalgae. These factors will depend on the source of wastewater effluent being used for cultivation. Abiotic contaminants in wastewater such as CO_2, NO_x (nitrogen oxides), SO_x (sulfur oxides), O_2, and NH_3 and heavy metals can also inhibit microalgae growth. Kumar et al. (2010) described that heavy metals can inhibit microalgae photosynthesis by blocking the prosthetic metal atoms in the active site of relevant enzymes which can make the cell physiologically incapable. In contrast, when essential nutrient concentrations in the wastewater are low specifically, trace mineral nutrients can result in poor growth, low biomass, and low lipid productivity (Christenson and Sims 2011). In such cases, it's needed to supplement these nutrients in wastewater to achieve high productivity. In an open pond system, carbon can be limited due to poor mass transfer. Some studies have suggested that bubbling of CO_2 can improve algae growth as well. For example, *Nannochloropsis oculata* growth and productivities were improved when aerated with 2% CO_2 in a semicontinuous cultivation system (Chiu et al. 2009). Also, high concentration of oxygen in wastewater can induce oxidative stress to microalgae cell and inhibit photosynthesis (Christenson and Sims 2011). Therefore, more research needs to be focused on how to optimize nutrient in wastewater for microalgae growth.

The cost and energy demand of harvesting microalgae from wastewater either by flocculation or centrifugation are still very high (Acién et al. 2012). This can be as high as 20–30% of the total energy and cost required for biofuel production (Rösch et al. 2009). Most of the current techniques for harvesting are yet to be demonstrated on a large scale. Therefore, there's a need for technological advancement in this area to make microalgae biotechnology more economically attractive.

Research in the use of wastewater for cultivating microalgae is still very limited as compared to research using synthetic inorganic medium. Lam and Lee (2012) found that only ~30% of published researches on microalgae cultivation are based on wastewater as a growth medium while the rest are using synthetic media, probably because the synthetic media are readily available and less contaminated and yield promising results. However, the use of synthetic media might be unsustainable in commercial terms.

2.3 Wastewater Effluent Supporting High Microalgae Productivity

Wastewater usually contains high concentration of nutrients in terms of nitrogen, phosphorus, carbon, and metals which are an essential requirement for microalgae growth in the presence of sunlight, CO_2, O_2, and optimal temperature and pH. Much of the nitrogen is in the form of ammonium, which is an available source of nitrogen for microalgae uptake, although this can potentially become toxic and inhibit growth at higher concentrations (Ip et al. 1982; Konig et al. 1987; Wrigley and Toerien 1990). Tolerance toward NH_3 is a very important criteria for the selection of microalgae to be grown in wastewater. For example, a comparative study of three green microalgae *Scenedesmus obliquus*, *Scenedesmus platensis*, and *Chlorella sorokiniana* growth in piggery wastewater showed that *Chlorella sorokiniana* has a high tolerance to high ammonium compared to the other species (de Godos et al. 2010). Furthermore, in a combined cultivation of these species, *Chlorella sorokiniana* was able to outgrow other species and become the dominant species in the continuous photobioreactor. The high concentration of N, P, and metals in wastewater effluent especially those from municipal and agriculture sources can potentially inhibit microalgae and in some cases will require dilution before inoculation. As shown in Table 2.1, a study reported that the use of a Biocoil helps in increasing biomass and lipid productivity of *Chlamydomonas reinhardtii* achieving 2000 mg/l/day of biomass and 505 mg/l/day of lipid when grown in municipal centrate (Kong et al. 2010). The high productivity was attributed to greater light exposure and intensity in the polyvinyl tubing because when the same microalgae species was grown in a flask, it produced 820 mg/l/d of biomass and 136 mg/l/d of lipid which are much lesser than in Biocoil. Another study reported that *Chlorella* grown in raw municipal centrate produced 1060 mg/l/d of biomass (Li et al. 2011). However, when the centrate was treated by autoclave, the productivity increased to 1175 mg/l/d providing biodiesel yield of 0.12 g-biodiesel/L-algae culture solution.

Whereas secondary-treated wastewater is generally low in nutrient concentration as compared to centrate or primary effluents, biomass and lipid productivities are usually lower. For example, *Botryococcus braunii* grown in municipal secondary-treated wastewater in a bioreactor yielded 345.6 mg/l/d and lipid of 62 mg/l/d (Orpez et al. 2009). In agricultural wastewater from piggery manure that contains high concentration of nutrient especially nitrogen, biomass and lipid productivities were recorded to be 700 mg/l/d and 69 mg/l/d, respectively (An et al. 2003a, b). Agriculture wastewater from diary manure has also been shown to promote high biomass and lipid productivities of *Chlorella* sp. attached to supporting polystyrene foam. Biomass productivity based on the algae biomass attached to the foam was 2.6 g m^{-2} day^{-1}, and the total fatty acid content was 230 mg m^{-2} day^{-1}.

In general, microalgae growth rates are lower in many industrial wastewaters, due to low nutrient and high toxin concentrations (Pittman et al. 2011). However, this

Table 2.1 Microalgae strains grown in various wastewaters and their biomass, lipid productivity, growth rate, and N and P removal efficiency

Wastewater type	Microalgae species	N (mg/l)	P (mg/l)	Growth rate	Biomass (mg L − 1 day−1)	Lipid %	Lipid productivity (mg L − 1 day−1)	Cultivation type	N removal	P removal	References
Municipal settle sewage effluent	Green algae, Blue green algae in low CO_2	40–50	7–8	na	25[b]	nd		Batch			Ip et al. (1982)
Municipal wastewater centrate	*Chlamydomonas reinhardtii*	128.6	120.6	0.564	2000	25.25	505	Batch in Biocoil photobioreactor	83	14.45	Kong et al. (2010)
Municipal wastewater centrate	*Chlamydomonas reinhardtii*	128.6	120.6	0.346	820	16.59	136	Batch in flask	55	15.4	Kong et al. (2010)
Municipal (secondary treated)	*Scenedesmus obliquus*	27.4[a]	11.8	0.0286[f]	26[c]	31.4[c]	8[d]	Batch bioreactor	94%[e]	98%[e]	Martinez et al. (2000)
Municipal (secondary treated)	*Botryococcus braunii*	11.9	11.5	0.14	345.6[g]	17.85	62	Batch bioreactor	na	na	Orpez et al. (2009)
Municipal (primary treated + CO_2)	Mix of *Chlorella* sp., *Micractinium* sp., and *Actinastrum* sp.	51	2.1	na	270.7[h]	9	24.4	Semicontinuous cultivation	99[i]	99[i]	Woertz et al. (2009)
Municipal (autocalved centrate)	*Chlorella* sp.	91[a]	212	0.466[j]	241.7[j]	30.9	74.7	Batch	na	na	Zhou et al. (2011)
Municipal (autocalved centrate)	*Hindakia* sp.	91[a]	212	0.498	275	28.3	77.8	Batch	na	na	Zhou et al. (2011)
Municipal (autocalved centrate)	*Scenedesmus* sp.	91[a]	212	0.472	247.5	30.1	74.5	Batch	na	na	Zhou et al. (2011)
Municipal (pretreated)[k]	*Chlorella* sp.	18.9	1.7	nr	74	31	22.9	Batch	92	86	Cho et al. (2011)

Agricultural (piggery manure with high NO3-N)	*B. braunii*	836	40	0.033	700[l]	Nr	69	Batch	80	82.5	An et al. (2003a, b)
Agricultural (dairy manure with polystyrene foam support)	*Chlorella* sp.	517	770	nr	2.6 g m^{-2} day^{-1}	9[m]	230[m] g m^{-2} day^{-1}	Batch	61–79%	62–93%	Johnson and Wen (2010)
Agricultural (fermented swine urine)	*Scenedesmus* sp.	86.4	20.2	0.12[n]	6[m]	0.9[m]	0.54[m]	Batch	nr	nr	Kim et al. (2007)
Agricultural (anaerobically digested dairy manure)	Mix of *Microspora willeana*, *Ulothrix zonata*, and *Ulothrix*	225	24.7	nr	5.5	nd	nd	Semi-batch	39	51	Wilkie and Mulbry (2002)
Agricultural (digested dairy manure, 20 × dilution)	*Chlorella* sp.	2232[a,p]	249.7[p]	0.407	81.4[o]	13.6[m]	11[m]	Batch	100	71.6	Wang et al. (2010)
Agricultural (dairy wastewater, 25% dilution)	Mix of *Chlorella* sp., *Micractinium* sp., and *Actinastrum* sp.	30.5[q]	2.6	nr	59	29	17	Continuous cultivation	96	>99	Woertz et al. (2009)
Industrial (carpet mill, untreated)	*B. braunii*	17.58–25.85[a]	5.47–13.83	nr	34	13.2	4.5	Batch	nr	nr	Chinnasamy et al. (2010)
Industrial (carpet mill, untreated)	*Chlorella saccharophila*	17.58–25.85[a]	5.47–13.83	nr	23	18.1	4.2	Batch	nr	nr	Chinnasamy et al. (2010)

(continued)

Table 2.1 (continued)

Wastewater type	Microalgae species	N (mg/l)	P (mg/l)	Growth rate	Biomass (mg L − 1 day−1)	Lipid %	Lipid productivity (mg L − 1 day−1)	Cultivation type	N removal	P removal	References
Industrial (carpet mill, untreated)	Consortium of 15 native algae isolated wastewaters	17.58–25.85[a]	5.47–13.83	nr	39	12	4.7	Batch	99.7[r]	99.1[r]	Chinnasamy et al. (2010)

nd not determined

DW dry weight

[a]Measured in form of Ammonium

[b]Estimated from biomass value of 1000 mg L^{-1} after 40 days

[c]Estimated from biomass value of 1.1 mg L^{-1} h^{-1}

[d]Fatty acid content and productivity determined rather than total lipid

[e]At a temperature of 20 °C without culture stirring

[f]Measured per hour

[g]Estimated from biomass value of 14.4 mg L^{-1} h-

[h]Estimated from biomass value of 812 mg L^{-1} after 3 days

[i]4 days growth with CO_2 supplement.

[j]Represent the highest value recorded

[k]Pretreated by filtration through 0.2 μm-pore size filter

[l]Estimated from biomass value of 7 g L − 1 after 10 days

[m]Fatty acid content and productivity determined rather than total lipid

[n]Pretreated by adding fermented swine urine (3%) (v/v) a control medium

[o]Estimated from biomass value of 1.71 g L − 1 after 21 days

[p]The actual value before dilution

[q]Estimated from lipid productivity and lipid content value

[r]Treated by bubbling with ambient air and incubated at 25 °C

will greatly vary depending on the industry. Industrial wastewater rich in nitrogen and phosphorus can be considered for the cultivation of microalgae. Though, some studies have used municipal wastewater centrate as a fertilizer to augment the low nutrients in industrial wastewater (Chinnasamy et al. 2010). Biofuel productivity of microalgae varies greatly in different wastewaters. Therefore, determining the best source of wastewater for microalgae cultivation will be governed by nutrient availability and level of growth-inhibiting factors.

2.4 Challenges of Pilot-Scale Biofuel Production from Wastewater Industries

Scaling up microalgae biofuel production from laboratory scale to commercial scale remains a challenge. Integrating microalgae biofuel in an existing wastewater treatment plant offers an inexpensive and economically feasible way of commercialization. However, this comes with some major technical and economic problems that still need to be improved.

2.4.1 Culture Instability

Wastewater provides a cheap source of growing microalgae; however, it comes with some challenges in terms of contamination and difficulty in maintaining a monoculture system. Using a natural strain that is indigenous to the cultivation pond might bring a solution to this. For example, in a previous report, consortia of natural strains grown in municipal wastewater has shown to produce lipid of up to 29% of its biomass and lipid productivity reaching 2800 mg/m^2/day (Woertz et al. 2009). Fungal, viral, and bacterial pathogenic activities can outcompete microalgae and disrupt the culture stability in large-scale cultivation. Hoffman et al. (2008) reported fungi found in cultures of *Haematococcus pluvialis* that grew epibiotically on the algal cells can cause the destruction of the host culture. So far little is known about pond speciation, ecological dynamics in wastewater, host specificity, and resistance mechanisms. In a US DOE (2010) report, three important facts that should be considered in large-scale cultivation of microalgae in wastewater were described. These include determining the extent of algae pathogen and predators in the wastewater pond. Macroalgae and other phytoplankton can dominate the pond or bioreactor. Furthermore, a dynamic pond monitoring technique and preventive measures needed to limit the take-over by a foreign organism.

2.4.2 Cultivation and Land Requirement

Techno-economically viable cultivation of microalgae by open or closed systems in an industrial scale is difficult to predict. Recent life cycle analysis of the production of biodiesel and biogas from *Haematococcus pluvialis* and *Nannochloropsis* has reported that a photobioreactor consumed about 56.5 MJ of electricity as compared to a raceway open pond that required 26.6 MJ (Razon and Tan 2011). The higher energy requirement can be because of culture mixing, gas exchange, and temperature maintenance. Bioreactors support higher growth and microalgae cell density than open cultivation systems and reduced evaporation which is suffered in an open pond system (Chisti 2007). Monoculture systems can easily be achieved in closed systems than in an open system. Land requirement for facilities is also a potential problem. Most wastewater treatment plants are usually sited close to the metropolitan areas with high land price and limited availability (U.S. DOE 2010). It will be expensive to transport wastewater over a long distance. Therefore, integrating wastewater into existing plants with a little modification is highly encouraged.

2.4.3 Microalgae Harvesting Challenges

While microalgae are capable of producing lipid and biomass suitable for biofuel, recovering and processing microalgae from the cultivation medium are still a challenge (De-Bashan et al. 2002). The effective harvesting method is the key to integrate this system in the wastewater industry as it costs more than 20–30% of biomass production (Mata et al. 2010). In older microalgal cultures, flocculation can occur, but, in some cases, flocculating agent such as alum ($Al_2(SO_4)_3$) can be added to the culture. Flocculation can also be triggered by manipulating culture pH or bioflocculation with other microorganisms that promote sedimentation (Sukenik and Shelaf 1984; Lavoie and de la Noue 1985). Flocculation increases particle size and promotes sedimentation. Among various harvesting techniques, flocculation has been found the easiest and economic. The use of chemically induced flocculation is currently expensive and results in chemical contamination of the harvested biomass, therefore, unsustainable for harvesting microalgae in a commercial scale. Numerous studies have been done on the coagulation and flocculation of various microalgal species with natural, inorganic, and synthetic chemicals. In a comparative study of flocculation efficiency, Gupta et al. (2014) reported excellent biomass recovery of *Scenedesmus* sp. from a polyamine-based synthetic polymer. It was also demonstrated that the microalgal biomass production could be as lowest as 0.5 USD with such flocculants, whereas the biomass harvesting through natural flocculants such as chitosan costs around 50 USD. Therefore, the selection of flocculent is very important for pilot-scale harvesting by flocculation, which also depends on the end use of the biomass. The use of chemical flocculants could be one of the most economical for low value products such as biofuels. Bioflocculation seemed to be a cheap option, but it is slower and not always reliable (Schenk et al. 2008). A comprehensive study on harvesting of various microalgae from different types of flocculent is recommended for further study (Gupta et al. 2017a, b, c).

Another method that is widely used in the wastewater industry is centrifugation and, in some cases, has been shown to be effective (Molina et al. 2003). Biomass recovery by centrifugation depends on the characteristic of the microalgae species such as size, settling rate, and speed of the centrifugation system. Molina et al. (2003) reported that cell harvest efficiency of >95% was obtained only at 13,000 × g and the harvest efficiency declined to 60% at 6000 × g and 40% at 1300 × g. Centrifugation can be rapid but consume more energy and can increase the cost of biofuel production. Another method is by filtration using pressure vacuum. This might not be suitable for some microalgae like *Scenedesmus* and *Chlorella* that are small in size (Brennan and Owende 2010). Filtrate recovery of microalgae may be unsatisfactory because filtration can be relatively slow and comes with the cost of replacing the filter membrane and energy cost of pumping (Pittman et al. 2011).

An alternative method of harvesting microalgae is by immobilization. It has been shown that immobilization is one of the feasible ways to achieve recovery of microalgae from wastewater (Moreno-Garrido 2008; De-Bashan et al. 2002). An immobilization technique offers a solution to these problems encountered during harvesting in a suspending growth system. This technique involves the absorption of microalgae when passed through the immobilized matrix system (Travieso et al. 1992; Zhang et al. 2008). Although this system is simple and allows the possibility to reuse the immobilized matrix system after recovery, there are still some constraints in terms of the cost of the polymer, effectiveness in removing some contaminants, and toxicity of some immobilizing techniques.

Being cost-effective and simple, filtration could be used as some microalgae harvesting technique. Various types of filtration techniques are also used for the harvesting of microalgae. The main advantage of this technique is that the harvested biomass is free from any chemical contaminants, and more importantly this can be used for the recycling of residual nutrient media (Ahmad et al. 2012). Uncontaminated biomass can be used for other purposes, while biomass harvested by coagulation has limited use. The major disadvantages of such techniques are the high energy consumption in pumping and frequent replacement of membrane due to fouling, choking, and wear and tear (Uduman et al. 2010). Hence, filtration cannot be used for pilot-scale harvesting purposes, especially for the low-value algal products. Therefore, development of cost-effective filtration techniques is the need of the day for improving economics of the harvesting. Sahoo et al. (2017) reported cost-effective and efficient dewatering and drying of *Scenedesmus* and *Chlorella* sp. using polypropylene non-woven fabric membrane (PNM). They demonstrated that dewatering with such type of membrane with natural drying could be achieved as economical as 0.048 USD per kg of dry algal biomass. Such type of arrangements could be a hope for the future.

2.4.4 Climatic Factors

One of the major parameters limiting commercialization of algae biofuel production is the climate change in terms of temperature and sunlight especially in an open cultivation system. Lower temperature below the optimal can reduce cell metabolism and result in low biomass and lipid productivities. The average suitable temperature for algae growth has been described to be in the region of 15 °C to 35 °C (Lundquist et al. 2010). Most regions of the earth in the upper hemisphere like Europe, where average temperature could be 15 °C or less at night, can be considered unsuitable region for maximum algae growth yield. One of the ways to ameliorate low temperature in those regions is by covering the pond with some plastic materials. Alternatively, in the night time, microalgae culture can be transferred into a settling pond that are normally deep and retain heat unlike the shallow growth ponds. The culture can then be circulated back to the growth pond during the day. However, this can potentially add to the cost and energy requirement.

2.4.5 Energy and Cost of Production: Life Cycle Analysis

The true potential of microalgae biofuel production using wastewater can be best determined by life cycle analysis of its cost and energy. Microalgae biofuel has been criticized for performing poorly than other plant-based biofuels in terms of energy requirement, greenhouse gas emission, and water usage (Clarens et al. 2010). Impacts associated with algae biofuel production were assessed using a stochastic life cycle model in comparison with switchgrass, canola, and corn. They concluded that major contributing factors are demand for CO_2 and fertilizer as a nutrient source. The use of wastewater for growing microalgae for biofuel can offset most of these factors and make microalgae biofuel more environmentally beneficial than the terrestrial crops that were accessed.

Cultivation of microalgae in either a photobioreactor or in an open system such as raceway pond requires considerable amounts of energy for mixing. The net energy requirement of a system has been defined as the ratio of the total energy output from the system over energy input for the entire production system. Net energy analysis of microalgae biofuel production is important as a diagnostic measure of identifying performance and improvement options that may be necessary. One of the potential positives for producing biofuel from microalgae is if microalgae can produce sustainable net surplus energy. However, a recent report has raised concerns that microalgae biodiesel may not deliver more energy than is required to produce it. Most of the energy demand has been attributed to the extraction process, which does require drying (Khoo et al. 2011). However, using wastewater as alternative to fertilizer and wet extraction instead of dry extraction can improve its energy performance.

Analysis of the microalgae biofuel production strategy have shown that the use of wastewater as cultivation medium could decrease water requirement by 90% and eliminate nutrient requirements except for phosphate (Clarens et al. 2010; Yang et al. 2011). Acién et al. (2012) performed a cost evaluation of a real microalgae plant; they found that the cost of running a 3 m^3 tubular photobioreactor operator in continuous mode for 2 years for producing biomass from *Scenedesmus almeriensis* would be 165,000.00 euros (47.41% of the total cost of production). Furthermore, a freeze dryer and storage of 70 kg/day were reported to cost 165,000.00 euros (31.6% of the total cost of production). They emphasized that it is important to reduce the cost of running a photobioreactor and use wastewater and flue gases in order to make microalgae biofuel production economically attractive. However, there is a need for more studies to access life cycle performance in terms of cost, energy, and environmental performance of microalgae in different wastewater scenarios.

2.5 Future of Microalgae-Based Bioenergy and Biofuels

Based on current knowledge, it is unlikely that microalgae will produce biofuel which is cost competitive with fossil fuels without major advances in technology. Most significant improvements are expected to be in strain selection, cultivation, harvesting, and oil extraction (Pittman et al. 2011). Integrating microalgae cultivation in wastewater for the dual purpose of remediation and biofuel production can potentially accelerate the commercialization of algae biofuel in the near term (Fig. 2.2). However, there are needs for better control of zooplankton and phytoplankton and development of lower cost ponds, clarifiers, and digesters (U.S. DOE 2010).

With the fast growing technology advancement in harvesting of microalgae, strain improvement, and selection, it is possible to break even in incorporating microalgae biofuel in wastewater soon. Coproducts, which are already being produced, can also be improved for combined production with biofuel. However, most of these coproducts require monocultures and high sterility which can be achieved only in a closed system. A single purpose commercial biofuel/biodiesel production from microalgae is likely to be a long-term goal owing to the current cost and energy demand for production (Harmelen and Oonk 2006). Research in algae biofuel needs to start focusing on the pond system rather than small-scale laboratory cultures to provide a robust data, minimize edge effects, and allow for extrapolating to commercial scale.

Fig. 2.2 Futuristic projection on algae biofuel based on the existing knowledge and technology

Acknowledgments Olumayowa Osundeko is grateful to the Sustainable Consumption Institute Doctoral Training Centre for providing a graduate studentship.

References

Abdel-Raouf N, Al-Homaidan AA, Ibraheem IBM (2012) Microalgae and wastewater treatment. Saudi J Biol Sci 19:257–275

Acién FG, Fernández JM, Magán JJ, Molina E (2012) Production cost of a real microalgae production plant and strategies to reduce it. Biotechnol Adv 30(6):1344–1353

Ahmad AL, Mat YNH, Derek CJC, Lim JK (2011) Microalgae as a sustainable energy source for biodiesel production: a review. Renew Sust Energ Rev 15(1):584–593

Ahmad A, Yasin NM, Derek C, Lim J (2012) Crossflow microfiltration of microalgae biomass for biofuel production. Desalination 302:65–70

Al-Jasser AO (2011) Saudi wastewater reuse standards for agricultural irrigation: Riyadh treatment plants effluent compliance. J King Saud Univ Eng Sci 23(1):1–8

Amaro HM, Guedes AC, Malcata FX (2011) Advances and perspectives in using microalgae to produce biodiesel. Appl Energy 88(10):3402–3410

Amin S (2009) Review on biofuel oil and gas production processes from microalgae. Energy Convers Manag 50(7):1834–1840

An JY, Sim SJ, Lee JS, Kim BW (2003a) Hydrocarbon production from secondarily treated piggery wastewater by the green alga Botryococcus braunii. J Appl Phycol 15:185–191

An JY, Sim SJ, Lee JS, Kim BW (2003b) Hydrocarbon production from secondarily treated piggery wastewater by the green alga Botryococcus braunii. J Appl Phycol 15:185–191

Ansari FA, Shriwastav A, Gupta SK, Rawat I, Guldhe A, Bux F (2015) Lipid extracted algae as a source for protein and reduced sugar: a step closer to the biorefinery. Bioresour Technol 179:559–564

Ansari FA, Shekh AY, Gupta SK, Bux F (2017a) Microalgae for biofuels: applications, process constraints and future needs. In: Algal Biofuels. Springer, Cham, pp 57–76

Ansari FA, Wahal S, Gupta SK, Rawat I, Bux F (2017b) A comparative study on biochemical methane potential of algal substrates: implications of biomass pre-treatment and product extraction. Bioresour Technol 234:320–326

Aresta M, Dibenedetto A et al (2005) Production of biodiesel from macroalgae by supercritical CO 2 extraction and thermochemical liquefaction. Environ Chem Lett 3(3):136–139

Aslan S, Kapdan IK (2006) Batch kinetics of nitrogen and phosphorus removal from synthetic wastewater by algae. Ecol Eng 28:64–70

Badawy MI, Wahaab RA, El-Kalliny Fenton AS (2009) Biological treatment processes for the removal of some pharmaceuticals from industrial wastewater. J Hazard Mater 167(1–3):567–574

Bajhaiya AK, Mandotra SK, Suseela MR, Toppo K, Ranade S (2010) Algal biodiesel: the next generation biofuel for India. Asian J Exp Biol Sci 1(4):728–739

Bajhaiya AK, Suseela MR, Ramteke PW (2012) Approaches and prospectives for algal fuel. In: The science of algal fuels. Springer, Dordrecht, pp 43–62

Balasubramani R, Gupta SK, Cho W, Kim J, Lee S, Jeong K, Lee D, Choi H (2016) Microalgae potential and multiple roles-current progress and future prospects-an overview. Sustainability 8 (12):1215

Belarbi EH, Molina E, Chisti Y (2000) A process for high yield and scaleable recovery of high purity eicosapentaenoic acid esters from microalgae and fish oil. Process Biochem 35:951–969

Bertoldi FC, Sant'Anna E, Da Costa Braga MV, Oliveira JLB (2006) Lipids, fatty acids composition and carotenoids of Chlorella vulgaris cultivated in hydroponic wastewater. Grasasy Aceites 57(3):270–274

Bich NN, Yaziz MI, Bakti NAK (1999) Combination of Chlorella vulgaris and Eichhornia crassipes for wastewater nitrogen removal. Water Res 33(10):2357–2362
Booth E (1966) Some properties of seaweed manures. In: Proceedings of 5th international seaweed symposium. Pergamon Press, London, pp 349–357
Brennan L, Owende P (2010) Biofuels from microalgae—a review of technologies for production, processing, and extractions of biofuels and co-products. Renew Sust Energ Rev 14(2):557–577
Certik M, Shimizu S (1999) Biosynthesis and regulation of microbial polyunsaturated fatty acid production. J Biosci Bioeng 87:1–14
Chabukdhara M, Gupta SK, Gogoi M (2017) Phycoremediation of heavy metals coupled with generation of bioenergy. In: Algal biofuels. Springer International Publishing, Cham, pp 163–188
Chan YJ, Mei FC, Chung LL, Hassell DG (2009) A review on anaerobic–aerobic treatment of industrial and municipal wastewater. Chem Eng J 155(1–2):1–18
Chapman VJ, Chapman DJ (1980) Seaweeds and their uses. Chapman & Hall, London, pp 30–42
Chen H, Zhang C, Han J, Yu Y, Zhang P (2012) PFOS and PFOA in influents, effluents, and biosolids of Chinese wastewater treatment plants and effluent-receiving marine environments. Environ Pollut 170:26–31
Chinnasamy S, Bhatnagar A, Hunt RW, Das KC (2010) Microalgae cultivation in a wastewater dominated by carpet mill effluents for biofuel applications. Bioresour Technol 101 (9):3097–3105
Chisti Y (2007) Biodiesel from microalgae. Biotechnol Adv 25(3):294–306
Chisti Y (2008) Response to Reijnders: do biofuels from microalgae beat biofuels from terrestrial plants? Trends Biotechnol 26(7):351–352
Chiu S, Kao C, Tsai M, Ong S, Chen C, Lin C (2009) Lipid accumulation and CO_2 utilization of Nannochloropsis oculata in response to CO_2 aeration. Bioresour Technol 100:833–838
Cho S, Luong TT, Lee D, Oh Y-K, Lee T (2011) Reuse of effluent water from a municipal wastewater treatment plant in microalgae cultivation for biofuel production. Bioresour Technol 102(18):8639–8645
Christenson L, Sims R (2011) Production and harvesting of microalgae for wastewater treatment, biofuels, and bioproducts. Biotechnol Adv 29(6):686–702
Clarens AF, Resurreccion EP, White MA, Colosi LM (2010) Environmental life cycle comparison of algae to other bioenergy feedstocks. Environ Sci Technol 44:1813–1819
Davis RD, Hall JE (1997) Production, treatment and disposal of wastewater sludge in Europe from a UK perspective. Eur Water Pollut Control 7(2):9–17
de Godos VVA, Blanco S, González MCG, Soto R, García-Encina PA, Becares E, Muñoz R (2010) A comparative evaluation of microalgae for the degradation of piggery wastewater under photosynthetic oxygenation. Bioresour Technol 101:5150–5158
De-Bashan LE, Bashan Y (2010) Immobilized microalgae for removing pollutants: review of practical aspects. Bioresour Technol 101:1611–1627
De-Bashan LE, Moreno M, Hernandez JP, Bashan Y (2002) Removal of ammonium and phosphorus ions from synthetic wastewater by the microalgae Chlorella vulgaris coimmobilized in alginate beads with the microalgae growth-promoting bacterium Azospirillum brasilense. Water Res 36(12):2941–2948
Demirbas A (2009) Potential resources of non-edible oils for Biodiesel. Energy Sources Part B Econ Plann Policy 4(3):310–314
Demirbas A (2010) Use of algae as biofuel sources. Energy Convers Manag 51(12):2738–2749
Doornbosch R, Steenblik R (2007). Biofuels: is the cure worse than the disease? Paris, France, September 11–12
Fitzgerald C, Gallagher E, Tasdemir D, Hayes M (2011) Heart health peptides from macroalgae: isolation and potential use in functional foods. J Agric Food Chem 59:6829–6836
González LE, Cañizares RO, Baena S (1997) Efficiency of ammonia and phosphorus removal from a colombian agroindustrial wastewater by the microalgae Chlorella vulgaris and Scenedesmus dimorphus. Bioresour Technol 60(3):259–262

Gupta SK, Kumar M, Guldhe A, Ansari FA, Rawat I, Kanney K, Bux F (2014) Design and development of polyamine polymer for harvesting microalgae for biofuels production. Energy Convers Manag 85:537–544

Gupta SK, Shriwastav A, Kumari S, Ansari FA, Malik A, Bux F (2015) Phycoremediation of emerging contaminants. In: Algae and environmental sustainability. Springer, New Delhi, pp 129–146

Gupta SK, Ansari FA, Shriwastav A, Sahoo NK, Rawat I, Bux F (2016) Dual role of Chlorella sorokiniana and Scenedesmus obliquus for comprehensive wastewater treatment and biomass production for bio-fuels. J Clean Prod 115:255–264

Gupta SK, Sriwastav A, Ansari FA, Nasr M, Nema AK (2017a) Phycoremediation: an eco-friendly algal technology for bioremediation and bioenergy production. In: Phytoremediation potential of bioenergy plants. Springer, Singapore, pp 431–456

Gupta SK, Ansari FA, Bauddh K, Singh B, Nema AK, Pant KK (2017b) Harvesting of microalgae for biofuels: comprehensive performance evaluation of natural, inorganic, and synthetic flocculants. In: Green technologies and environmental sustainability. Springer International Publishing, pp 131–156

Gupta SK, Ansari FA, Nasr M, Rawat I, Nayunigari MK, Bux F (2017c) Cultivation of Chlorella sorokiniana and Scenedesmus obliquus in wastewater: fuzzy intelligence for evaluation of growth parameters and metabolites extraction. J Clean Prod 147:419–430

Harmelen T, Oonk H (2006) Microalgae biofixation processes. Application and potential contributions to greenhouse gas mitigation options. TNO Built Environment and Geosciences http://www.fluxfarm.com/uploads/3/1/6/8/3168871/biofixation.pdf. Accessed on 31 July 2012

Hoffman Y, Aflalo C, Zarka A, Gutman J, James TY, Boussiba S (2008) Isolation and characterization of a novel chytrid species (phylum Blastocladiomycota), parasitic on the green alga Haematococcus. Mycol Res 112(1):70–81

Hoffmann JP (1998) Wastewater treatment with suspended and nonsuspended algae. J Phycol 34:757–763

Ip SY, Bridger JS, Chin CT, Martin WRB, Raper WGC (1982) Algal growth in primary settled sewage – the effects of five key variables. Water Res 16:621–632

Johnson MB, Wen ZY (2010) Development of an attached microalgal growth system for biofuel production. Appl Microbiol Biotechnol 85:525–534

Kagami M, Bruin AD, Ibelings B, Donk EV (2007) Parasitic chytrids: their effects on phytoplankton communities and food-web dynamics. Hydrobiologia 578:113–129

Khan SA, Rashmi HMZ, Prasad S, Banerjee UC (2009) Prospects of biodiesel production from microalgae in India. Renew Sust Energ Rev 13(9):2361–2372

Khoo HH, Sharratt PN, Das P, Balasubramanian RK, Naraharisetti PK, Shaik S (2011) Life cycle energy and CO_2 analysis of microalgae-to-biodiesel: preliminary results and comparisons. Bioresour Technol 102(10):5800–5807

Kim MK, Park JW, Park CS, Kim SJ, Jeune KH, Chang MU, Acreman J (2007) Enhanced production of Scenedesmus spp. (green microalgae) using a new medium containing fermented swine wastewater. Bioresour Technol 98:2220–2228

Kong QX, Li L, Martinez B, Chen P, Ruan R (2010) Culture of microalgae Chlamydomonas reinhardtii in wastewater for biomass feedstock production. Appl Biochem Biotechnol 160:9–18

Konig A, Pearson HW, Silva SA (1987) Ammonia toxicity to algal growth in waste stabilization ponds. Water Sci Technol 19:115–122

Kovacevic V, Wesseler J (2010) Cost-effectiveness analysis of algae energy production in the EU. Energy Policy 38(10):5749–5757

Kumar M, Sharma MK, Kumar A (2005) Spirulina fusiformis: a food supplement against mercury induced hepatic toxicity. J Health Sci 51:424–430

Kumar A, Ergas S, Yuan X, Sahu A, Zhang Q, Dewulf J, Malcata FX, Langenhove HV (2010) Enhanced CO_2 fixation and biofuel production via microalgae: recent developments and future directions. Trends Biotechnol 28:371–380

Lam MK, Lee KT (2012) Microalgae biofuels: a critical review of issues, problems and the way forward. Biotechnol Adv 30(3):673–690
Lavoie A, de la Noue J (1985) Hyperconcentrated cultures of Scenedesmus obliquus. A new approach for wastewater biological tertiary treatment. Water Res 19:1437–1442
Li Y, Chen YF, Chen P, Min M, Zhou W, Lanca BM, Zhu J, Ruan R (2011) Characterization of a microalga chlorella sp. well adapted to highly concentrated municipal wastewater for nutrient removal and biodiesel production. Bioresour Technol 102(8):5138–5144
Lohrey C, Kochergin V (2012) Biodiesel production from microalgae: co-location with sugar mills. Bioresour Technol 108:76–82
Louime C, Marshall RW, Vasanthaiah HKN, Onokpise O (2012) Genomics and potential bioenergy applications in the developing world. Genomics Appl Dev World 6:263–272
Lundquist TJ, Woertz IC, Quinn NWT, Benemann JR (2010). A realistic technology and engineering assessment of algae biofuel production. Ener Biosci Inst Berkeley California: 1–178
Martinez ME, Sanchez S, Jimenez JM, Yousfi FE, Munoz L (2000) Nitrogen and phosphorus removal from urban wastewater by the microalga Scenedesmus obliquus. Bioresour Technol 73:263–272
Mata TM, Martins AA, Caetano NS (2010) Microalgae for biodiesel production and other applications: a review. Renew Sust Energ Rev 14(1):217–232
McGriff CE Jr, McKinney RE (1972) The removal of nutrients and organics by activated algae. Water Res 6(10):1155–1164
Melo JJD, Câmara AS (1994) Models for the optimization of regional wastewater treatment systems. Eur J Oper Res 73(1):1–16
Metcalf, Eddy I (1991) Wastewater engineering: treatment, disposal, and reuse, 4th edn. McGraw-Hill, Inc, New York
Mohamed M, Ahmed T, El-Gohary F (2011) Use of down-flow hanging sponge (DHS) reactor as a promising post-treatment system for municipal wastewater. Chem Eng J 168(2):535–543
Molina GE, Belarbi EH, Fernández FGA, Medina AR, Chisti Y (2003) Recovery of microalgal biomass and metabolites: process options and economics. J Biotechnol Adv 20:491–515
Moreno-Garrido I (2008) Microalgae immobilization: current techniques and uses. Bioresour Technol 99(10):3949–3964
Olguin EJ (2012) Dual purpose microalgae-bacteria-based systems that treat wastewater and produce biodiesel and chemical products within a biorefinery. Biotechnol Adv 30(5):1031–1046
Orpez R, Martinez ME, Hodaifa G, El Yousfi F, Jbari N, Sanchez S (2009) Growth of the microalga Botryococcus braunii in secondarily treated sewage. Desalination 246:625–630
Oswald WJ (1988) Micro-algae and wastewater treatment. In: Borowitzka MA, Borowitzka LJ (eds) Micro-algal biotechnology. Cambridge University Press, New York, pp 305–328
Oswald WJ, Gotaas HB, Golueke CG, Kellen WR (1957) Algae in waste treatment. Sewage Ind Wastes 29:437–457
Oswald WJ, Golueke CG, Gee HK (1959) Wastewater reclamation through production of algae. Contribution 22, Water Resources Center, University of California, Berkley
Park BK, Craggs RJ, Shilton AN (2011) Wastewater treatment high rate algal ponds for biofuel production. Bioresour Technol 102:35–42
Pittman JK, Dean AP, Osundeko O (2011) The potential of sustainable algal biofuel production using wastewater resources. Bioresour Technol 102(1):17–25
Rawat I, Kumar RR, Mutanda T, Bux F (2011) Dual role of microalgae: Phycoremediation of domestic wastewater and biomass production for sustainable biofuels production. Appl Energy 88(10):3411–3424
Rawat I, Gupta SK, Shriwastav A, Singh P, Kumari S, Bux F (2016) Microalgae applications in wastewater treatment. In: Algae biotechnology. Springer, Cham, pp 249–268
Razon LF, Tan RR (2011) Net energy analysis of the production of biodiesel and biogas from the microalgae: Haematococcus pluvialis and Nannochloropsis. Appl Energy 88(10):3507–3514
Riaño B, Molinuevo B, MC G-G (2011) Treatment of fish processing wastewater with microalgae-containing microbiota. Bioresour Technol 102(23):10829–10833

Riley H (2002) Effects of algal fibre and perlite on physical properties of various soils and on potato nutrition and quality on a gravelly loam soil in southern Norway. Acta Agric Scandin B – Plant Soil Sci 52:86–95

Robins JG (2010) Cool-season grasses produce more total biomass across the growing season than do warm-season grasses when managed with an applied irrigation gradient. Biomass Bioenergy 34(4):500–505

Rösch C, Skarka J, Patyk A (2009) Microalgae – opportunities and challenges of an innovative energy source. In: 17th European biomass conference and exhibition, Hamburg, Germany; June 29–July 3

Ruiz-Marin A, Mendoza-Espinosa LG, Stephenson T (2010) Growth and nutrient removal in free and immobilized green algae in batch and semi-continuous cultures treating real wastewater. Bioresour Technol 101:58–64

Sahoo NK, Gupta SK, Rawat I, Ansari FA, Singh P, Naik SN, Bux F (2017) Sustainable dewatering and drying of self-flocculating microalgae and study of cake properties. J Clean Prod 159:248–256

Schenk P, Thomas-Hall S, Stephens E, Marx U, Mussgnug J, Posten C, Kruse O, Hankamer B (2008) Second generation biofuels: high-efficiency microalgae for biodiesel production. Bioenergy Res 1:20–43

Shiklomanov IA (1999) World water resources: modern assessment and outlook for the 21st century, (Summary of world water resources at the beginning of the 21st century, prepared in the framework of the IHP UNESCO). Federal Service of Russia for Hydrometeorology & Environment Monitoring, State Hydrological Institute, St. Petersburg

Shriwastav A, Gupta SK, Ansari FA, Rawat I, Bux F (2014) Adaptability of growth and nutrient uptake potential of Chlorella sorokiniana with variable nutrient loading. Bioresour Technol 174:60–66

Sialve B, Bernet N, Bernard O (2009) Anaerobic digestion of microalgae as a necessary step to make microalgal biodiesel sustainable. Biotechnol Adv 27(4):409–416

Singh J, Gu S (2010) Commercialization potential of microalgae for biofuels production. Renew Sust Energ Rev 14(9):2596–2610

Sonune A, Ghate R (2004) Developments in wastewater treatment methods. Desalination 167:55–63

Spolaore P, Joannis-Cassan C, Duran E, Isambert A (2006) Commercial applications of microalgae. J Biosci Bioeng 101(2):87–96

Sujatha M, Reddy TP, Mahasi MJ (2008) Role of biotechnological interventions in the improvement of castor (Ricinus communis L.) and Jatropha curcas L. Biotechnol Adv 26(5):424–435

Sukenik A, Shelaf G (1984) Algal autoflocculation-verification and proposed mechanism. Biotechnol Bioeng 26:142

Tam NFY, Wong YS (1989) Wastewater nutrient removal by Chlorella and Scenedesmus sp. Environ Pollut 58:19–34

Ternes TA (1998) Occurrence of drugs in German sewage treatment plants and rivers. Water Res 32 (11):3245–3260

Travieso L, Benitez F, Dupeiron R (1992) Sewage treatment using immobilized microalgae. Bioresour Technol 40:183–187

U.S. DOE (2010) National algal biofuels technology roadmap. Report no.: DOE/EE-0332, U.S. Department of Energy, Office of Energy Efficiency and Renew Energy, Biomass Program

Uduman N, Qi Y, Danquah MK, Forde GM, Hoadley A (2010) Dewatering of microalgal cultures: a major bottleneck to algae-based fuels. J Renew Sust Energ 2:012701

Wang LA, Min M, Li YC, Chen P, Chen YF, Liu YH (2010) Cultivation of green algae Chlorella sp. in different wastewaters from municipal wastewater treatment plant. Appl Biochem Biotechnol 162:1174–1186

Wei Y, Van Houten RT, Borger AR, Eikelboom DH, Fan Y (2003) Minimization of excess sludge production for biological wastewater treatment. Water Res 37(18):4453–4467

Wilde EW, Benemann JR (1993) Bioremoval of heavy metals by the use of microalgae. Biotechnol Adv 11(4):781–812

Wilkie AC, Mulbry WW (2002) Recovery of dairy manure nutrients by benthic freshwater algae. Bioresour Technol 84:81–91

Woertz I, Feffer A, Lundquist T, Nelson Y (2009) Algae grown on dairy and municipal wastewater for simultaneous nutrient removal and lipid production for biofuel feedstock. J Environ Eng 135:1115–1122

Wrigley TJ, Toerien DF (1990) Limnological aspects of small sewage ponds. Water Res 24:83–90

Wu LF, Chen PC, Huang AP, Lee CM (2012) The feasibility of biodiesel production by microalgae using industrial wastewater. Bioresour Technol 113:14–18

Yamaguchi K (1997) Recent advances in microalgal bioscience in Japan, with special reference to utilization of biomass and metabolites: a review. J Appl Phycol 8:487–502

Yang J, Xu M, Zhang X, Hu Q, Sommerfeld M, Chen Y (2011) Life-cycle analysis on biodiesel production from microalgae: water footprint and nutrients balance. Bioresour Technol 102(1):159–165

Zhang ED, Wang B, Wang QH, Zhang SB, Zhao BD (2008) Ammonia-nitrogen and orthophosphate removal by immobilized Scenedesmus sp isolated from municipal wastewater for potential use in tertiary treatment. Bioresour Technol 99:3787–3793

Zhou W, Li Y, Min M, Hu B, Chen P, Ruan R (2011) Local bioprospecting for high-lipid producing microalgal strains to be grown on concentrated municipal wastewater for biofuel production. Bioresour Technol 102(13):6909–6919

Chapter 3
Risk of Metal Contamination in Agriculture Crops by Reuse of Wastewater: An Ecological and Human Health Risk Perspective

Sanjay Kumar Gupta, Sayanti Roy, Mayuri Chabukdhara, Jakir Hussain, and Manish Kumar

Abstract Agriculture sector is one of the major users of water resources. Due to limited availability of freshwater resources, domestic and industrial wastewater is being used in agriculture. Such water and wastewater contain varying number of micronutrients such as carbon and nitrogen as well as other toxic elements. Continuous irrigation with such type of water results overloading of these nutrients and some of the times pathogens, if not treated, in agricultural top soils. Heavy metals are nonbiodegradable and cumulative in nature. The accumulation and bioavailability of the metals depend on various environmental factors such as climatic conditions, temperature, rain pattern, and physicochemical properties of the soil, i.e., organic contents, pH, cationic exchange capacity, etc., which regulate accumulation of metals in soil and its bioavailability. Therefore, such toxic elements once enter in the food chain, get accumulated in various trophic levels, and exert undesirable effects to the flora and fauna. The major concern is its accumulation of toxic metal in agricultural crops from the wastewater-irrigated topsoil and associated health risk to the end-use consumers. Other than ingestion, there are various other routs of heavy metal exposure to the human beings. Therefore, for effective use and management of the wastewater in agriculture, periodic monitoring and risk assessment of heavy metal contamination are very important. This book chapter deals with the

S. K. Gupta (✉) · S. Roy
Department of Civil Engineering, Indian Institute of Technology, Delhi, India

M. Chabukdhara
Institute for Water and Wastewater Technologies, Durban University of Technology, Durban, South Africa

J. Hussain
Central Water Commission, Government of India, Delhi, India

M. Kumar
Department of Biology, Texas State University, San Marcos, TX, USA

R. P. Singh et al. (eds.), *Water Conservation, Recycling and Reuse: Issues and Challenges*, https://doi.org/10.1007/978-981-13-3179-4_3

comprehensive evaluation of pros and cons of reuse of wastewater in agricultural with special reference to heavy metal contamination and associated human health risk.

Keywords Heavy metal contamination · Agricultural soil · Wastewater · Fruits · Human health risk

3.1 Introduction

The ever-increasing population has led to exponential increase in growth of urban areas and industries. In the process, there is an increase in consumption of resources and generation of wastes, which have reduced the environment's assimilation capacity and lead to accumulation of the wastes in the environment. Thus, the wastes, which comprise of various compositions of several noxious substances, are released into the environment. Considering the wastewater being released from several sectors of human society, it is supposed to contain organic matter, deposition of nutrients, pathogens, and innumerable heavy metals (Ping et al. 2011; Khan et al. 2013). The surge in the water scarcity is pushing the envelope of wastewaters for irrigation purpose (Khan et al. 2013) and, often, without prior treatment. It has been estimated that 20 million hectares of worldwide agricultural land are irrigated with wastewater (Wuana and Okieimen 2011). Another major problem particularly in developing countries is the improper discharge of industrial effluents into adjacent water bodies or into groundwater by reverse boring. Water from these sources is also used for irrigation which introduces the contaminants from the water into the agricultural soil. Due to this long-term and continuous use of the wastewaters in agricultural fields, the soil saturates with the heavy metals and leaches them into soil solution (Sharma et al. 2007). These soluble and bioavailable toxic heavy metals in the wastewater are absorbed by the crops, thereby, posing serious risks for contaminating the food chain and the environment. This eventually becomes another major concern, when the crops are food crops and are being consumed by humans in their daily diet, resulting in biomagnification. Besides, these heavy metals also tend to bind to the soil particles in the irrigated agricultural fields and pose threat when gets dislodged due to wind and suspends into the air to enter human body systems through inhalation exposure pathway. Another way this practice of reusing wastewater in agriculture results in heavy metal exposure to human beings is when the contaminated soil comes in contact to skin and adheres to it; however, absorption through this pathway is most likely to affect the farmers, since they are the ones coming in direct contact with the wastewater-irrigated soil. Apart from this, depending upon the soil properties and components, the heavy metals in soil solution may also leach down to the aquifers, if present, which contributes to the contamination of the groundwater. Further, governed by the type of aquifer and the underground topology, the groundwater is subjected to relocate to nearby areas. In that way, the ecosystem as well as humans is directly exposed to these heavy metals while utilizing the water for consumption. Since these numerous exposures and co-exposures of heavy metals produce a string of additive, antagonistic, or synergistic effects to human health (Wang and Bruce 2008; Tchounwou et al. 2012), including effect on circulatory, nervous, endocrine, pulmonary, renal, skeletal, enzymatic, and immune systems (Żukowska and Biziuk 2008; Zhang et al. 2012), it has become a matter of concern. Pertaining to this, several studies have been reported on the input of heavy metals in soil and edible plants as a result of irrigation using wastewater and the ecological and health risks associated with this. Considering the water availability of different countries, the practice of reusing wastewater in agricultural sector varies. Moreover, the concept of water footprint of a country (Hoekstra and Mekonnen 2012) also plays a role that determines the extent of use of wastewater in the country. Water footprint is the amount of water consumed for the production of commodities in a country. According to a report, China, India, and the United States contribute to the largest water footprint (1207, 1182, and 1053 Gm^3 /year, respectively) within their territory. It has also been reported that the water footprint in agricultural sector occupies the maximum share within all the three countries. Among these three countries, largest blue water footprint (24%) has been reported to be in India, where irrigation in wheat cultivation requires the largest share, followed by irrigation of rice and sugarcane, i.e., 33%, 24%, and 16%, respectively. However, the water availability per capita of the country is projected to be decreasing as reported by the Ministry of Water Resources, GOI (2009). Therefore, this chapter focuses on heavy metal concentrations in the agricultural soil and the comprehensive assessment of ecological risk as well as human health risk related to the reuse of wastewater in agriculture in different regions of India.

3.2 Heavy Metal Contamination of Soil due to Reuse of Wastewater

The pollution due to heavy metals has become a major concern, as the metals tend to become persistent in the environment and find their way into the other components of the environment through several biological and physiological processes. Once the heavy metals enter the food chain, they bioaccumulate in the living tissues, that is, the concentrations of heavy metals within a biological organism increase over a long time, and along with their magnification to higher trophic levels, than that in the environment (Du et al. 2013). Moreover, even a very low concentration of most heavy metals is toxic, and often a carcinogenic effect is produced in humans (Dockery and Pope 1996; Willers et al. 2005). While certain heavy metals are known to serve as essential elements to plants and humans at trace amount, a rise above the threshold concentration leads to adverse impacts on the living systems. The significant presence of the toxic heavy metals has also been held responsible for the inhibition of natural biodegradation of organic pollutants (Maslin and Maier

2000). The major concern for heavy metals is, thus, attributed to their high-level toxicity, long biological residence time, solubility, and potential of bioaccumulation (Arora et al. 2008).

In wastewater, the most significant sources of heavy metals are industries including effluents from power plants; metallurgical, chemical, and inorganic pesticide manufacturing plants; automobiles; pigment and dyes; textile; tannery; electroplating; galvanizing; cement; paint and asbestos industries; etc. (Ahluwalia and Goyal 2007), along with mine wastewater containing tailings (Moore and Ramamoorthy 1984; Dudka and Adriano 1997; Navarro et al. 2008). Besides, domestic sources may include corrosion of sewerage pipe and plumbing equipment, laundry detergents, cosmetic ingredients, and preservatives (Aonghusa and Gray 2002; Sharma et al. 2007; Sahu et al. 2014). Other major sources of wastewater containing significant amount of heavy metals are laboratories of educational, scientific, and medical institutions, which disposed chemicals, antibiotics, cancer therapeutics, anti-inflammatory drugs, contraceptives, and other hormones (Hernando et al. 2006; Nikolaou et al. 2007). While certain sources like metallurgical, chemical, and electroplating industries, etc. release innumerable heavy metals, effluents from industries producing dyes, textile, and tannery mainly comprise of chromium, zinc, iron, calcium, etc. (APHA 1995). Zinc, cadmium, and copper have been reported to be significant in laundry wastewater (Aonghusa and Gray 2002), whereas chromium, nickel, and mercury concentration dominates in cosmetic products such as lipsticks and fairness creams (Sahu et al. 2014). In most countries including India, the irrigation of vegetable crops with domestic or industrial wastewater has become a regular practice (Gupta et al. 2008a, b; Garg et al. 2014; Singh et al. 2010). The heavy metals present in such type of wastewater get accumulated in the agricultural soil, from which it is passed and accumulated in the agricultural crops (Krishna and Govil 2005; Godson et al. 2002; Barman et al. 2000; Fazeli et al. 1991).

Singh et al. 2009 reported the following ranges of heavy metals (mg/mL), i.e., Cd (0.03–0.04), Cr (0.05–0.147), Cu (0.043–0.053), Zn (0.093–0.117), Pb (0.043–0.063), Mn (0.077–0.11), and Ni (0.02–0.05), in the wastewater used for irrigation in Dinapur and Lohta sites of Varanasi district in India. The higher ranges of Ni, Cr, Pb, and Mn at the Lohta site were attributed to the untreated industrial effluents discharged from several industries where these metals were used for making of metal alloys, metal plating, and coloring (Singh et al. 2009). Dyeing and paint industries in Varanasi contribute to high concentrations of heavy metals especially Cd, Cr, and Pb to wastewater (Sharma et al. 2007), whereas Ni and Pb are added by battery- and metal-plating industries (Sharma et al. 2006, 2007). It has been suggested that substantial reduction of heavy metal concentration is possible by screening of sewage and other types of wastewater (Panicker 1995). However, certain metals in traces remain in screened wastewater which may get accumulated in the soil and agricultural crops over long-term use and can cause phytotoxicity (Ghafoor et al. 1999). Once the toxic metals get accumulated in plants, they induce physiological stress and subsequent changes in biochemical composition of the plants (Gupta et al. 2010). Various studies have reported decreased chlorophyll concentrations in vascular plants due to the physiological stress of toxic metals (Monni et al. 2001; Patsikka et al. 2002).

3.3 Heavy Metal Contamination of Food Crops Irrigated with Wastewater

The reuse of wastewater containing innumerable heavy metals for irrigation exposes them into the soil, then to the crops, and finally to the consumers. However, the dissolution, uptake, and bioaccumulation of heavy metals in edible crops, cereals, and vegetables are governed by various factors (Fig. 3.1). This may include climatic conditions; the nature and composition of soil, i.e., concentrations of organic matter and pH; and the presence and concentrations of various anions and the concentrations and solubility sequences of heavy metals in soil, assimilative capacity of the soil, atmospheric depositions, the plant species, and the degree of maturity of the plants during the harvesting period (Lake et al. 1984; Scott et al. 1996; Voutsa et al. 1996; Kafka and Kuras 1997). The extent of heavy metal binding to the soil particles is reliant on the pH and ion properties, as the binding forces of heavy metals are inversely related to the soil pH. Moreover, the metal ions with higher charges have higher tendency to adhere to soil particles than the ions with less charges (Dobrzanski and Zawadzki 1993). The sorption of heavy metals in the soil is also established to be influenced by the presence of humus in the soil which has a significant role in metal adsorption (Stevenson 1992). According to Schulten and Leinweber (2000), the heavy metal content reduces from clay to silt due to high surface area of clay, and soil containing organic matter and higher clay fractions can have higher concentration of heavy metals. Soil pH influences the solubility of the heavy metals, and it decreases with increase of soil pH (i.e., alkaline range). The increase in the organic contents of soil which facilitates more binding of metals to the soil therefore increases the metal solubility and adsorption in soil (Hough et al.

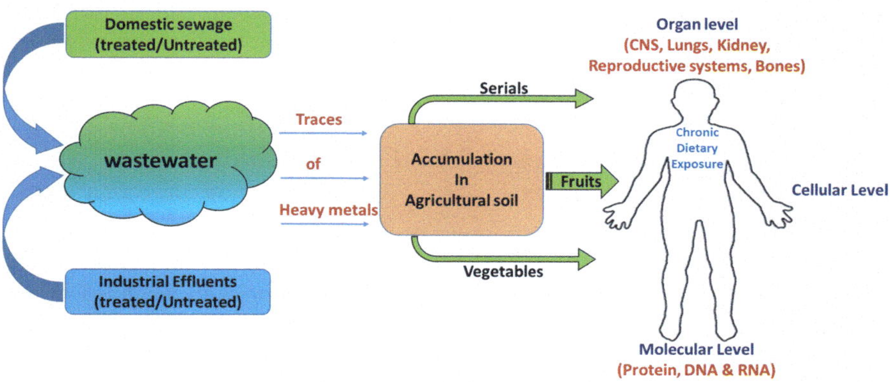

Fig. 3.1 Dietary exposure of heavy metals from agricultural products contaminated by wastewater

2003). The pH also affects the bioavailability of metals present in the wastewater-irrigated soils. It has been established that the hydrogen ions have a greater affinity for competing with metal ions; therefore, at lower pH metal ions are easily released from the soil and become available for the plant uptake (Singh et al. 2009). Rupa et al. 2003 reported the increased uptake of heavy metals in wheat plants at comparably higher levels of organic matters. Similarly, pH and organic contents enhance the solubility and availability of heavy metals and facilitate higher availability of metal ions to the plant from the soil (Ram et al. 2006).

Several studies suggest that heavy metal accumulation in plants varies in different species and different parts of plants. Mostly, leafy edible parts of vegetables are reported to contain high heavy metal concentrations than fruit crops or grain crops (Chopra and Pathak 2015). However, the leaves and roots of crops show a greater accumulation of heavy metals than the storage organs or fruits (Jinadasa et al. 1997; Lehoczky et al. 1998; Mapanda et al. 2005; Sharma and Agrawal 2006). Nevertheless, researchers (Davis et al. 1994; Traina and Laperche 1999; Violante et al. 2010) have suggested that bioavailability of heavy metals is majorly controlled by the metal species, affinity of the heavy metal for plant roots, the existing forms of metals in soil, and the properties of soil, instead of the total heavy metal concentrations. Even minute concentrations of heavy metals get adsorbed on soil particles and are retained for a long time as colloidal association (Sauve et al. 2000).

Through a study on *Beta vulgaris* receiving wastewater for irrigation, Sharma et al. (2007) have reported that concentrations of metals such as cadmium, zinc, mercury, and chromium were higher during summer, but in winter season, higher concentrations were measured in case of metals like copper, lead, and nickel. In another study by Sharma et al. (2008), the highest concentration of lead was observed in *Beta vulgaris*, whereas zinc and copper were highest in *Brassica oleracea*, and cadmium was highest in *Abelmoschus esculentus* and *Beta oleracea*. Rice grain has been reported to accrue very high concentrations of lead and cadmium (Zhuang et al. 2009). Fiber crops like flax and cotton, when cultivated in heavily contaminated soils, have also been detected to take up heavy metals (Angelova et al. 2004). It is evident that different species accumulate heavy metals. Various studies indicated that the duration of use of wastewater also correlated with heavy metal accumulation in vegetables (Sinha et al. 2005; Sharma et al. 2006, 2007).

3.4 Ecological Risk of Heavy Metal Contamination

In nature, soil acts as sink and filters for the heavy metals, by the process of binding and immobilization. Nevertheless, the continuous input of heavy metals in the soil by the means of wastewater irrigation alters the soil's capacity to retain the heavy metals, hence, leading to the consequent release of heavy metals (Sharma et al. 2007). This may also lead to imbalance in the essential trace metal composition of the soil, which is likely to further adverse ecological conditions in the soil microenvironment as well as to the plants growing on it. Increase in heavy metal concentration in soil results in increase in uptake of the metals by crops growing on the soil (Whatmuff 2002 and McBride 2003), which further creates and stimulates stress conditions in plants by impeding physiological and metabolic functioning of the plants. In plants, heavy metals can cause structural disorganization of organelles, disrupt cell membranes, and retard normal growth rate (Long et al. 2003; Zhang et al. 2002; Chien and Kao 2000; Kimbrough et al. 1999). Apart from these, the elevated concentration of heavy metals in soil also produces toxic effect on the soil microorganisms. Studies have reported that the heavy metals tend to alter the microbial processes in the soil ecosystem, which can be attributed to the physiological stress caused to the soil microbes by the activities of heavy metals. The metals in soil are also responsible for hindering enzymatic and metabolic activities of the soil microbiota (Giller et al. 1998; Wang et al. 2007). This further disrupts the microbial activities in the soil that are essentials for plants such as nitrogen fixation, other nutrient cycles, etc. Apart from directly posing enormous threat to the soil quality, and crops and vegetables cultivated in the contaminated soil, the heavy metals may get introduced to nearby surface water bodies through runoffs and threatens the aquatic ecosystem (Gupta et al. 2014). Fishes have been reported to accumulate significant quantities of toxic contaminants in their tissues on exposure to polluted aquatic bodies (Lewis et al. 2007; Yılmaz 2010; Chabukdhara and Nema 2012, 2013; Leung et al. 2014; Gupta et al. 2015). Therefore, it is well conformed from several studies that heavy metal contamination in soil initiates the interaction of the heavy metals with the other components of the ecosystem. When the contamination level and the load of heavy metal pollution exceed the threshold, limits pose risk to the environment.

3.4.1 *Quantification and Assessment of Potential Ecological Risk of Heavy Metal Contaminations in Soils*

Based on reported heavy metal concentration in the soil of the different regions of India, a cumulative assessment has been done to determine the level of pollution and associated ecological risk in particular due to reuse of wastewater for irrigation. The heavy metal concentration in the soil is equally important as that of their concentrations in the edible products because it is from the soil that the metals find their way into the plant tissues and then to the consumers.

3.4.1.1 Potential Ecological Risk

The potential ecological risk index (*RI*) proposed by Hakanson (1980) and Zhu et al. (2008) is one of the most common methods of the quantification of the potential ecological risk of the heavy metals, which can be calculated by contamination factor (C^i_f) and the "toxic-response" factor. The potential risk index can be obtained as:

$$E^i_r = T^i_r C^i_f \tag{3.1}$$

$$C^i_{f=Cn/Cnr} \tag{3.2}$$

$$C_{deg} = \sum C^i_f \tag{3.3}$$

$$RI = \sum E^i_r \tag{3.4}$$

where E^i_r is the potential ecological risk index of an individual metal, C^i_f is single-metal pollution factor, C_n is the concentration of the metal in samples, and C_{nr} is a background value for metal. The chemical compositions of the continental crust were used as the background values in this chapter which are 46,700 mg/kg for Fe, 95 mg/kg for Zn, 20 mg/kg for Pb, 68 mg/kg for Ni, 850 mg/kg for Mn, 45 mg/kg for Cu, 90 mg/kg for Cr, and 0.3 mg/kg for Cd (Turekian and Wedepohl 1961). Loska et al. (2004) classified the metal contamination levels as follows: low (Cif <1), moderate (1 $\leq$ Cif < 3), considerable (3 $\leq$ Cif < 6), and very high (6 $\leq$ Cif) contamination levels. The degree of contamination (C_{deg}) is the sum of contamination factors for all of the metals. Based on the value of C_{deg}, metal contamination levels are categorized as follows: low ($C_{deg} < 5$), moderate ($5 \leq C_{deg} < 10$), considerable ($10 \leq C_{deg} < 20$), and very high ($20 \leq C_{deg}$) degree of contamination (Duong and Lee, 2011). If the C_{deg} values exceeded 20, then necessary measures are required to reduce heavy metal contamination (Abdel-Latif and Saleh 2012). T^i_r denotes the "toxic-response" factor for heavy metals. The Tr values of Cu, Cr, Pb, Cd, Zn, Mn, and Ni are 5, 2, 5, 30, 1,1, and 5, respectively (Xu et al. 2008; Hakanson 1980). The scale of ecological risk can be categorized as follows: $E^i r < 40$, low risk; $40 \leq Er^i < 80$, moderate risk; $80 \leq Er^i < 160$, considerable risk; $160 \leq Er^i < 320$, high risk; and $Er^i \geq 320$, very high risk (Islam et al. 2015).

Based on the estimation, the degree of contamination and potential ecological risk index due to metal contamination in agricultural soils are presented in Table 3.1. As can be seen, most of the wastewater-irrigated sites showed very high degree of contamination due to heavy metals. The highest degree of contamination and ecological risk is found at Durgapur and Burdwan region of West Bengal (Gupta et al. 2008a, b) that were irrigated with wastewater, effluents, or effluent-contaminated water. Based on the study done in Delhi, where the major source of irrigation is groundwater, the soil showed least degree of contamination and risk (Kaur and Rani 2006). This indicates that the effluents or wastewater discharges are not safe for use in irrigation and these need proper treatment prior to disposal at different sites. Agricultural sites in Kanpur, Uttar Pradesh, and Delhi showed low risk (Sinha et al. 2006; Kaur and Rani 2006), Ghaziabad showed moderate risk (Chabukdhara et al. 2016), and Hyderabad showed considerable risk (Chary et al. 2008). All other sites showed high to very high risk.

Among metals, Pb and Cd showed the higher levels of contamination as compared to other metals in the wastewater-irrigated soil. As expected, the highest levels of ecological risks are also associated with Pb and Cd. While considering the degree of contamination to assess the contamination level in the affected soils of the

Table 3.1 Degree of contamination (C_{deg}) and ecological risk index (*RI*) due to heavy metals in Indian agricultural soils irrigated with wastewater

Agricultural soil site	C^i_f							C_{deg}	E^i_r							*RI*	References
	Cu	Cr	Pb	Cd	Zn	Mn	Ni		Cu	Cr	Pb	Cd	Zn	Mn	Ni		
Durgapur, West Bengal	0.00	0.61	44.80	20.00	0.00	0.10	0.00	**65.51**	0.00	1.22	224.00	600.00	0.00	0.10	0.00	**825.32**	Gupta et al. (2010)
Titagarh, West Bengal	2.00	1.65	6.52	102.40	2.29	0.00	1.52	**116.38**	10.00	3.30	32.61	3072.00	2.29	0.00	7.62	**3127.82**	Gupta et al. (2008a)
Ghaziabad, India	0.41	0.31	1.53	1.47	0.90	0.36	0.85	**5.83**	1.97	0.65	7.45	42.70	0.89	0.36	4.80	**58.82**	Chabukdhara et al. (2016)
Hyderabad, Andhra Pradesh	0.71	0.37	25.60	0.00	4.06	0.00	0.81	**31.55**	3.56	0.73	128.00	0.00	4.06	0.00	4.04	**140.40**	Chary et al. (2008)
Kanpur, Uttar Pradesh	1.58	1.87	0.00	0.00	3.23	0.37	0.62	**7.67**	7.89	3.75	0.00	0.00	3.23	0.37	3.08	**18.32**	Sinha et al. (2006)
Varanasi, Uttar Pradesh	0.71	1.05	0.89	8.61	0.79	0.19	0.21	**12.45**	3.55	2.09	4.45	258.33	0.79	0.19	1.07	**270.48**	Sharma et al. (2007)
Varanasi, Uttar Pradesh	0.47	0.21	1.07	10.40	0.61	0.00	0.35	**13.11**	2.35	0.43	5.35	312.00	0.61	0.00	1.74	**322.47**	Singh et al. (2010)
Varanasi, Uttar Pradesh	0.32	0.00	0.32	4.23	1.15	0.00	0.00	**6.03**	1.62	0.00	1.60	127.00	1.15	0.00	0.00	**131.37**	Singh et al. (2009)
Varanasi, Uttar Pradesh	0.32	0.20	0.90	14.22	0.26	0.49	0.20	**16.60**	1.59	0.41	4.51	426.60	0.26	0.49	1.02	**434.88**	Singh et al. (2009)
Burdwan, West Bengal	1.43	7.76	0.00	105.90	1.52	0.17	0.00	**116.78**	7.15	15.51	0.00	3177.00	1.52	0.17	0.00	**3201.35**	Gupta et al. (2008b)

(continued)

Table 3.1 (continued)

Agricultural soil site	C^i_f							C_{deg}	E^i_r							*RI*	References
Delhi	1.53	0.00	2.32	7.80	1.39	0.00	0.00	**13.04**	7.65	0.00	11.61	234.00	1.39	0.00	0.00	**254.64**	Singh and Kumar (2006)
Kanpur, Uttar Pradesh	1.97	0.21	0.76	6.97	0.30	0.00	1.19	**11.39**	9.85	0.43	3.78	209.00	0.30	0.00	5.93	**229.29**	Sanghi and Sasi (2001)
Varanasi, Uttar Pradesh	0.91	0.01	0.95	7.72	0.95	0.00	0.00	**10.54**	4.57	0.03	4.74	231.50	0.95	0.00	0.00	**241.79**	Rai and Tripathi (2008)
Varanasi, Uttar Pradesh	0.45	0.59	3.96	33.27	0.00	0.00	1.24	**39.51**	2.26	1.18	19.79	998.00	0.00	0.00	6.22	**1027.44**	Pandey et al. (2012)
Delhi	0.03	0.00	0.04	0.00	0.01	0.00	0.00	**0.09**	0.16	0.00	0.20	0.00	0.01	0.00	0.02	**0.40**	Kaur and Rani (2006)
Durgapur, India	0.00	0.61	44.80	20.00	0.00	0.10	0.00	**65.51**	0.00	1.22	224.00	600.00	0.00	0.10	0.00	**825.32**	Gupta et al. (2010)

country, there is a very high level of contamination estimated to have been persisting. In addition to the elevated contamination level, the pollution load index also depicts a deteriorating quality of the environment, where wastewater irrigation of agricultural soil is prevalent. However, the presence of cadmium poses a very high ecological risk in the environment where wastewater is being used for irrigation. In contrary to this, the ecological risk due to other metals considered for the evaluation has been determined to be low, except for lead, which has been detected to pose moderate ecological risk. Thereby, it is quite clear that the cumulative presence of cadmium in the soil as a result of wastewater irrigation in the studied areas of the country is beyond the safe limit and, therefore, requires intensive remediation measures.

3.5 Human Health Risk of Heavy Metals

In several studies, heavy metals have been accounted to interrupt the normal functioning of cellular organelles such as endoplasmic reticulum, lysosome, mitochondria, certain enzymes, nuclei, and cell membrane. This results in conformational changes in cellular structure and functions, leading to variation in cell cycle, apoptosis, and carcinogenic and teratogenic effects (Chang et al. 1996; Wang and Shi 2001; Beyersmann and Hartwig 2008). The production of reactive oxygen species (ROS) leads the subsequent oxidative stress in human bodies (Coman and Draghici 2011). Researchers have accounted that intake of heavy metal-contaminated food is capable of reducing the immunological defenses by depleting certain essential nutrients from the body. Several other health effects such as impaired fetus development, psychosociological behavior, gastrointestinal cancer, etc. are also associated with undernourishment. Various scientific literatures have established the disorders likely to occur in human bodies in relation to the dietary intake of food contaminated with heavy metals. While lead and cadmium have been held responsible for upper gastrointestinal cancer, breast cancer mortalities have been related to chromium intake (Iyengar and Nair 2000; Jarup 2003; Turkdogan et al. 2003; Pasha et al. 2010). Lead has also been recognized to cause encephalopathy in children, improper hemoglobin synthesis, renal infections, high blood pressure, and reproductive system disruption (Kanwal and Kumar 2011; Sanders et al. 2009; UNEP 2006; Fewtrell et al. 2003). The intake of excess cadmium through ingestion causes adverse health effects such as prostate and breast cancer; kidney, bone, and pregnancy disorders; as well as disturbances of male fertility (Kippler et al. 2012; Julin et al. 2012; Thomas et al. 2011; Godt et al. 2006). Cancer, fatigue, headache, skin rashes, dizziness, heart problems, and respiratory illness are also related with high concentration of nickel in food. However, the effects of heavy metals to human system are governed by age group, gender, prevalent health status of an individual, etc. Therefore, evaluation of health risk requires the consideration of these factors.

3.6 Exposure and Risk Assessment

An attempt has been made to summarize the health risk due to heavy metals via crop and vegetable consumption in India. The reported risk due to metals in crops and vegetables has been included as such in this chapter, and for others, the calculation of daily intake of metals (DIM) for adults was determined using the following equation:

$$\mathrm{DIM} = \left(C_{\mathrm{metal}}\ \mathrm{X}\ C_{\mathrm{factor}}\ \mathrm{X}\ D_{\mathrm{food\ intake}}/\mathrm{BW}_{\mathrm{averageweight}}\right.$$

where C_{metal} stands for the heavy metal concentrations in plants/crops (mg/kg), C_{factor} stands for conversion factor (0.085) (Rattan et al. 2005), $D_{food\ intake}$ stands for daily intake of vegetables, and $BW_{averageweight}$ stands for average body weight. The average daily intakes of food crops and vegetables for adult were considered to be 0.345 kg/person/day (Ge 1992; Wang et al. 2005).

3.6.1 Health Risk Index (HRI)

In this review, we assessed the possible potential health risk of heavy metals for India which was based on reported heavy metal data in crops and vegetables. The health risk index was computed for Cd, Pb, Cr, Cu, Zn, and Ni as the ratio of average daily intake of metals to oral reference dose through dietary intake of crops/vegetables as food following the method of Cui et al. (2004).

$$\mathrm{HRI} = \mathrm{DIM}/\mathrm{RfD}$$

where DIM represents the daily exposure of metals and RfD represents reference oral dose. RfD value for Cu, Cr, Pb, Cd, Zn, and Ni is 0.04, 1.5, 0.004, 0.001, 0.003, and 0.02 (mg/kg bw/day), respectively (USEPA 2001, 1997; USEPA IRIS 2006).

3.6.2 Noncarcinogenic Risk of Heavy Metals for Adults Through Contaminated Cereals and Vegetables

Exposure of heavy metals to the human being generally occurs through ingestion of the food crops and vegetables cultivated in the agricultural lands irrigated with wastewater. As summarized in Table 3.2, HRI for Zn, Pb, and Cd exceeded the safe limit for many vegetables. This clearly revealed that dietary intake of such metal-contaminated vegetables is likely to induce serious health hazard to the consumers, i.e., human beings, if such vegetable is regularly consumed due to chronic exposure. Some of the heavy metals such as lead and cadmium are potential carcinogens as these metals are associated with aetiology of a number of diseases.

Table 3.2 Noncarcinogenic risk due to heavy metal exposures through vegetables grown in soil irrigated with wastewater (Based on the available data)

Agricultural soil site	Common name	Botanical name	Heavy metals						References
			Cu	Cr	Pb	Cd	Zn	Ni	
Hyderabad, Andhra Pradesh	Spinach	*Beta vulgaris*	1.18E-03	1.01E-03	4.07E-01		**1.75**	8.39E-02	Chary et al. (2008)
Titagarh, West Bengal	Spinach	*Beta vulgaris*	4.52E-01	3.37E-02	**6.53**	**7.65**	**2.70E + 01**	**1.82**	Gupta et al. (2008a)
Sri Ganganagar, Rajasthan	Spinach	*Beta vulgaris*	2.16E-01				**5.79**		Arora et al. (2008)
Varanasi, Uttar Pradesh	Spinach	*Beta vulgaris*	2.66E-01		1.31E-01	5.14E-01	**6.71**		Sharma et al. (2009)
Hyderabad, Andhra Pradesh	Spinach	*Beta vulgaris*		4.71E-03	**1.96**	**3.36**			Srikanth and Reddy (1991)
Delhi	Spinach	*Beta vulgaris*	3.11E-01		6.09E-01	**2.21**	**1.57E + 01**		Singh and Kumar (2006)
Durgapur, West Bengal	Spinach	*Beta vulgaris*	7.75E-01	3.60E-03	**6.36**	**2.62**	**2.57E + 01**	2.89E-01	Kisku et al. (2000)
Titagarh, West Bengal	Spinach	*Beta vulgaris*	4.21E-01	3.35E-02	**6.25**	**6.80**	**2.59E + 01**	**1.80**	Gupta et al. (2012)
Ludhiana, Punjab	Spinach	*Beta vulgaris*		2.36E-03	7.59E-01	2.36E-01		2.21E-01	Dheri and Brar (2007)
Varanasi, Uttar Pradesh	Spinach	*Beta vulgaris*	**1.04**		2.36E-01	**3.54**	5.40E-03		Sharma et al. (2008)
Varanasi, Uttar Pradesh	Spinach	*Beta vulgaris*	2.40E-02	2.40E-04	**2.91**	**4.62**	5.10E-03	**1.31**	Singh et al. (2010)
Varanasi, Uttar Pradesh	Okra	*Abelmoschus esculentus*	**2.68**		5.08E-01	**9.07**	1.86E-02		Sharma et al. (2008)
Hyderabad, Andhra Pradesh	Okra	*Abelmoschus esculentus*	7.87E-03	4.90E-04	4.72E-01		6.47E-01	6.30E-02	Chary et al. (2008)

(continued)

Table 3.2 (continued)

Agricultural soil site	Common name	Botanical name	Heavy metals						References
Varanasi, Uttar Pradesh	Okra	*Abelmoschus esculentus*	2.11E-01		1.15E-01	4.72E-01	**6.07**		Sharma et al. (2009)
Varanasi, Uttar Pradesh	Okra	*Abelmoschus esculentus*	6.00E-03	2.90E-04	**1.52**	**5.08**	5.70E-03	4.3E-01	Singh et al. (2010)
Varanasi, Uttar Pradesh	Cauliflower	*Brassica oleracea*	**7.49**		**1.33**	**2.28**	4.36E-02		Sharma et al. (2008)
Titagarh, West Bengal	Cauliflower	*Brassica oleracea*	2.00E-01	3.01E-02	**3.89**	**6.54**	**1.59E + 01**	**1.55**	Gupta et al. (2012)
Titagarh, West Bengal	Cauliflower	*Brassica oleracea*	2.05E-01	3.04E-02	**4.07**	**7.24**	**1.69E + 01**	**1.55**	Gupta et al. (2008a)
Sri Ganganagar, Rajasthan	Cauliflower	*Brassica oleracea*	6.86E-02				**7.03**		Arora et al. (2008)
Varanasi, Uttar Pradesh	Cauliflower	*Brassica oleracea*	2.18E-01	0.00E + 00	1.34E-01	6.61E-01	**9.01**		Sharma et al. (2009)
Varanasi, Uttar Pradesh	Cauliflower	*Brassica oleracea*	3.10E-03	2.00E-04	**7.49**	**3.88**	7.40E-03	**1.8**	Singh et al. (2010)
Varanasi, Uttar Pradesh	Amaranthus	*A. retroflexus*	1.60E-02	4.20E-04	**4.1**	**5.32**	3.40E-03	**1.61**	Singh et al. (2010)
Hyderabad, Andhra Pradesh	Amaranthus	*A. retroflexus*	1.84E-02	8.39E-04	3.80E-01		**1.40**	8.13E-02	Chary et al. (2008)
Hyderabad, Andhra Pradesh	Amaranthus	*A. retroflexus*		3.54E-03	**1.60**	5.77E-01			Srikanth and Reddy (1991)
Durgapur, West Bengal	Amaranthus	*Amaranthus viridis*	6.64E-01	3.15E-03	**7.58**	**1.57**	**2.55E + 01**	2.96E-01	Kisku et al. (2000)
Varanasi, Uttar Pradesh	Brinjal	*Solanum melongena*	8.00E-03	8.00E-05	**1.11**	**1.48**	1.10E-03	4.7E-01	Singh et al. (2010)

Hyderabad, Andhra Pradesh	Brinjal	*Solanum melongena*	9.18E-03	3.85E-04	3.93E-01		7.87E-01	8.13E-02	Chary et al. (2008)
Sri Ganganagar, Rajasthan	Brinjal	*Solanum melongena*	1.34E-01				**3.93**		Arora et al. (2008)
Durgapur, West Bengal	Brinjal	*Solanum melongena*	7.11E-01	3.81E-03	**8.60**	**2.47**	**2.43E + 01**	2.94E-01	Kisku et al. (2000)
Varanasi, Uttar Pradesh	Tomato	*Solanum lycopersicum*	5.00E-03	5.00E-05	4.3E-01	4.9E-01	6.00E-04	1.8E-01	Singh et al. (2010)
Varanasi, Uttar Pradesh	Bottle gourd	*Lagenaria siceraria*	6.00E-03	2.50E-04	8.2E-01	**1.07**	8.00E-04	3.7E-01	Singh et al. (2010)
Varanasi, Uttar Pradesh	Sponge gourd	*Luffa aegyptiaca*	9.00E-03	2.30E-04	9.5E-01	**2.1**	1.20E-03	5.4E-01	Singh et al. (2010)
Varanasi, Uttar Pradesh	Bitter gourd	*Momordica charantia*	3.00E-03	3.00E-04	**1.09**	**1.61**	1.00E-03	3.9E-01	Singh et al. (2010)
Varanasi, Uttar Pradesh	Pumpkin	*Cucurbita argyrosperma*	4.00E-03	7.00E-05	9.6E-01	**1.14**	1.00E-03	3.6E-01	Singh et al. (2010)
Varanasi, Uttar Pradesh	Pointed gourd	*Luffa acutangula*	4.00E-03	1.30E-04	3.1E-01	7.6E-01	5.00E-04	2.1E-01	Singh et al. (2010)
Varanasi, Uttar Pradesh	Radish	*Raphanus raphanistrum*	4.00E-03	4.00E-04	4.1E-01	4.9E-01	7.00E-04	2.1E-01	Singh et al. (2010)
Varanasi, Uttar Pradesh	Wheat	*Triticum aestivum*	5.16E-02	2.97E-04	**4.37**	**5.86**	4.81E-03	**2.4**	Singh et al. (2010)
Varanasi, Uttar Pradesh	Rice	*Oryza sativa*	4.39E-02	2.17E-04	**6.83**	**9.15**	8.40E-03	**1.32**	Singh et al. (2010)
Hyderabad, Andhra Pradesh	Cabbage	*Brassica oleracea*		2.17E-03	9.86E-01	**1.51**			Srikanth and Reddy (1991)
Durgapur, West Bengal	Cabbage	*Brassica oleracea*	5.72E-01	4.13E-03	**7.9**	**2.99**	**2.47E + 01**	2.12E-01	Kisku et al. (2000)
Titagarh, West Bengal	Radish	*Raphanus raphanistrum*	3.68E-01	2.73E-02	**7.56**	**9.33**	**2.43E + 01**	**1.64**	Gupta et al. (2008a)

(continued)

Table 3.2 (continued)

Agricultural soil site	Common name	Botanical name	Heavy metals						References
Sri Ganganagar, Rajasthan	Radish	*Raphanus raphanistrum*	7.82E-02				**3.93**		Arora et al. (2008)
Titagarh, West Bengal	Radish	*Raphanus raphanistrum*		2.87E-02	**2.89**	**3.41**			Gupta et al. (2008a)
Durgapur, West Bengal	Radish	*Raphanus raphanistrum*	7.78E-01	2.94E-03	**5.59**	**3.04**	**2.26E + 01**	1.86E-01	Kisku et al. (2000)
Titagarh, West Bengal	Radish	*Raphanus raphanistrum*	3.54E-01	2.67E-02	**6.79**	**8.48**	2.39E + 01	**1.58**	Gupta et al. (2012)
Titagarh, West Bengal	Lettuce	*Lactuca sativa*	3.27E-01	2.15E-02	**4.58**	**7.02**	**2.99E + 01**	**1.37**	Gupta et al. (2008a)
Titagarh, West Bengal	Celery	*Apium graveolens*	2.70E-01	1.22E-02	**3.18**	**6.31**	**1.63E + 01**	**1.12**	Gupta et al. (2008a)
Hyderabad, Andhra Pradesh	Coriander	*Coriandrum sativum*	1.57E-02	7.34E-04	3.54E-01	–	9.44E-01	7.08E-02	Chary et al. (2008)
Titagarh, West Bengal	Coriander	*Coriandrum sativum*	3.29E-01	1.69E-02	**4.08**	**7.37**	**2.38E + 01**	**1.35**	Gupta et al. (2008a)
Sri Ganganagar, Rajasthan	Coriander	*Coriandrum sativum*	1.59E-01	–	–	–	**5.40**	–	Arora et al. (2008)
Durgapur, West Bengal	Coriander	*Coriandrum sativum*	8.41E-01	2.69E-03	**9.46**	**3.88**	**2.59E + 01**	2.26E-01	Kisku et al. (2000)
Titagarh, West Bengal	Coriander	*Coriandrum sativum*	3.15E-01	1.54E-02	**4.04**	**6.92**	**2.35E + 01**	**1.34**	Gupta et al. (2012)
Ludhiana, Punjab	Coriander	*Coriandrum sativum*	–	1.14E-03	5.80E-01	1.52E-01	–	6.51E-02	Dheri and Brar (2007)
Hyderabad, Andhra Pradesh	Mint	*Mentha spicata*	1.44E-02	4.90E-04	2.89E-01	–	**1.14**	6.30E-02	Chary et al. (2008)

Titagarh, West Bengal	Mint	*Mentha spicata*	3.44E-01	2.37E-02	**2.83**	**5.44**	**2.43E + 01**	**1.41**	Gupta et al. (2008a)
Sri Ganganagar, Rajasthan	Mint	*Mentha spicata*	1.67E-01	–	–	–	**7.87**	–	Arora et al. (2008)
Titagarh, West Bengal	Mint	*Mentha spicata*	3.28E-01	2.29E-02	**2.74**	**4.93**	**2.35E + 01**	**1.46**	Gupta et al. (2012)
Ludhiana, Punjab	Mint	*Mentha spicata*	–	1.24E-03	5.50E-01	1.36E-01	–	6.06E-02	Dheri and Brar (2007)
Durgapur, West Bengal	Mustard	*Brassica nigra*	–	1.26E-03	**4.07**	**3.41**	–	–	Gupta et al. (2010)
Durgapur, West Bengal	Taro	*Colocasia esculenta*	–	9.55E-03	**3.15**	**3.30**	–	–	Gupta et al. (2010)
Burdwan, West Bengal	Tomato	*Solanum lycopersicum*	–	9.90E-03	–	**4.15**	**6.73**	–	Gupta et al. (2008b)
Varanasi, Uttar Pradesh	Tomato	*Solanum lycopersicum*	1.34E-02	–	3.15E-02	–	–	–	Rai and Tripathi (2008)
Delhi	Okra	*Hibiscus esculentus*	2.69E-01	–	3.36E-01	**1.94**	**1.95E + 01**		Singh and Kumar (2006)
Durgapur, West Bengal	Okra	*Hibiscus esculentus*	6.98E-01	3.22E-03	**7.11**	**2.89**	–	2.31E-01	Kisku et al. (2000)
Titagarh, West Bengal	Chinese onion	*Allium chinense*	2.30E-01	1.62E-02	**4.49**	**6.03**	**2.19E + 01**	**1.24**	Gupta et al. (2008a)
Titagarh, West Bengal	Parsley	*Petroselinum crispum*	3.88E-01	2.67E-02	**4.10**	**6.48**	**1.95E + 01**	**1.47**	Gupta et al. (2008a)
Titagarh, West Bengal	Parsley	*Petroselinum crispum*	3.73E-01	2.64E-02	**3.99**	**6.44**	**1.87E + 01**	**1.47**	Gupta et al. (2012)
Sri Ganganagar, Rajasthan	Turnip	*Brassica rapa*	2.11E-01	–	–	–	**5.12**	–	Arora et al. (2008)
Varanasi, Uttar Pradesh	Turnips	*Brassica rapa*	3.67E-02	–	8.79E-01	9.97E-01	–	–	Rai and Tripathi (2008)

(continued)

Table 3.2 (continued)

Agricultural soil site	Common name	Botanical name	Heavy metals						References
Sri Ganganagar, Rajasthan	Carrot	*Daucus carota*	2.20E-01	–	–	–	8.11	–	Arora et al. (2008)
Varanasi, Uttar Pradesh	Carrot	*Daucus carota*	1.61E-02	–	7.11E-01	–	–	–	Rai and Tripathi (2008)
Varanasi, Uttar Pradesh	Fenugreek	*Trigonella foenum-graecum*	2.39E-01	–	–	–	**4.74**	–	Arora et al. (2008)

The list may include diseases of the nervous system, kidney, blood, cardiovascular, and many others (Jarup 2003; WHO/FAO 2007). Among rooted vegetables, radish showed higher risk due to Pb, Cd, and Zn, while among leafy vegetables, coriander, mint, cauliflower, parsley, and onion showed higher risk for consumption. Cu and Cr showed no risk for consumption of vegetables. HRI values for Ni also exceeded the safe limit (HRI > 1) in some vegetables, but it was comparatively lower than those due to Zn, Pb, and Cd.

3.7 Conclusions

The scarcity of precious freshwater and groundwater initiated the search of alternative water resources for the agriculture crops. Recycling and reuse of wastewater seem the most suitable option among all others for the sustainable management of water resources. Reuse of domestic or treated industrial wastewater for irrigation is often viewed as the most economic and environmental-friendly option. However, such wastewater contains variety of chemicals including traces of heavy metals. Prolonged use of such wastewater for the production of crop and vegetables leads gradual accumulation of trace elements in the agricultural lands. Various environmental and geochemical factors often moderate the leachability and bioavailability of such metals from the soil which accumulate in the growing crops. Therefore the chronic exposure of such metals to the humans through dietary intake poses serious threat. An assessment of human health risk of heavy metals through dietary intake was comprehensively assessed on the basis of available literature in Indian scenarios. The results showed that the metal concentrations in agricultural soils in India are categorized as high to very high risk in most of the wastewater-irrigated sites. Furthermore, in Indian scenario, the potential health risk index exceeded the safe limit (HRI > 1) for some of the metals such as Cd, Pb, Ni, and Zn. The observed health risk clearly indicated the poor quality of wastewater due to the presence of some of the heavy metals, and irrigation of crops/vegetables with such wastewater could pose serious health risk to the consumers. Immediate action of regulatory authorities is recommended to regulate the use of such type of wastewater contaminated with traces of the selected heavy metals for safeguarding the health of the general public.

References

Abdel-Latif NM, Saleh IA (2012) Heavy metals contamination in roadside dust along major roads and correlation with urbanization activities in Cairo, Egypt. J Am Sci 8:379–389

Ahluwalia SS, Goyal D (2007) Microbial and plant derived biomass for removal of heavy metals from wastewater. Bioresour Technol 98(12):2243–2257

American Public Health Association (APHA) (1995) Standard methods for the examination for water and wastewater, 19th edn. Byrd Prepess Springfield, Washington, DC

Angelova V, Ivanova R, Delibaltova V, Ivanov K (2004) Bio-accumulation and distribution of heavy metals in fiber crops (flax, cotton and hemp). Ind Crop Prod 19(3):197–205
Aonghusa CN, Gray NF (2002) Laundry detergents as a source of heavy metals in Irish domestic wastewater. J Environ Sci Health Part A 37(1):1–6
Arora M, Kiran B, Rani S, Rani A, Kaur B, Mittal N (2008) Heavy metal accumulation in vegetables irrigated with water from different sources. Food Chem 111:811–815
Barman SC, Sahu RK, Bhargava SK, Chaterjee C (2000) Distribution of heavy metals in wheat, mustard, and weed grown in field irrigated with industrial effluents. Bull Environ Contam Toxicol 64(4):489–496
Beyersmann D, Hartwig A (2008) Carcinogenic metal compounds: recent insight into molecular and cellular mechanisms. Arch Toxicol 82(8):493–512
Chabukdhara M, Nema AK (2012) Assessment of heavy metal contamination in Hindon River sediments: a chemometric and geochemical approach. Chemosphere 87(8):945–953
Chabukdhara M, Nema A (2013) Heavy metals assessment in urban soil around industrial clusters in Ghaziabad, India: probabilistic health risk approach. Ecotoxicol Environ Saf 87:57–64
Chabukdhara M, Munjal A, Nema AK, Gupta SK, Kaushal RK (2016) Heavy metal contamination in vegetables grown around peri-urban and urban-industrial clusters in Ghaziabad, India. Hum Ecol Risk Assess Int J 22(3):736–752
Chang LW, Magos L, Suzuki T (eds) (1996) Toxicology of metals. CRC Press, Boca Raton
Chary NS, Kamala CT, Raj DSS (2008) Assessing risk of heavy metals from consuming food grown on sewage irrigated soils and food chain transfer. Ecotoxicol Environ Saf 69(3):513–524
Chien HF, Kao CH (2000) Accumulation of ammonium in rice leaves in response to excess cadmium. Plant Sci 156:111–115
Chopra AK, Pathak C (2015) Accumulation of heavy metals in the vegetables grown in wastewater irrigated areas of Dehradun, India with reference to human health risk. Environ Monit Assess 187:445. https://doi.org/10.1007/s10661-015-4648-6
Coman G, Draghici C (2011) Heavy metals activity mechanisms at cellular level and possible action on children's bodies. Environmental heavy metal pollution and effects on child mental development. Springer, Dordrecht, pp 145–158
Cui YJ, Zhu YG, Zhai RH, Chen DY, Huang YZ, Qiu Y, Ling JN (2004) Transfer of metals from soil to vegetables in an area near a smelter in Nanning. China Environ Int 31:785–791
Davis A, Ruby MV, Bergstrom PD (1994) Factors controlling lead bioavailability in the Butte mining district, Montana, USA. Environ Geochem Health 3(4):147–157
Dheri GS, Brar MS (2007) Heavy-metal concentration of sewage-contaminated water and its impact on underground water, soil, and crop plants in alluvial soils of northwestern India. Commun Soil Sci Plant Anal 38:1353–1370
Dobrzanski B, Zawadzki S (1993) Gleboznawstwo – Praca Zbiorowa. Pwril, Warszawa
Dockery D, Pope A (1996) Epidemiology of acute health effects: summary of time-series studies. In: Wilson R, Spengler JD (eds) Particles in our air. Concentration and health effects. Harvard University Press, Cambridge, MA, pp 123–147
Du Y, Gao B, Zhou H, Ju X, Hao H, Yin S (2013) Health risk assessment of heavy metals in road dusts in urban parks of Beijing, China. Procedia Environ Sci 18:299–309
Dudka S, Adriano DC (1997) Environmental impacts of metal ore mining and processing: a review. J Environ Qual 26:590–602
Duong TT, Lee BK (2011) Determining contamination level of heavy metals in road dust from busy traffic areas with different characteristics. J Environ Manag 92:554–562
Environmental site assessment guideline (2009) DB11/T 656-2009
Fazeli MS, Sathyanarayan S, Satish PN, Mutanna L (1991) Effects of paper mill effluents on the accumulation of heavy metals in coconut trees near Nanjangud, Mysore District, Karnataka, India. Environ Geol Water Sci 17:47–50
Fewtrell L, Kaufmann R, Prüss-Üstün A (2003) Lead assessing the environmental burden of disease at national and local levels. World Health Organization, Protection of the Human Environment, Geneva

Garg VK, Yadav P, Mor S, Singh B, Pulhani V (2014) Heavy metals bioconcentration from soil to vegetables and assessment of health risk caused by their ingestion. Biol Trace Elem Res 157:256. https://doi.org/10.1007/s12011-014-9892-z

Ge KY (1992) The status of nutrient and meal of Chinese in the 1990s. Beijing People's Hygiene Press, 414e434

Ghafoor A, Ahmed S, Qadir M, Hussain SI, Murtaz G (1999) Formation and leaching of lead species from a sandy loam alluvial soil as related to pH and Cl:SO_4 ratio of leachates. J Agric Res 30:391–401

Giller KE, Witter E, McGrath SP (1998) Toxicity of heavy metals to microorganisms and microbial processes in agricultural soils: a review. Soil Biol Biochem 30:1389–1414

Godson RE, Ana E, Sridhar MKC (2002) Soil quality near a chemical fertilizer industry at Port Harcourt, Nigeria. AJEAM/RAGEE 4(2):50–57

Godt J, Scheidig F, Ch G–S, Esche V, Brandenburg P, Reich A, Groneberg DA (2006) The toxicity of cadmium and resulting hazards for hu-man health. J Occup Med Toxicol 1(1):22

Gupta N, Khan DK, Santra SC (2008a) An assessment of heavy metal contamination in vegetables grown in wastewater-irrigated areas of Titagarh, West Bengal, India. Bull Environ Contam Toxicol 80:115–118

Gupta S, Nayek S, Saha RN, Satpati S (2008b) Assessment of heavy metal accumulation in macrophyte, agricultural soil, and crop plants adjacent to discharge zone of sponge iron factory. Environ Geol 55:731–739

Gupta S, Satpati S, Nayek S, Garai D (2010) Effect of wastewater irrigation on vegetables in relation to bioaccumulation of heavy metals and biochemical changes. Environ Monit Assess 165:169–177

Gupta N, Khan DK, Santra SC (2012) Heavy metal accumulation in vegetables grown in a long-term wastewater-irrigated agricultural land of tropical India. Environ Monit Assess 184:6673–6682

Gupta SK, Chabukdhara M, Pandey PK, Singh J, Bux F (2014) Evaluation of potential ecological risk of metal contamination in the Gomti River: a biomonitoring approach. Ecotoxicol Environ Saf 110:49–55

Gupta SK, Chabukdhara M, Singh J, Bux F (2015) Evaluation and potential health hazard of selected metals in water, sediments, and fish from the Gomti River. Hum Ecol Risk Assess 21 (1):227–240

Hakanson L (1980) An ecological risk index for aquatic pollution control: A sedimentological approach. Water Res 14:975–1001

Hernando MD, Mezcua M, Fernandez-Alba AR, Barcelo D (2006) Environmental risk assessment of pharmaceutical residues in wastewater effluents, surface waters and sediments. Talanta 69:334–342

Hoekstra AY, Mekonnen MM (2012) The water footprint of humanity. Proc Natl Acad Sci 109 (9):3232–3237

Hough RL, Young SD, Crout NMJ (2003) Modelling of Cd, Cu, Ni, Pb, and Zn uptake, by winter wheat and forage maize, from a sewage disposal farm. Soil Use Manag 19:19–27

Islam MS, Ahmed MK, Habibullah-Al-Mamun M, Masunaga S (2015) Potential ecological risk of hazardous elements in different land-use urban soils of Bangladesh. Sci Total Environ 512–513:94–102

Iyengar V, Nair P (2000) Global outlook on nutrition and the environment: meeting the challenges of the next millennium. Sci Total Environ 249:331–346

Jarup L (2003) Hazards of heavy metal contamination. Br Med Bull 68:167–182

Jinadasa KBPN, Milham PJ, Hawkins CA, Cornish PSD, Williams PA, Kaldor CJ, Conroy JP (1997) Survey of Cd levels in vegetables and soils of greater Sydney, Australia. J Environ Qual 26:924–933

Julin B, Wolk A, Johansson JE, Andersson SO, Andrén O, Akesson A (2012) Dietary cadmium exposure and prostate cancer incidence: a population-based prospective cohort study. Br J Cancer 107(5):895–900

Kafka Z, Kuras M (1997) Heavy Metals in soils contaminated from different sources. In: Cheremissinoff Paul N (ed) Ecological issues and environmental impact assessment. Gulf Publishing Company, Houston, pp 175–180
Kanwal SK, Kumar V (2011) High prenatal and postnatal lead exposure associated lead encephalopathy in an infant. Indian J Pediatr 78(11):1420–1423
Kaur R, Rani R (2006) Spatial characterization, and prioritization of heavy metal contaminated soil-water resources in peri-urban areas of national capital territory (nct), Delhi. Environ Monit Assess 123:233–247
Khan MU, Malik RN, Muhammad S (2013) Human health risk from heavy metal via food crops consumption with wastewater irrigation practices in Pakistan. Chemosphere 93:2230–2238
Kimbrough DE, Cohen Y, Winer AM, Creelman L, Mabuni C (1999) A critical assessment of chromium in the environment. Crit Rev Environ Sci Technol 29:1–46
Kippler M, Tofail F, Gardner R et al (2012) Maternal cadmium exposure during pregnancy and size at birth: a prospective cohort study. Environ Health Perspect 120(2):284–289
Kisku GC, Barman SC, Bhargava SK (2000) Contamination of soil and plants with potentially toxic elements irrigated with mixed industrial effluent and its impact on the environment. Water Air Soil Pollut 120:121–137
Krishna AK, Govil PK (2005) Heavy metal distribution and contamination in soils of Thane-Belapur industrial development area, Mumbai, Western India. Environ Geol 47:1054–1061
Lake DL, Kirk PWW, Lester JN (1984) The fractionation, characterization, and speciation of heavy metals in sewage sludge and sewage sludge amended soils: a review. J Environ Qual 13:175–183
Lehoczky E, Szabo L, Horvath S (1998) Cadmium uptake by lettuce in different soils. Commun Soil Sci Plant Anal 28:1903–1912
Leung DY, Caramanna G, Maroto-Valer MM (2014) An overview of current status of carbon dioxide capture and storage technologies. Renew Sust Energy Rev 39:426–443
Lewis MA, Neighbors C, Oster-Aaland L, Kirkeby BS, Larimer ME (2007) Indicated prevention for incoming freshmen: personalized normative feedback and high-risk drinking. Addict Behav 32(11):2495–2508
Long XX, Yang XE, Ni WZ, Ye ZQ, He ZL, Calvert DV, Stoffella JP (2003) Assessing zinc thresholds for phytotoxicity and potential dietary toxicity in selected vegetable crops. Commun Soil Sci Plant Anal 34:1421–1434
Loska K, Wiechula D, Korus I (2004) Metal contamination of farming soils affected by industry. Environ Int 30:159–165
Mapanda F, Mangwayana EN, Nyaman Gara J, Giller KE (2005) The effect of long form irrigation using waste water an heavy metal contents of soils under vegetables in Marare, Zimbabwe. Ecosyst Environ 107:151–165
Maslin P, Maier RM (2000) Rhamnolipid-enhanced mineralization of phenanthrene in organic-metal co-contaminated soil. Biorem J 4(4):295–308
McBride MB (2003) Toxic metals in sewage sludge-amended soils: has promotion of beneficial use discounted the risks? Adv Environ Res 8:5–19
Monni S, Uhlig C, Hansen E, Magel E (2001) Ecophysiological responses of Empetrum nigrum to heavy metal pollution. Environ Pollut 112:121–129
Moore JW, Ramamoorthy S (1984) Aromatic hydrocarbons—polycyclics. In: Organic chemicals in natural waters. Springer, New York, pp 67–87
Natori S, Lai S, Finn JP, Gomes AS, Hundley WG, Jerosch-Herold M et al (2006) Cardiovascular function in multi-ethnic study of atherosclerosis: normal values by age, sex, and ethnicity. Am J Roentgenol 186(6_supplement_2):S357–S365
Navarro MC, Pérez-Sirvent C, Martínez-Sánchez MJ, Vidal J, Tovar PJ, Bech J (2008) Abandoned mine sites as a source of contamination by heavy metals: a case study in a semi-arid zone. J Geochem Explor 96:183–193
Nikolaou A, Meric S, Fatta D (2007) Occurrence patterns of pharmaceuticals in water and wastewater environments. Anal Bioanal Chem 387(4):1225–1234

Pandey R, Shubhashish K, Pandey J (2012) Dietary intake of pollutant aerosols via vegetables influenced by atmospheric deposition and wastewater irrigation. Ecotox Environ Safety 76:200–208

Pasha Q, Malik SA, Shaheen N, Shah MH (2010) Investigation of trace metals in the blood plasma and scalp hair of gastrointestinal cancer patients in comparison with controls. Clin Chim Acta 411(7–8):531–539

Patsikka E, Kairavuo M, Sersen F, Aro EM, Tyystjarvi E (2002) Excess copper predisposes photosystem II to photoinhibition in vivo by outcompeting iron and causing decrease in leaf chlorophyll. Plant Physiol 129:1359–1367

Ping L, Hai-jun Z, Li-li W, Zhao-hui L, Lin JW, Yan-qin W, Li-hua J, Liang D, Yu-feng Z (2011) Analysis of heavy metal sources for vegetable soils from Shandong Province, China. Agric Sci China 10:109–119

PVRC P (1995) Recycling of human waste in agriculture. In: Tandon HLS (ed) Recycling of waste in agriculture. Fertiliser Development and Consultation Organisation, New Delhi, pp 68–90

Rai PK, Tripathi BD (2008) Heavy metals in industrial wastewater, soil and vegetables in Lohta village, India. Toxicol Environ Chem 90(2):247–257

Ram LC, Srivastava NK, Tripathi RC, Jha SK, Sinha AK, Singh G, Manoharans V (2006) Management of mine spoil for crop productivity with lignite fly ash and biological amendments. J Environ Manag 79:173–187

Rattan RK, Dutta SP, Chhonkar PK, Suribabu K, Singh AK (2005) Long-term impact of irrigation with sewage effluents on heavy metal content in soil crops and ground water—a case study. Agric Ecosyst Environ 109:310–322

Rupa TR, Sinivas RC, Subha RA, Singh M (2003) Effect of farmyard manure and phosphorus on Zn transformation and phytoavailability in two altisol of India. Bioresour Technol 87 (3):279–288

Sahu R, Saxena P, Johnson S (2014) Heavy metals in cosmetics. Centre for Science and Environment and Pollution Monitoring Laboratory. http://www.cseindia.org/node/5293

Sanders T, Liu Y, Buchner V, Tchounwou PB (2009) Neurotoxic effects and biomarkers of lead exposure: a review. Rev Environ Health 24(1):15–45

Sanghi R, Sasi KS (2001) Pesticides and heavy metals in agricultural soil of Kanpur, India. Bull Environ Contam Toxicol 67:446–454

Sauve S, Norvell WA, Mcbride M, Hendershot W (2000) Speciation and complexation of cadmium in extracted soil solutions. Environ Sci Technol 34:291–296

Schulten HR, Leinweber P (2000) New insights into organic-mineral particles: composition, properties and models of molecular structure. Biol Fertil Soils 30(5–6):399–432

Scott D, Keoghan JM, Allen BE (1996) Native and low input grasses – a New Zealand high country perspective. N Z J Agric Res 39:499–512

Sharma RK, Agrawal M (2006) Effects of single and combined treatment of Cd and Zn on carrots: uptake and bioaccumulation. J Plant Nutr 29:1791–1804

Sharma RK, Agrawal M, Marshall FM (2006) Heavy metals contamination in vegetables grown in wastewater irrigated areas of Varanasi. Ind Bull Environ Contam Toxicol 77:311–318

Sharma RK, Agrawal M, Marshall F (2007) Heavy metal contamination of soil and vegetables in suburban areas of Varanasi, India. Ecotoxicol Environ Saf 66:258–266

Sharma RK, Agrawal M, Marshall F (2008) Heavy metal (Cu, Zn, Cd and Pb) contamination of vegetables in urban India: a case study in Varanasi. Environ Pollut 154:254–263

Sharma RK, Agrawal M, Marshall FM (2009) Heavy metals in vegetables collected from production and market sites of a tropical urban area of India. Food Chem Toxicol 47(3):583–591

Singh S, Kumar M (2006) Heavy metal load of soil, water, and vegetables in peri-urban Delhi. Environ Monit Assess 120:79–91

Singh A, Sharma RK, Agrawal M, Marshall M (2009) Effects of wastewater irrigation on physicochemical properties of soil and availability of heavy metals in soil and vegetables. Commun Soil Sci Plant Anal 40:3469–3490

Singh A, Sharma RK, Agrawal M et al (2010) Health risk assessment of heavy metals via dietary intake of foodstuffs from the wastewater irrigated site of a dry tropical area of India. Food Chem Toxicol 48:611–619

Sinha S, Pandey K, Gupta AK, Bhatt K (2005) Accumulation of metals in vegetables and crops grown in the area irrigated with river water. Bull Environ Contam Toxicol 74:210–218

Sinha S, Gupta AK, Bhatt K, Pandey K, Rai UN, Sinh KP (2006) Distribution of metals in the edible plants grown at Jajmau, Kanpur (India) receiving treated tannery wastewater: relation with physico-chemical properties of the soil. Environ Monit Assess 115:1–22

Srikanth R, Reddy SRP (1991) Lead, cadmium and chromium levels in vegetables grown in urban sewage sludge Hyderabad, India. Food Chem 40:229–234

Stevenson FJ (1992) Humus chemistry: genesis, composition and reactions. Wiley Intersc. Publ, New York

Tchounwou PB, Yedjou CG, Patlolla AK, Sutton DJ (2012) Heavy metals toxicity and the environment. NIH Public Access Author Manuscript, EXS 101:133–164

Thomas LD, Michaëlsson K, Julin B, Wolk A, Åkesson A (2011, July 1) Dietary cadmium exposure and fracture incidence among men: a population-based prospective cohort study. J Bone Miner Res 26(7):1601–1608

Traina SJ, Laperche V (1999) Contaminant bioavailability in soils, sediments, and aquatic environments. Proc Natl Acad Sci USA 96(7):3365–3371

Turekian KK, Wedepohl KH (1961) Distribution of the elements in some major units of the earth's crust. Geol Soc Am Bull 72(2):175–192

Turkdogan MK, Fevzi K, Kazim K, Ilyas T, Ismail U (2003) Heavy metals in soil, vegetables and fruits in the endemic upper gastrointestinal cancer region of Turkey. Environ Toxicol Pharmacol 13:175–179

United Nations Environment Programme (2006) Division of Technology, Industry and Economics (UNEP. DTIE/CHEMICALS). Interim review of scientific information on lead. UNEP, Geneva

USEPA (1997) Exposure factors handbook EPA/600/P-95/002F

USEPA (2001) Baseline Human Health Risk Assessment. Vasquez Boulevard and I-70 superfund site Denver, Denver (Co)

USEPA IRIS (2006) United States, Environmental Protection Agency, Integrated Risk Information System. http://www.epa.gov/iris/substS

Violante A, Cozzolino V, Perelomov L, Caporale AG, Pigna M (2010) Mobility and bioavailability of heavy metals and metalloids in soil environments. J Soil Sci Plant Nutr 10(3):268–292

Voutsa D, Grimanis A, Samara C (1996) Trace elements in vegetables grown in an industrial area in relation to soil and air particulate matter. Environ Pollut 94:325–335

Wang G, Bruce F (2008) Roles of biomarkers in evaluating interactions among mixtures of lead, cadmium and arsenic. Toxicol Appl Pharmacol 233:92–99

Wang S, Shi X (2001) Molecular mechanisms of metal toxicity and carcinogenesis. Mol Cell Biochem 222:3–9

Wang X, Sato T, Xing B, Tao S (2005) Health risks of heavy metals to the general public in Tianjin, China via consumption of vegetables and fish. Sci Total Environ 350:28–37

Wang YP, Shi JY, Wang H, Lin Q, Chen XC, Chen YX (2007) The influence of soil heavy metals pollution on soil microbial biomass, enzyme activity, and community composition near a copper smelter. Ecotoxicol Environ Safe 67:75–81

Wei B, Jiang F, Li X, Mu S (2010) Heavy metal induced ecological risk in the city of Urumqi, NW China. Environ Monit Assess 160(1):33–45

Whatmuff MS (2002) Applying biosolids to acid soil in New South Wales: are guideline soil metal limits from other countries appropriate? Aust J Soil Res 40:1041–1056

WHO/FAO/UNU (2007) Protein and amino acid requirements in human nutrition report of a joint WHO/FAO/UNU Expert Consultation, 935

Willers S, Gerhardsson L, Lundh T (2005) Environmental tobacco smoke (ETS) exposure in children with asthma-relation between lead and cadmium, and nicotine concentrations in urine. Respir Med 99:1521–1527

Wuana RA, Okieimen FE (2011) Heavy metals in contaminated soils: a review of sources, chemistry, risks and best available strategies for remediation. Int Sch Res Netw (ISRN) Ecol 2011:402647

Xu ZQ, Ni SJ, Tuo XG et al (2008) Calculation of heavy metals` toxicity coefficient in the evaluation of potential ecological risk index. Environ Sci Technol 31(2):112–115

Yilmaz I (2010) Comparison of landslide susceptibility mapping methodologies for Koyulhisar, Turkey: conditional probability, logistic regression, artificial neural networks, and support vector machine. Environ Earth Sci 61(4):821–836

Zhang GP, Fukami M, Sekimoto H (2002) Influence of cadmium on mineral concentrations and yield components in wheat genotypes differing in cd tolerance at seedling stage. Field Crop Res 77:93–98

Zhang C, Qiao Q, Appel E, Huang B (2012) Discriminating sources of anthropogenic heavy metals in urban street dusts using magnetic and chemical methods. J Geochem Explor 119:60–75

Zhu W, Bian B, Li L (2008) Heavy metal contamination of road-deposited sediments in a medium size city of China. Environ Monit Assess 147(1–3):171–181

Zhuang P, McBride MB, Xia H, Li N, Li Z (2009) Health risk from heavy metals via consumption of food crops in the vicinity of Dabaoshan mine, South China. Sci Total Environ 407 (5):1551–1561

Żukowska J, Biziuk M (2008) Methodological evaluation of method for dietary heavy metal intake. J Food Sci 00:R1–R9

Chapter 4
Biological Wastewater Treatment for Prevention of River Water Pollution and Reuse: Perspectives and Challenges

N. K. Singh, G. Gupta, A. K. Upadhyay, and U. N. Rai

Abstract Wastewater discharge with high biological oxygen demand (BOD) and high nutrient levels (e.g., nitrate, phosphate) affects water quality and is a major reason for degradation of water bodies, including rivers. In addition, metals and other toxic elements are also concentrated in aquatic bodies due to the continuous disposal of wastewater that is treated or partially treated. In many developing countries, wastewater treatment facilities are not fully operational due to energy crises and improper maintenance. However, under the provisions of the Environmental Protection Act, maximum permissible limits have been established for the disposal of different pollutants into surface water bodies. Therefore, the appropriate treatment of wastewater containing various pollutants is mandatory before its disposal into a body of water. Conventional methods of wastewater treatment use sewage treatment plants; however, they may be unable to treat wastewater properly and completely due to their higher cost and maintenance requirements. In this case, green plant-based technologies such as phytoremediation, the development of constructed wetlands, and algal pond systems may perform key roles in treating wastewater by removing nutrients and toxic metals before their discharge into rivers. By implementing plant-based, low-cost, and eco-friendly technologies for the treatment of wastewater at the source of origin up to a permissible level of discharge, we can prevent the pollution of surface water bodies and recycle the treated water in agriculture for irrigation, gardening, and other purposes.

Keywords Wastewater treatment · Water pollution · Reuse · Plant · Phytoremediation

---

N. K. Singh (✉) · G. Gupta
Environmental Science discipline, Department of Chemistry, Manipal University Jaipur, Jaipur, Rajasthan, India

A. K. Upadhyay · U. N. Rai
Plant Ecology and Environmental Science Division, CSIR-National Botanical Research Institute, Lucknow, Uttar Pradesh, India

R. P. Singh et al. (eds.), *Water Conservation, Recycling and Reuse: Issues and Challenges*, https://doi.org/10.1007/978-981-13-3179-4_4

4.1 Introduction

Increasing overpopulation, urbanization, and modernization are creating problems for sewage discharge and thus the surface water contamination of lakes and rivers. Rivers are the major dumping site for highly polluted wastewaters emerging from different sources (Georgetti et al. 2011; Guittonny-Philippe et al. 2014). For example, the Ganga River became one of the most highly polluted rivers in the world in recent decades (Wong et al. 2007) because of sewage containing high BOD and toxic elements that was discharged without treatment from urban areas (Purushothaman and Chakrapani 2007; Rai et al. 2011). The most problematic source of water contamination is sewage that is discharged directly into the rivers without any kind of treatment (CPCB 2005). Therefore, the safe discharge of wastewater, including sewage, is essential to protect water resources and the environment in accordance with the Environmental Protection Act, which established maximum permissible limits for wastewater discharged into aquatic bodies, particularly rivers (EPA 1986). Therefore, the proper treatment of sewage and wastewater is mandatory prior to its final discharge into water bodies, including rivers. However, wastewater treatment facilities are not fully operational in many developing countries because of power shortages or inadequate maintenance.

The deteriorating water quality in rivers from anthropogenic and industrial activities has led to the need to establish efficient water management systems. Conventional methods of wastewater treatment are based on physicochemical methods and have limitations due to their high cost of implementation and operation (Carty et al. 2008). However, green plant-based technologies are innovative solutions for environmental protection due their low cost and ecofriendly nature. Plants and algae growing in the littoral zones of rivers are of particular importance in wastewater remediation because of their efficiency in removing accumulated metals and nutrients from wastewater. These efficient metal-accumulating plants and algae have water-purifying potential that can be used in phytoremediation technology, either individually or as a part of a constructed wetland. Plant-based technologies are solar driven, ecofriendly, and cost-effective because they require no energy and comparatively lower costs for their implementation and operation. After biological treatment, wastewater may be used for irrigation or gardening, or it may be recycled for other purposes. Therefore, treating wastewater by growing efficient aquatic plants and algae in a designed plant-based system could be used not only as an alternative to conventional methods, but also to prevent and control river water pollution.

4.2 Wastewater Sources and Characteristics

4.2.1 Sources of Wastewater

The direct discharge of untreated or partially treated sewage into rivers leads to pollution and affects water quality. According to Singh et al. (2004), on average more than 1.3 billion liters of sewage per day is discharged directly into rivers without any treatment. Out of 38,254 MLD (Million Litre per Day) sewage generated from 498 class I cities and 410 class II towns in India, only 11,787 MLD (35%) was treated in existing treatment facilities; therefore, 26,467 MLD (65%) of untreated sewage was discharged directly into the river system (CPCB 2007; CPCB 2009). Furthermore, other sources of contamination include effluent treatment plants and industrial operations in the watersheds of river, which deteriorate water quality by discharging wastewater with high BOD and nutrients into rivers (Rai et al. 2011; Markandya and Murty 2004). When wastewater containing heavy metals such as Hg, Cd, Zn, Ni, Pb, Cr, Co and Cu is discharged into rivers, it leads to bioaccumulation and biomagnification in the food chain, which affects aquatic biota because of their persistent nature (Gochfeld 2003).

4.2.2 Characteristics of Wastewater

Wastewater is defined as water used by different communities that contains a variety of impurities mixed with either suspended or dissolved solids. Generally, wastewater contains 99.9% water and 0.1% solids. The wastewater can be characterized as high or low pH, with high total suspended solids (TSS) and total dissolved solids (TDS), low dissolved oxygen (DO), high concentrations of nitrates and phosphates, high BOD, and high metal content. Wastewater includes sewage and effluent from different industries that have low DO, high BOD, phosphate, nitrate, and metals.

4.2.3 Effects of Wastewater

The water quality of rivers and lakes, which are the main source of drinking water, is continuously degrading because of discharge of wastewater. This can lead to a variety of waterborne diseases, including cholera, renal disease, heart diseases, diarrhea, and dysentery (Gray 2008). Furthermore, wastewater that is discharged on land can leach into underground water tables and contaminate groundwater, making it unsafe for use (Table 4.1). The discharge of sewage causes eutrophication of aquatic bodies, which damages the flora and fauna that grow in the water, as well as disturbs the food chain and food web of the ecosystem. Similarly, wastewater used in irrigation contains high nutrients, toxic elements, and organic contaminants that may affect the production of crops (Birol and Das 2010).

4.2.4 Wastewater Quality Parameters and Permissible Limits

The water quality parameters of wastewater and allowable levels of pollution and metal content are given in Table 4.2.

Table 4.1 Wastewater characteristics and their sources

Characteristics	Source
Color	Domestic sewage, industrial effluents, undecomposed organic materials
Odor	Wastewater, industrial waste
Solids	Municipal water supply, industrial wastewater, soil erosion, inflow infiltration
pH and temperature	Urban and industrial wastewater
Organic contents	Commercial and industrial wastewater
Oils and grease	Sewage and industrial wastewater
Pesticides	Agricultural runoff
Phenols	Industrial effluents
Proteins and fats	Sewage, commercial, and industrial wastes
Alkalinity	Domestic water supply, groundwater infiltration
Chlorides	Domestic wastes, domestic water supply
Heavy metals	Industrial effluent, agriculture runoff, groundwater infiltration
Nitrogen and phosphorus	Sewage, commercial and industrial wastes, natural runoff
Animals, plants, and microbes	Open watercourses, treatment plants, and domestic waste

Table 4.2 Water quality parameters of wastewater and their maximum permissible limits for discharge

Water quality parameter	Standards for wastewater discharge (EPA 1986)
pH	5.5–9.0
Temperature (°C)	Not to exceed 5 °C above the receiving water temperature
Conductivity (μs cm^{-1})	500
Total dissolved solids (TDS)	500
Dissolved oxygen (DO)	4–6
Biological oxygen demand (BOD)	30
Total suspended solids (TSS)	100
Nitrates (NO_3-N)	10
Ammoniacal nitrogen (NH_4-N)	50
Phosphates (PO_4-P)	5
Chromium (Cr)	2
Manganese (Mn)	2
Cobalt (Co)	3
Nickel (Ni)	3
Copper (Cu)	3
Zinc (Zn)	5
Arsenic (As)	0.2
Lead (Pb)	0.1

Values are in mg L^{-1} unless otherwise noted.

4.3 Biological Wastewater Treatment Technologies

The conventional approach of wastewater treatment includes sewage treatment plants based on physico-chemical processes (e.g., activated sludge, wastewater maturation ponds, trickling filters, high-rate stabilization ponds), anaerobic processes (e.g., anaerobic ponds, upflow anaerobic sludge blanket reactors), or a combination of both. However, because of the higher costs of operation, equipment corrosion, high requirements for electricity supply, and the non-reusability of treated water, these conventional methods are not more feasible and or more successful than biological methods. Biological wastewater treatment technologies are comparatively cheap and more ecofriendly. They often apply two methods: phytoremediation (i.e., the use of green plants to remove, degrade, and detoxify the pollutants) and bioremediation (i.e., the use of microorganisms to degrade, convert, transform, and detoxify the pollutants). In conventional sewage treatment plants, there is a possibility of operational failure due improper maintenance and accidents, thus risking a huge capital and harm to workers. However, plant- and microbe-based eco-technology have zero risk of hazards, create zero residues, and use zero energy.

4.3.1 Algae-Based Wastewater Oxidation Pond System

Algae-based wastewater oxidation pond system (AIWPS) technology is a very efficient way to use solar energy for algal growth. AIWPS releases photosynthetic oxygen from the supporting water and uses a special design to foster pond methane formation. This technology has been successfully used at many locations in California and elsewhere, and it has now been applied in Varanasi, India.

AIWPS has a series of four ponds: (i) an advance facultative pond, (ii) a high-rate pond, (iii) an algal settling pond, and (iv) a maturation pond. The wastewater is first screened out and then allowed to pass through the AIWPS. In this technology, produced methane gas is purified up to 85–88% and collected in the gas collectors for combustion. Greenhouse gas emissions are minimized as the carbon dioxide generated during fermentation and electricity generation is absorbed in the pond. The system is designed to grow crops of algae and release a maximum amount of free molecular oxygen as dissolved oxygen (DO) to the surrounding water under controlled conditions. The DO in soluble form goes to bacteria, which break down organic waste and purify water.

Algal systems can treat sewage, livestock waste, agro-industrial waste, and industrial waste (Shelef et al. 1980; Ibraheem 1998; Kaplan et al. 1988; Ma et al. 1990). The most commonly used arrangements for the treatment of wastewater are high-rate algal ponds (Oswald 1988) and the patented Algal Turf Scrubber (Craggs et al. 1996), which employs common green algae (*Chlorella, Scenedesmus, Cladophora*), cyanobacteria (*Spirulina, Oscillatoria, Anabaena*), or both. Various species of green algae, such as *Enteromorpha* and *Cladophora,* have been used to monitor heavy metal concentrations around the world (Al-Homaidan et al. 2011).

The ability of algae to accumulate and biotransform metals in their tissues is the key factor for their widespread use in the biomonitoring of different ecosystems (Mehta and Gaur 2005). The blue-green algae *Phormidium* has been reported to accumulate metals (Cd, Zn, Pb, Ni and Cu) from wastewater, while *Caulerpa racemosa* has been used for the removal of boron species from water (Bursali et al. 2009). Metal accumulation by microalgae and macroalgae can be used in phytoremediation methods, which are not costly and environmental friendly. Algae have different mechanisms for sequestering and synthesizing phytochelatins and metallothioneins, which bind with metals and translocate them into vacuoles (Suresh and Ravishankar 2004). The uptake of toxic elements by algae generally depends on the process of adsorption and metabolism-dependent active uptake (Lomax et al. 2011). Furthermore, the role of algae in arsenic transformation and the reduction of toxicity have been reported by many researchers (Wang et al. 2013; Bahar et al. 2013; Upadhyay et al. 2016).

4.3.2 Wastewater Treatment by Plants

The solar-driven and cost-effective technology for wastewater treatment, popularly known as phytotechnology or green technology, is based on the use of efficient metal-accumulating plants to remove pollutants, including metals and radionuclides, from soil and water. A variety of water-purifying plants may be used, depending on the local climate and geographical location. The chosen plants are usually indigenous to a specific location for ecological reasons and to optimize the functioning of the system.

Several aquatic plants have been found to be more efficient at utilizing solar energy than many terrestrial plants and hence show high growth rates. More than 450 angiospermic plants have been identified and reported as metal hyperaccumulators (Rascioa and Navari-Izzo 2011), which are able to accumulate potentially phytotoxic elements to concentrations up to 1000 times higher than average plants (Cherian and Oliveira 2005).

In phytoremediation processes, plants perform different mechanisms, such as phytoextraction, phytodegradation, rhizofiltration, phytostabilization, and phytovolatilization to remove toxic elements from wastewater. Several investigations have demonstrated that aquatic plants are quite effective at removing metals from wastewater without any visible injury (Lesage et al. 2007; Mishra and Tripathi 2009). Metal removal by the use of aquatic macrophytes is a cost-effective and eco-friendly technique for wastewater treatment (Tangahu et al. 2011; Rai et al. 2013). Submerged macrophytes have shown great potential to accumulate metals (Peng et al. 2008).

Furthermore, aquatic plants absorb metals and nutrients from the water and sediment to which they are exposed (Fritioff and Greger 2003; Upadhyay et al. 2014, 2017). High concentrations of arsenic and other heavy metals in naturally grown plants and algae have been reported in arsenic-affected areas of West Bengal, India, suggesting that macrophytes such as *Eichhornia crassipes, Lemna minor,* and

Spirodela polyrhiza may be used for removing metals from contaminated water in a plant-based treatment system (Singh et al. 2016).

4.3.3 Wastewater Treatment by Constructed Wetlands

Aquatic plants are used in constructed wetlands, which are designed wastewater treatment systems based on biological, chemical, and physical processes for treating wastewater (Rai 2004). Constructed wetlands mimic treatment processes that occur in natural wetlands to remove pollutants from the water (USEPA 1993). The treatment potential and efficiency of constructed wetlands can vary depending on plant combinations and their potential to accumulate and translocate metals in different parts of the plant (Sune et al. 2007). Constructed wetlands have been successfully used for the reduction of water contaminants by treating a wide variety of wastewaters, including industrial effluents, urban and agricultural runoff, sewage, leachates, pharmaceutical waste, and mine drainage (Scholz and Lee 2005; Hadad et al. 2006; Sheoran and Sheoran 2006; Zhang et al. 2012).

A constructed wetlands treatment system is composed of a bed substrate, efficient aquatic plants, and a microbial population. The substrate may be sand, gravel, or soil for anchoring wetland plants. There are two types of constructed wetland systems: horizontal flow systems and vertical flow systems. Most constructed wetland treatment systems are surface-flow or free-water surface systems. In surface-flow wetlands, the water flow on the surface of the wetland above the substrate has a variety of vegetation, such as emergent, floating, rooted, and submerged (Reed et al. 1995). In a subsurface flow system, also called a root-zone system, water flows below the substrate from one end to the other through a permeable bed due to gravitational force.

Various aquatic macrophyte species can be used in a constructed wetland system. However, naturally occurring emergent aquatic plants, such as common reed, adapt well to the local climate and soil conditions and provide adequate treatment. The plants supply oxygen and other nutrients that promote microbial growth in the substrate. The microbial population is largely responsible for treatment in the constructed wetland. In addition, the plants growing in constructed wetlands may provide a habitat for wildlife through remediation of pollution.

The planting and raising of reed beds are very popular in European countries for constructed wetlands, using plants such as cattails (*Typha* spp.), water hyacinth (*Eichhornia crassipes*), sedges, and *Pontederia* sp. Recent research in use of constructed wetlands for subarctic regions has shown that buckbeans (*Menyanthestrifoliata*) and pendant grass (*Arctophilafulva*) are also useful for metal uptake (Rai et al. 2010). Aquatic vegetation may play an important role in removal of phosphorus from water (Breen 1990; Guntensbergen 1989; Rogers et al. 1991). Phosphorus removal in a surface flow wetland treatment system has been reported using the aquatic plants *Scirpus* sp., *Phragmites* sp., and *Typha* sp. (Finlayson and Chick 1983). More often, the macrophyte *Phragmites australis* is used in sewage treatment systems to treat sewage. Constructed wetlands have been used for metal and metalloid removal from wastewater. Water-purifying plants, which supply oxygen and shade, are also added to the complete ecosystem to improve wastewater treatment. The aquatic plants *Phragmites australisa* and *Typha latifolia* have commonly been used for their metal tolerance, uptake, and filtration ability for wastewater treatment.

Over time, the bacterial suspension forms a biofilm around the roots of the treating plants (Peterson and Teal 1996). These micro-organisms assimilate the nutrients from the water and supply them to the aquatic plants for energy to grow and develop. Furthermore, macrophytes can transport atmospheric oxygen to into wastewater through their leaves, stems, and roots (Howes and Teal 1994). Some portion of transported oxygen is consumed for root respiration, with the remaining oxygen dissolved in the water column of the wetland, which leads to oxidation of organic carbon by facultative bacteria (Abbasi and Ramasami 1999). Moreover, anoxic conditions promote the growth of denitrifying micro-organisms. The plants not only assimilate nitrogen as a key nutrient in the life cycle for growth and development (Ellis et al. 1994) but also release oxygen and provide a suitable condition for purification reactions by enhancing a variety of chemical and microbial processes in constructed wetlands (Jenssen et al. 1993).

The key mechanisms of phosphorus removal from wastewater in constructed wetlands are physicochemical processes, such as the fixation of phosphate by iron and aluminum in the substrate (Arias et al. 2001). Aquatic plants accumulate metals in their tissue from water, which augments their use in designing a plant-based treatment system coupled with mechanical skills (Rai et al. 2012). A reduction of more than 90% of BOD and an increase in DO levels in sewage has been reported in constructed wetlands that treat water using the aquatic plants *Typha latifolia, Phragmites australis,* and *Colocasia esculenta* (Rai et al. 2013). Similarly, the metal removal potential of these aquatic plants was also reported after 36 h of sewage treatment (Rai et al. 2015). Masi et al. (2013) reported an average of 86% removal for the organic load, 60% for total nitrogen, 43% for total phosphorus, 89% for total suspended solids, and 76% for ammonium from sewage in constructed wetlands. The hardiness, ability to survive under adverse environmental conditions, and high productivity of vascular aquatic weeds can also be utilized to make them efficient bio-agents for treating wastewater (Abbasi and Ramasami 1999). These spent plants can be used as an energy feedstock. Recently, simulated constructed wetlands planted with *Typha latifolia* and *Polygonum hydropiper* were reported to efficiently treat sewage containing metals before its discharge into freshwater to prevent water pollution (Upadhyay et al. 2017).

4.4 Treatment Processes

A number of conventional approach for wastewater remediation have been put forth, but they have been unable to reach the desired safe disposal limit. In addition, different investment priorities, a large gap between treated and generated sewage, high costs, and maintenance requirements also contribute significantly to the disposal of untreated sewage. Plant-based technology, including constructed wetlands designed for the secondary treatment of wastewater, use physical, chemical, and biological processes with plants and microbes to reduce the toxic level of different pollutants to a safer level (Breitholtz et al. 2012).

Wetland microbes have the ability to remove excessive amounts of nutrient runoff from agricultural/human sources (Wu et al. 2011). It has been observed that the suspended solids and oxidized nutrients present in wastewater are readily used by wetland organisms for their growth and metabolism. As the water passes through the wetland system, contaminants are removed from the water through adsorption by the beds (phosphates and large organic compounds), microbially mediated removal (biochemical reactions), or uptake into plants (heavy metals and some organic compounds).

The phosphorus present in wastewater can be removed by different processes, including mineralization, absorption, uptake by the plants, dissolution of insoluble P to soluble by microbial secretion of the phosphatase enzyme, and precipitation through the formation of a complex with metals present in the wastewater. Similarly, nitrogen in the wastewater can be removed by the microbial actions of ammonification, nitrification, and denitrification (Stewart 1966). Nitrogen can be easily taken up by the plants (Kartal et al. 2010).

Metal accumulation processes involve the following steps: (1) binding to the substrate, particulates, and soluble organics; (2) precipitation as insoluble salts; and (3) uptake by plants and micro-organisms (Gill et al. 2014). Furthermore, enhancements in the metal uptake of plants can be regulated (increased or decreased) by the increased bioavailability of metals in solution, the addition of chelating molecules and the genetic modification of transporters because these proteins can directly control the uptake, distribution, and accumulation of various elements in plants (Ovecka and Takac 2014; Salt et al. 1995).

The key step in the hyperaccumulation of metals depends on genetic regulation and overexpression of different types of genes and transporters present in the plants (Raskin 1996). Metals are either absorbed in the roots or translocated to the aboveground parts of the plant through xylem, then deposited into vacuoles to remove excess metal ions (Assunção et al. 2003). Metal uptake is governed by a variety of molecules, such as the the transporter protein Zip family protein (zinc iron permease), natural resistance-associated macrophage, metallothionein, phytochelatins, Lsi 1 and 2, aquaporins, multidrug and toxin efflux, and heavy metal-transporting ATPases (Talke et al. 2006; Fulekar et al. 2009; Pilon et al. 2009). After treatment through the plant system, the wastewater could be recycled and reused for agriculture and industrial processes.

4.5 Conclusions

Green plant-based technologies such as phytoremediation, constructed wetlands, and algal pond systems play an important role in treating wastewater by removing nitrogen, phosphorus, and toxic metals before their discharge into rivers, thus preventing water pollution. Aquatic plants with high metal tolerances and accumulation potentials can be used to establish constructed wetlands for treating wastewater by removing metals and nutrients. Furthermore, planting soil binders and pollution abator plants along the banks of rivers may play an important role in CO_2 mitigation in the atmosphere, the settlement of suspended particulate matter from water, and the control of flooding in the river streams. Therefore, plant-based wastewater biological treatment may serve as an eco-friendly and cost-effective technology for prevention of river water pollution and conservation of their ecosystem compared with conventional methods. Biological wastewater treatment may be supplemented with conventional wastewater treatment for the proper and complete treatment of wastewater, filling the gap between sewage generation and existing treatment facilities to conserve the water quality and ecological entities of rivers.

Acknowledgements The author is thankful to the Dean and Director of the School of Basic Sciences at Manipal University, Jaipur, India, for providing facilities.

References

Abbasi SA, Ramasami EV (1999) Biotechnological methods of pollution control. University Press, Hyderabad

Al-Homaidan AA, Al-Ghanayem AA, Areej AH (2011) Int J Water Resour Arid Environ:1–10

Arias CA, Bubba MD, Brix H (2001) Phosphorus removal by sands for use as media in sub-surface flow constructed reed beds. Water Res 35:1159–1168

Assunção AG, Schat H, Aarts MG (2003) *Thlaspi caerulescens*, an attractive model species to study heavy metal hyper accumulation in plants. *New Phytol* 159(2):351–360

Bahar MM, Megharaj M, Naidu R (2013) Toxicity transformation and accumulation of inorganic arsenic species in microalga *Scendesmus* sp. isolated from soil. J Appl Phycol 25:913–917

Birol E, Das S (2010) Estimating the value of improved wastewater treatment: the case of River Ganga, India. J Environ Manag 91:2163–2171

Breen PF (1990) A mass balance method for assessing the potential of artificial wetlands for wastewater treatment. Wat Res 24:689–697

Breitholtz M, Naslund M, Strae D, Borg H (2012) An evaluation of free water surface wetlands as tertiary sewage water treatment of micro-pollutants. Ecotoxi Environ Saf 78:63–71

Bursali EA, Cavas L, Seki Y, Bozkurt SS, Yurdakoc M (2009) Chem Eng J 150:385

Carty A, Scholz M, Heal K, Gouriveau F, Mustafa A (2008) The universal design, operation and maintenance guidelines for farm constructed wetlands (FCW) in temperate climates. Bioresour Technol 99:6780–6792

Cherian S, Oliveira MM (2005) Transgenic plants in phytoremediation: recent advances and new possibilities. Environ Sci Technol 39(24):9377–9390

CPCB (2005) Performance status of common effluent treatment plants in India. Central Pollution Control Board, India

CPCB (2007) Advance methods for treatment of textile industry effluents, Resource Recycling Series: RERES/&/2007. Central Pollution Control Board, India

CPCB (2009) Forth report on status of water supply, sewage generation and treatment series in India. Central Pollution Control Board, India

Craggs RJ, Adey WH, Jenson KR, St. John MS, Green GF, Oswald WJ (1996) Phosphorous removal from wastewater using and algal turf scrubber. Water Sci Technol 33:191–198

Ellis JB, Shutes RB, Revitt DM, Zhang TT (1994) Use of macrophytes for pollution treatment in urban wetlands. Resour Conserv Recycl 11:1–12

EPA, Environmental (Protection) Act, India (1986)

Finlayson MC, Chick AJ (1983) Testing the significance of aquatic plants to treat abattoir effluent. Wat Res 17:15–422

Fritioff A, Greger M (2003) Aquatic and terrestrial plant species with potential to remove heavy metals from storm water. Int J Phytoremediation 5:211–224

Fulekar MH, Singh A, Bhaduri AM (2009) Genetic engineering strategies for enhancing phytoremediation of heavy metals. Afr J Biotechnol 8:1–4

Gill LW, Ring P, Higgins NM, Johnston PM (2014) Accumulation of heavy metals in a constructed wetland treating road runoff. Ecol Eng 70:133–139

Giorgetti L, Talouizte H, Merzouki M, Caltavuturo L, Geri C, Frassinetti S (2011) Genotoxicity evaluation of effluents from textile industries of the region Boulmane, Morocco: acase study. Ecotoxicol Environ Saf 74:2275–2283

Gochfeld M (2003) Cases of mercury exposure, bioavailability, and adsorption. Ecotoxicol Environ Saf 56:174–179

Gray NF (2008) Drinking water quality: problems and solutions. Cambridge University Press. EPA, Environmental (Protection) Act, India, 1986

Guittonny-Philippe A, Masotti V, Höhener P, Boudenne JL, Viglione J, Laffont-Schwob I (2014) Constructed wetlands to reduce metal pollution from industrial catchments in aquatic Mediterranean ecosystems: a review to overcome obstacles and suggest potential solutions. Environ Int 64:1–16

Guntensbergen GR, Stearns F, Kadlec JA (1989) Wetland vegetation. In: Hammer DA (ed) Constructed wetlands for wastewater treatment. Lewis publishers, Chelsea, pp 73–88

Hadad HR, Maine MA, Bonetto CA (2006) Macrophytes growth in a pilot-scale constructed wetland for industrial wastewater treatment. Chemosphere 63:1744–1753

Howes BL, Teal JM (1994) Oxygen loss from *Spartina alterniflora* and its relation to salt marsh oxygen balance. Oecologia 97:431–438

Ibraheem IBM (1998) Utilization of certain algae in the treatment of wastewater. Ph.D. Thesis, Faculty of Science Al-Azhar University, Cairo, Egypt, pp 197

Jenssen PD, Maehlum T, Krogstad T (1993) Potential use of constructed wetlands for wastewater treatment in northern environments. Water Sci Technol 28:149–157

Kaplan D, Christiaen D, Arad S (1988) Binding of heavy metals by algal polysaccharides. In: Stadler T, Mollion J, Verdus MC, Karamanos Y, Morvan H, Christiaen D (eds) Algal biotechnology. Elsevier Applied Science, London, pp 179–187

Kartal B, Kuenen JV, Van Loosdrecht MCM (2010) Sewage treatment with anammox. Science 328:702–703

Lesage E, Mundia C, Rosseau DPL, Van de Moortel AMK, Du Laing G, Meers E, Tack FMG, De Pauw N, Verloo MG (2007) Sorption of Co, Cu, Ni, and Zn from industrial effluents by the submerged aquatic macrophyte *Myriophyllum spicatum* L. Ecol Eng 30:320–325

Lomax C, Liu WJ, Wu L, Xue K, Xiong J, Zhou J, McGrath SP, Meharg AA, Miller AJ, Zhao FJ (2011) Methylated arsenic species in plants originate from soil microorganisms. New Phytol 193:665–672

Ma AN, Cheah SC, Chow MC (1990) Current status on treatment and utilization of palm oil industrial wasters in Malaysia. Special coordination meeting of the working group on environmental biotechnology, Kuala Lumpur

Markandya A, Murty MN (2004) Cost benefit analysis of cleaning the Ganges: some emerging environment and development issues. Environ Dev Econ, Cambridge University Press 9 (01):61–81

Masi F, Caffaz S, Ghrabi A (2013) Multi-stage constructed wetland systems for municipal wastewater treatment. Water Sci Technol 67:1590–1598

Mehta SK, Gaur JP (2005) Use of algae for removing heavy metals ions from wasterwater, progress and prospects. Crit Rev Biotechnol 25:113–152

Mishra VK, Tripathi BD (2009) Accumulation of chromium and zinc from aqueous solutions using water hyacinth (*Eichhornia crassipes*). J Hazard Mater 164:1059–1063

Oswald WJ (1988) Micro-algae and wastewater treatment. In: Borowitzka MA, Borowitzka LJ (eds) Micro-algal biotechnology. Cambridge University Press, New York, pp 305–328

Ovecka M, Takac T (2014) Managing heavy metal toxicity stress in plants: biological and biotechnological tools. Biotechnol Adv 32:73–86

Peng KJ, Luo CL, Lou LQ, Li XD, Shen ZG (2008) Bioaccumulation of heavy metals by the aquatic plants *Potamogeton pectinatus* L. and *Potamogeton malaianus* Miq. and their potential use for contamination indicators and in wastewater treatment. Sci Total Environ 392:22–29

Peterson SB, Teal JM (1996) The role of plants in ecologically engineered wastewater treatment systems. Ecol Eng 6:137–148

Pilon M, Cohu CM, Ravet K, Abdel-Ghany SE, Gaymard F (2009) Essential transition metal homeostasis in plants. Curr Opin Plant Biol 12:347–357

Purushothaman P, Chakrapani GJ (2007) Heavy metals fractionation in Ganga River sediments. India Environ Monit Assess Sep 132:475–489

Rai UN (2004) Plantation on the banks of river ganga. Gandhi, Ganga, Giriraj. Navjivan Publishing House, Ahmedabad, pp 1–20

Rai UN, Singh NK, Shukla MK, Verma S, Prasad D, Khan SR (2010) Plant based Management of Ganga Water Pollution: constructed wetland for waste treatment and river front plantation. In: Pandey DD (ed) Environmental pollution: a threat to living world. Jaspal Prakashan, Patna, pp 1–9

Rai UN, Singh NK, Verma S, Prasad D, Upadhyay AK (2011) Perspectives in plant based management of ganga water pollution: a negative carbon technique to rehabilitate river ecosystem. Appl Bot Abs 31:64–81

Rai UN, Prasad D, Verma S, Upadhyay AK, Singh NK (2012) Bio monitoring of metals in ganga water at different Ghats of Haridwar: implications of constructed wetland for sewage detoxification. Bull Environ Contam Toxicol 89:805–810

Rai UN, Tripathi RD, Singh NK, Upadhyay AK, Dwivedi S, Shukla MK, Mallick S, Singh SN, Nautiyal CS (2013) Constructed wetland as an Ecotechnological tool for pollution treatment for conservation of Ganga River. Bioresour Technol 148:535–541

Rai UN, Upadhyay AK, Singh NK, Dwivedi S, Tripathi RD (2015) Seasonal applicability of horizontal sub-surface flow constructed wetland for trace elements and nutrient removal from urban wastes to conserve Ganga River water quality at Haridwar, India. Ecol Eng 81:115–122

Rascioa N, Navari-Izzo F (2011) Heavy metal hyperaccumulating plants: how and why do they do it? And what makes them so interesting? Plant Sci 180:169–181

Raskin I (1996) Plant genetic engineering may help with environmental cleanup. Proc Natl Acad Sci U S A 93:3164–3166

Reed SC, Crites RW, Middlebrooks EJ (1995) Natural Systems for Waste Management and Treatment 2nded. McGraw Hill, New York, pp 173–284

Rogers KH, Breen PF, Chick AJ (1991) Nitrogen removal in experimental wetland treatment systems: evidence for the role of aquatic plants. Res J Water Polit Control Fed 63:934–941

Salt DE, Blaylock M, Kumar PBAN, Dushenkov V, Ensley BD, Chet I (1995) Phytoremediation: a novel strategy for removal of toxic metals from the environment using plant. Nat Biotechnol 13:468

Scholz M, Lee BH (2005) Constructed wetlands: a review. Int J Environ Stud 62(4):421–447

Shelef G, Azov Y, Moraine R, Oron G (1980) In: Shelef G, Soeder CJ (eds) Algal mass production as an integral part of wastewater treatment and reclamation system in algal biomass. Elsevier, pp 163–190

Sheoran AS, Sheoran V (2006) Heavy metal removal mechanism of acid mine drainage in wetlands: a critical review. Miner Eng 19(2):105–116

Singh KP, Mohan D, Sinha S, Dalwani R (2004) Impact assessment of treated/untreated wastewater toxicants discharged by sewage treatment plants on health, agricultural, and environmental quality in the wastewater disposal area. Chemosphere 55:227–255

Singh NK, Raghubanshi AS, Upadhyay AK, Rai UN (2016) Arsenic and other heavy metal accumulation in plants and algae growing naturally in contaminated area of West Bengal, India. Ecotoxicol Environ Saf 130:224–232

Stewart WDP (1966) Nitrogen fixation in plants. Athlene Press, London

Sune N, Sanchez G, Caffarattia S, Maine MA (2007) Cadmium and chromium removal kinetics from solution by two aquatic macrophytes. Environ Pollut 145:467–473

Suresh B, Ravishankar GA (2004) Phytoremediation: a novel and promising approach for environmental cleanup. Crit Rev Biotechnol 24:97–124

Talke IN, Hanikenne M, Krämer U (2006) Zinc-dependent global transcriptional control, transcriptional deregulation, and higher gene copy number for genes in metal homeostasis of the hyperaccumulator *Arabidopsis halleri*. Plant Physiol 142:148–167

Tangahu BV, Abdullah SRS, Basri H, Idris M, Anuar N, Mukhlisin M (2011) A review on heavy metals (As, Pb, and Hg) uptake by plants through phytoremediation. Int J Chem Eng 2011:31. https://doi.org/10.1155/2011/939161

U.S. Environmental Protection Agency (1993) Subsurface flow constructed wetlands for wastewater treatment. A Technology Assessment. Office of Water, Washington, DC

Upadhyay AK, Singh NK, Rai UN (2014) Comparative metal accumulation potential of Potamogeton pectinatus and Potamogeton crispus: role of enzymatic and non-enzymatic antioxidants in tolerance and detoxification of metals. Aquat Bot 117:27–32

Upadhyay AK, Singh NK, Singh R, Rai UN (2016) Amelioration of arsenic toxicity in rice: comparative effect of inoculation of *Chlorella vulgaris* and *Nannochloropsis* sp. on growth, biochemical changes and arsenic uptake. Ecoto Environ Saf 124:68–73

Upadhyay AK, Singh NK, Bankoti NS, Rai UN (2017) Designing and construction of simulated constructed wetland for treatment of sewage containing metals. Environ Technol 38:2691–2699

Wang NX, Li Y, Deng XH, Miao AJ, Ji R, Yang LY (2013) Toxicity and bioaccumulation kinetics of arsenate in two freshwater green algae under different phosphate regimes. Water Res 47:2497–2506

Wong CM, Williams CE, Pittock J, Collier U, Schelle P (2007) World's top 10 rivers at risk. WWF International, Gland

Wu H, Zhang J, Li P, Zhang J, Xie H, Zhang B (2011) Nutrient removal in constructed microcosm wetlands for treating polluted river water in northern China. Ecol Eng 37:560–568

Zhang DQ, Gersberg RM, Hua T, Zhu J, Tuan NA, Tan SK (2012) Pharmaceutical removal in tropical sub-surface flow constructed wetlands at varying hydraulic rates. Chemosphere 87:273–277

Chapter 5
Climate Change and Sustainable Management of the Rivers System with Special Reference to the Brahmaputra River

Pallavi Das and Manish Kumar

Abstract Climate change is one of the biggest challenges and likely to have significant impact on the hydrology. Due to the increase in urbanization, industrialization and climate change, availability and requirement scenario of freshwater have been changing. Water availability and water security are becoming more uncertain through changes in temperature and precipitation, shifts in the timing and intensity of the monsoon, increased frequency of extreme events such as droughts and floods and accelerated melting of the Himalayan glaciers resulting in changes in short- and long-term runoff, snow cover and melting. The Brahmaputra River basin is one of the most vulnerable areas in the world as it is subject to the combined effects of glacier melt, extreme monsoon rainfall and sea level rise. For years, the river has been serving as one of the most important freshwater resources for agriculture, irrigation, transportation and electricity and habitat for aquatic organisms in the north-eastern India. People residing along its banks are heavily dependent on the river for their livelihood, thus making them a highly vulnerable riverine community. As climate change is a major concern, we should reduce both greenhouse gas emissions and develop effective management strategy for freshwater resource. Climate change adds uncertainty in controlling core issues of water management due to lack of capacity to address climate change. Therefore to address this problem, we need resilient institutions. At the same time, we need political process, involvement of diverse array of actors and foreign policies of riparian countries for transboundary water management.

Keywords Climate change · Sustainable management · Brahmaputra River

---

P. Das
Department of Environmental Science, Indira Gandhi National Tribal University, Amarkantak, Madhya Pradesh, India

M. Kumar (✉)
Department of Earth Sciences, Indian Institute of Technology Gandhinagar, Ahmedabad, Gujarat, India

R. P. Singh et al. (eds.), *Water Conservation, Recycling and Reuse: Issues and Challenges*, https://doi.org/10.1007/978-981-13-3179-4_5

5.1 Introduction

Climate change is one of the biggest challenges having significant impact on the hydrology. Due to increase in urbanization, industrialization and climate change, availability and requirement scenario of freshwater have been changing. According to the IPCC's 2007 report, rising temperature and increase in carbon dioxide concentration that contributed to the climate change would have direct impacts on river system. This would subsequently cause melting of glaciers, change in rainfall pattern and loss of biodiversity (Bates et al. 2008). Change in temperature and rainfall pattern led to shift in the timing and intensity of the monsoon, increase in the extreme event frequency such as droughts and floods and accelerated melting of the Himalayan glaciers, causing changes in the short- and long-term runoff, snow cover and melting (Rasul 2015; Eriksson et al. 2009; Shrestha and Aryal 2011). Climate change causes increase in temperature of the river system and affects aquatic organisms. The higher temperatures and lower rainfall has impact on riverine flow, which in turn affects the water quality of the river. Reduce flow causes higher concentrations of pollution due to less dilution. Higher rainfall intensity could lead to sediment deposition and other contaminants entering rivers leading to more silt deposition.

All of these, melting of glaciers, extreme rainfall during monsoon season and rise in sea level along with geographical location, have combined impact on the Brahmaputra River. Hence this river basin is considered as one of the most vulnerable areas in the world. In the north-eastern region of India, the Brahmaputra River is one of the most important resources for freshwater. For years the river water has been used for agriculture, irrigation, transportation and electricity besides serving as the natural habitat for the aquatic organisms. People residing along its banks are heavily dependent on the river for their livelihood and thus making them a highly vulnerable riverine community. As climate change is a major concern, we should reduce both greenhouse gas emissions and develop effective management strategy for freshwater resources. This chapter highlights the impact of climate change on the river system with special emphasis on sustainable management of the Brahmaputra River. The study also made an attempt to correlate weathering and CO_2 consumption rate with the extent of heavy metal pollution in the environment.

5.2 Materials and Method

The river water samples ($n = 54$) were collected in three different seasons during 2011 and 2014, and nine surface sediment samples ($n = 27$) were collected during 2011, 2012 and 2013 from upstream to downstream of the Brahmaputra River. The collection areas which covered nine sites starting from upstream to downstream of the Brahmaputra River, i.e. Guijan, Romeria, Dibrugarh, Jorhat, Dhansiri Mukh, Tezpur, Guwahati, Jogighopa and Dhubri, are shown in Fig. 5.1. Water samples were analyzed for major ions as per standard methods prescribed by the American

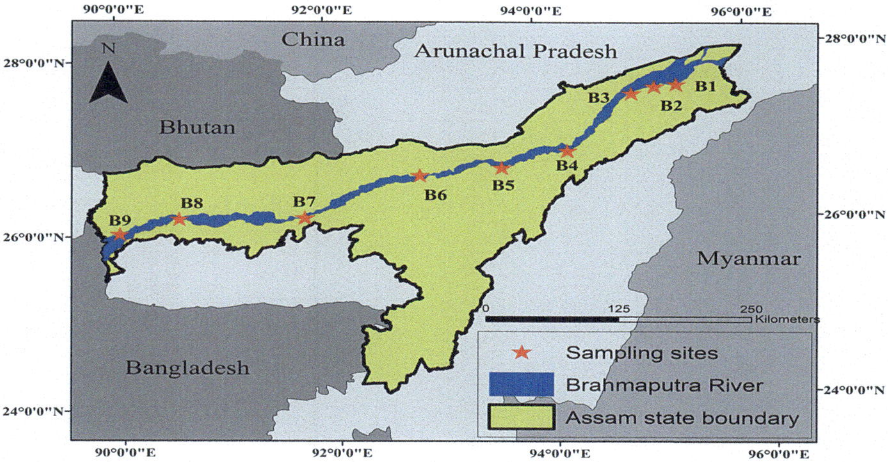

Fig. 5.1 Map of the study area illustrating the Brahmaputra River with sampling locations. Guijan to Dhubri (upstream to downstream) is assigned sample ID B1 to B9 serially

Public Health Association: {calcium (Ca^{2+}) and magnesium (Mg^{2+}) ICP-OES}; {sodium (Na^{+}) and potassium (K^{+}), flame photometer};{phosphate (PO_4^{3-})}, {dissolved silica (H_4SiO_4), sulphate (SO_4^{2-}), nitrate (NO_3^{-}), UV spectrophotometer}; and {chloride (Cl^{-}), titration method}. Heavy metals (Sr, Ti, Fe, Mn, Cu, Zn, Pb, Ni, Co, Cr) were analysed in ICP-OES (2100 DV Perkin Elmer). A statistical analysis (PCA) was performed using SPSS Version 22.0.

5.3 CO_2 Consumption Rates of the Brahmaputra River

Chemical weathering of rock involves consumption of CO_2, a greenhouse gas that exerts strong influence on climate (Berner et al. 1983; Das et al. 2005). Weathering of silicates and carbonates represents an important carbon sink at various scales. The carbon dioxide in the atmosphere dissolves in rainwater forming carbonic acid, which, once in contact with rocks, slowly dissolves them. The CO_2 consumption rate due to silicate weathering (W_{Sil}-CO_2) and carbonate weathering (W_{Carb}-CO_2) of the Brahmaputra River for 2011–2012 and 2013–2014 is shown in Table 5.1 and compared with the world's major rivers. From the value of TDS flux and CO_2 consumption of the world's major rivers (Table 5.1), it was observed that TDS flux of the Brahmaputra River was found to be higher than the Indus River (Himalayan River), the Yellow River (Tibetan River) and the Amazon River (global river). The present result showed that TDS yield was lower than that of the major Indian rivers (Ganga, Godavari, Krishna) and other non-Indian rivers like Mekong and the Amazon but higher than that of the Indus and the Yellow Rivers. As compared with the world's major rivers, the CO_2 consumption rate due to carbonate weathering of the Brahmaputra River was found to be higher than that of the Ganga, Indus,

Table 5.1 TDS flux and CO_2 consumption of the Brahmaputra River and other major rivers of the world

Rivers	TDS flux (tkm^2yr^{-1})	W Sil-CO_2	WCarb-CO_2	References
Brahmaputra	57	0.52	0.57	**Present study**
Brahmaputra	26	0.55	0.58	
Ganges	171	0.45	0. 24	Gaillardet et al. (1999)
Indus	10	0.06	0.09	
Godavari	102	0.29	NA	Jha et al. (2009)
Krishna	28	0.36	NA	Das et al. (2005)
Huang He	NA	0.09	0.27	Wu et al. (2008)
Amazon	147	0.05	0. 11	Gaillardet et al. (1999)
Mekong	56	0.28	0. 58	
Ganga	72	0.38	NA	Singh et al. (2005)
Indus	42	0.06	NA	
Mekong	72	0.24	NA	
Yellow	25	0.08	NA	
Amazon	35	0.05	NA	
World avg	36	0.09	NA	

WCarb-CO_2 - (carbonate weathering $x10^6$ mol Km^{-2} yr^{-1}). Source: Das et al. (2016)
WSil-CO_2 - (silicate weathering $x10^6$ mol Km^{-2} yr^{-1})
NA Not available

Amazon and Huang He Rivers and lower than the Mekong River. Our result shows that high CO_2 consumption rate was associated with carbonate weathering.

5.4 Heavy Metal Distribution in Surface Sediment

Heavy metals (Sr, Ti, Fe, Mn Cu, Zn, Ni, Pb, Co and Cr) in distribution in surface sediments of the Brahmaputra River are shown in Fig. 5.2. Figure 5.2 shows that there was considerable variation in heavy metal distribution among the nine sampling stations of the Brahmaputra River. Distribution of Fe, Mn, Sr, Ti, Cr and Co in surface sediments shows decreasing trend from upstream to downstream, whereas spatial distribution of Cu, Zn, Ni and Pb shows increasing trend from upstream to downstream. In some sites metal content was exceptionally high, which may be due to several anthropogenic activities such as urbanization, dumping of solid waste, industrial effluents, etc., and taking place in the catchment area. At site 7 (Guwahati) concentrations of Cu, Zn, Ni and Pb were comparatively higher. Site 7 is an important river port which receives high volume of industrial effluent and urban runoff. All the drainage outlets in the Guwahati city discharge into the Brahmaputra, thereby polluting the river. Moreover, extensive agricultural practices and high dependence on chemical fertilizers and pesticides are possible anthropogenic sources that contribute to the increase in metal concentration in the river.

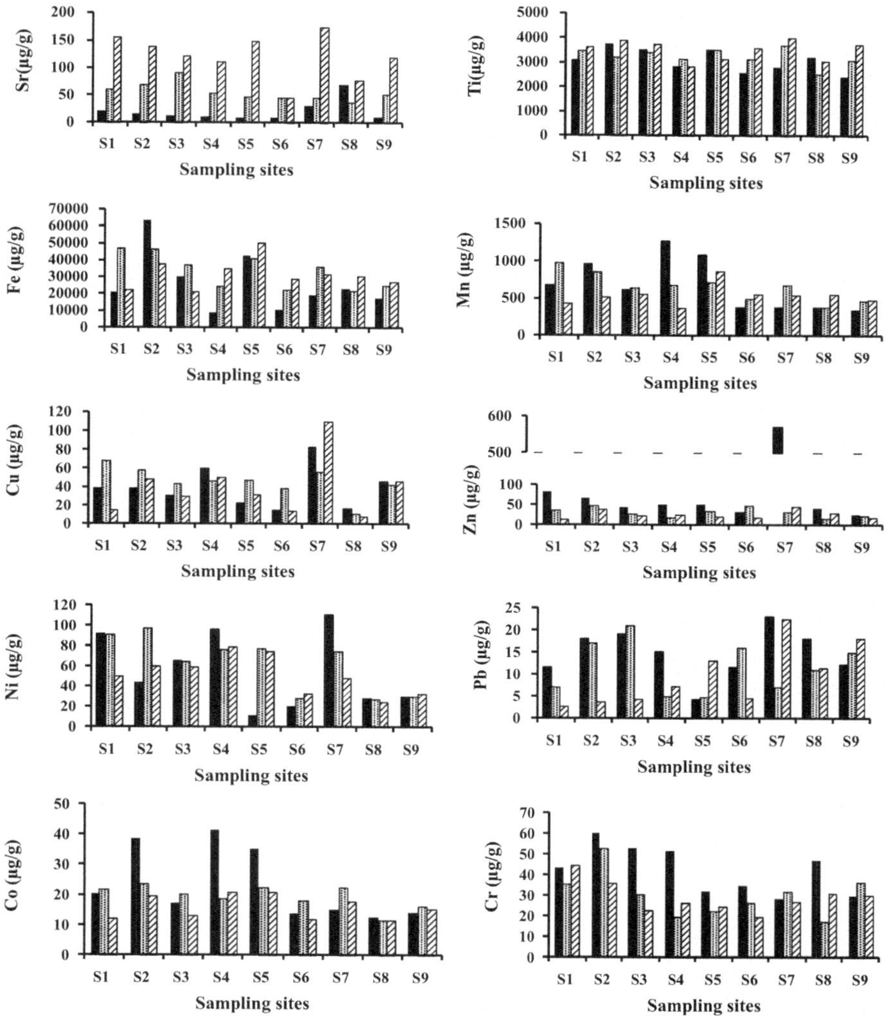

Fig. 5.2 Spatial and temporal variation of heavy metal (Sr, Ti, Fe, Mn, Cu, Zn, Ni, Pb, Co, Cr) in surface sediment of the Brahmaputra River

5.5 Statistical Analysis

Principal Component Analysis (PCA) in Surface Sediment

PCA was performed on heavy metal content in surface sediment samples in order to identify the major process that affects the water chemistry of the river and possible source of different variables. PCA for 13 variables was performed, and 4 components were extracted with eigenvalue >1.

Table 5.2 Factor loading of principal component analysis (PCA) of heavy metal, pH and organic matter in surface sediment samples

	Component			
	F1	F2	F3	F4
Fe	**0.578**	**0.504**	0.002	0.113
Sr	−0.214	**0.774**	0.177	−0.254
Ti	0.249	**0.810**	−0.007	0.115
Mn	**0.900**	0.050	0.037	0.072
Cu	0.143	0.177	**0.849**	0.089
Zn	−0.198	−0.339	**0.713**	0.175
Ni	0.292	0.060	**0.742**	−0.073
Pb	−0.161	−0.120	0.423	**0.712**
Co	**0.935**	−0.162	0.127	0.124
Cr	0.437	−0.080	−0.064	**0.749**
pH	−0.188	**0.772**	−0.129	−0.094
OM	**0.636**	−0.184	0.326	−0.451
Eigenvalue	2.93	2.34	2.13	1.43
% of variance	24.4	19.5	17.8	12.0
Cumulative %	24.4	44.0	61.7	73.6

The results of PCA for heavy metal concentrations, pH and organic matter in surface sediment samples are shown in Table 5.2. The first factor (F1) accounts for 24.4% variance with Fe, Mn, Co and OM as dominant variables. These elements mainly originate from lithogenic sources such as rock weathering and soil erosion. High positive loading of Mn and Co with OM indicates complexation of organic matter with lithophile elements or participation in biological activity with the mobility of these metals being controlled by organic colloids. Association of Fe with Co and Mn indicates that during weathering, Co and Mn are released as Co^{2+} and Mn^{2+}, respectively, from ferromagnesium minerals. It was reported that Co^{2+} and Mn^{2+} were retained in sediments in association with silicate layers and organic colloids and mostly in oxides of Mn (Alloway 2010).

The second factor (F2) accounts for 19.5% variance. In factor 2 high loading of Ti, Sr and pH indicates that pH controls the distribution of Sr and Ti. pH is a vital factor that regulates concentration of heavy metal in sediment. With decrease in pH, sorption of heavy metal decreases, and at high pH, sorption of heavy metal in sediment increases (Christensen 1989). The third factor (F3) accounts for 17.8% variance. The dominant variables were Cu, Zn and Ni indicating metal released from similar anthropogenic source. Industrial waste and sewage in urban areas also add to ubiquitous distribution of Cu and Zn (Kumar et al. 2013a, b). The use of anti-corrosion paints on fishing boats may add to the occurrence of Zn in sediment (Aris et al. 2009). Agricultural runoff containing chemical fertilizers is another important source of these metals in the river basin. The fourth factor (F4) accounts for 12.0% variance with dominant variables, Pb and Cr, indicating strong association with vehicular emission. Common sources of Pb are sewage sludge, vehicle exhaust, lead arsenate pesticide and pesticides (Lei et al. 2008).

5.6 Assessment Imprints of Climate Change on Weathering Rate and Subsequent Metal Release

Table 5.3 shows the effect of climate change on weathering. It shows that factors like temperature, rainfall, runoff, partial pressure of CO_2 (P_{CO2}) and increase in sea level lead to an increase in chemical weathering rate. The carbon dioxide in the atmosphere dissolves in rainwater forming carbonic acid, which, once in contact with rocks, slowly dissolves them. Eventually CO_2 that was present in the atmosphere gets introduced to the river through wet precipitation and then transported into the oceans, where it is trapped for several thousand years, before returning to the atmosphere again or alternatively being stored in river sediment.

In the present study partial pressure of CO_2 (log PCO_2) was calculated from CO_2 concentration recorded in IPCC reports from 2000 to 2010 using the following equations:

$$\log (PCO_2) = -\log KH + \log (H_2CO_3^-) \tag{5.1}$$

Table 5.3 Effect of climate change on weathering

Effects of climate change	Impacts on weathering	Verdict
Increase in temperature	High levels of CO_2 cause high temperatures through the greenhouse effect. Increase of temperature speeds up chemical weathering. Silicate weathers faster at high temperature, consuming larger quantities of CO_2. This causes CO_2 levels and global temperature to drop. At low temperatures, silicate weathering and CO_2 consumption are minimal, causing CO_2 to accumulate in the atmosphere	Favours weathering Positive
Increase in rainfall	When CO_2 dissolves in rainwater, carbonic acid is produced. As the rainwater moves through soil, the carbonic acid dissolves calcite	High rainfall favours weathering Positive
Increase in runoff	Climate change will lead to increased runoff. High runoff increases chemical weathering rate. Runoff – warmer = more precipitation => weathering increases	Increased runoff favours weathering Positive
Increase in partial pressure of CO_2	Increased partial pressure of CO_2 near the surface of Earth results in increase in CO_2 dissolution in water	Increase partial pressure, CO_2 consumption increases. Positive
Glacier melt and rising of sea level	Increase in sea level leads to increase in runoff. As a result the amount of suspended solids, sediment load and associated contaminant metal flux increases	Positive

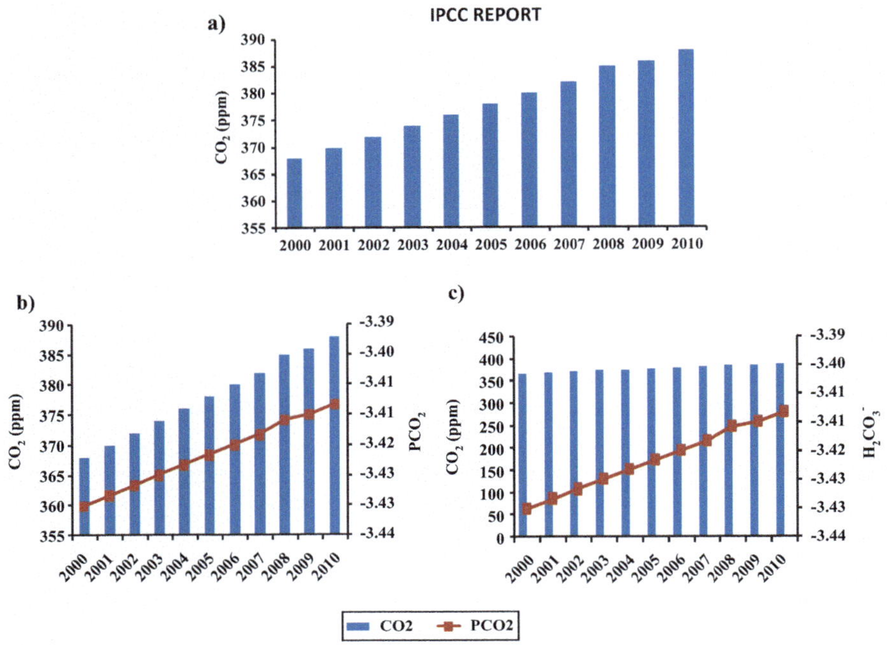

Fig. 5.3 (**a**) Atmospheric CO_2 concentration of IPCC report from 2000 to 2010. (**b**) Partial pressure of CO_2 (log PCO_2) calculated from atmospheric CO_2. (**c**) Carbonic acid (H_2CO_3) calculated from log PCO_2

$$\log (H_2CO_3^-) = -\log K1 - pH + \log (HCO_3^-) \tag{5.2}$$

River alkalinity is an indicator of climate change. For every 1 mole of mineral reacting with carbonated water, there is proportional release of dissolved silica and bicarbonate (alkalinity). The exact amount will depend on the type of minerals being degraded by water, the atmospheric CO_2 concentration and temperature conditions. Figure 5.3 shows that with increase of CO_2 concentration in the atmosphere, partial pressure of CO_2 (logPCO_2) in the atmosphere will increase. In turn $H_2CO_3^-$ and HCO_3^- concentration will increase.

Figure 5.4 (a) shows variations of CWR and CO_2 consumption rate. As silicate weathering gives net sink of CO_2, significant correlation between SWR and CO_2 consumption was drawn, (b) indicating that with increase of weathering rate, CO_2 consumption rate increases.

Weathering is a key natural process that releases most of the metals. Overall findings in the present study suggest that input of heavy metals was mainly through geogenic source. Multivariate statistical analyses suggest that metals were mostly released from terrigenous source. Therefore chemical weathering is one of the key

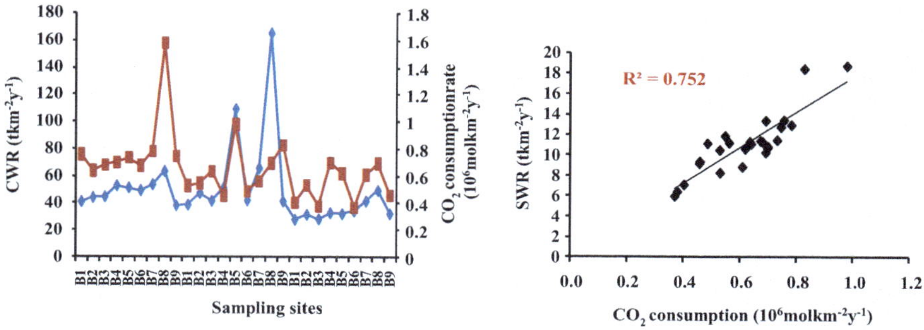

Fig. 5.4 (**a**) Variation of chemical weathering rate (CWR) and CO_2 consumption rate. (**b**) Plot of CO_2 consumption v/s SWR indicate significant correlation between SWR and CO_2

processes that control the concentrations of metals in the bed sediment of the Brahmaputra River. As the weathering rate increases, concentration of metal release through geogenic source increases. Hence metal behaviour is indirectly related with climate change.

5.7 Sustainable Management of the River System

The Brahmaputra River is a major Asian river cutting across international boundaries as it traverses four very distinct regions: the Tibetan Plateau, the Eastern Himalayan Mountains, the Assam plains and the delta in Bangladesh (Singh et al. 2005). The Brahmaputra River is characterized by frequent channel pattern change and shift, due to which it has high seasonal discharge and sediment load and regarded as one of the largest sandbar braided rivers in the world (Thorne et al. 1993). Geologically the Brahmaputra River is the youngest among major rivers of the world.

The Brahmaputra River is known to be the lifeline of Assam, but at the same time it creates havoc through flood and erosion. In India, among the eight north-eastern states, Assam faces the most severe brunt of flood and erosion. Both flood and erosion have severely impacted the economy as well as the political, social and cultural life of the people in Assam.

In order to reduce the hazard of flood and bank erosion, we need proper planning, community participation and application of contemporary technologies for identification of needs and opportunities and for understanding the governing cause-effect relationships and the related management options. Climate change and sustainable development maintain dual relationship (Cohen and Waddell 2009), and not only does climate change affect development, but development affects climate too.

Sustainable Development Policies

1. Application of alternative development pathways: Alternative development pathways such as modification in building infrastructure can save biodiversity loss and degradation of ecosystem services. Managing water resources.
2. Sectoral environmental/economic policies: Economic policies should include economic analyses of regulatory policy instruments such as emission trading, estimating greenhouse gas reduction, benefitting the role of uncertainty and modelling the economic impacts.
3. Institutional/managerial changes: Institution should involve in planning and implementation of water resource management. Promote hazard and risk mapping, disaster management.
4. Innovation/technological changes and their application: Spread the benefits of industrialization worldwide and without unsustainable impacts on water and other natural resources.

Climate Change Policies

1. Avoid climate change damage: Implementing adaptation strategies, risk reduction, financial insurance, micro-insurance, microfinance and risk-pooling transfer are some of the policies that can minimize the risk of loss and damage and financial burden of adapting to climate change.
2. Secondary benefits/costs: Policies to reduce emissions of greenhouse gases also have impacts on other policy goals, especially health.
3. Spill over/trade effects management: Impact of mitigation actions by the industrialized countries on the level of greenhouse gas emissions in the developing countries.
4. Encourage innovation/technological change: Innovation technology like low-carbon technologies can increase welfare.
5. Direct national/sectoral costs.

5.8 Transboundary Water Management

Figure 5.5 shows the interrelation of different factors and parties in transboundary water management. The Brahmaputra River is a transboundary river shared by different countries and managed differently based on the interest of the respective countries. Due to this, conflict arises between states in India and neighbouring countries, and also lack of consistent cohesive cooperation among the countries to address the core issues of water management and climate change adds further uncertainties in controlling the same. A resilient cohesive workable institutional framework is required to drive and resolve the issues. At the same time, we need political initiative and process, state policies, foreign policies and active and efficient actors in governing the system.

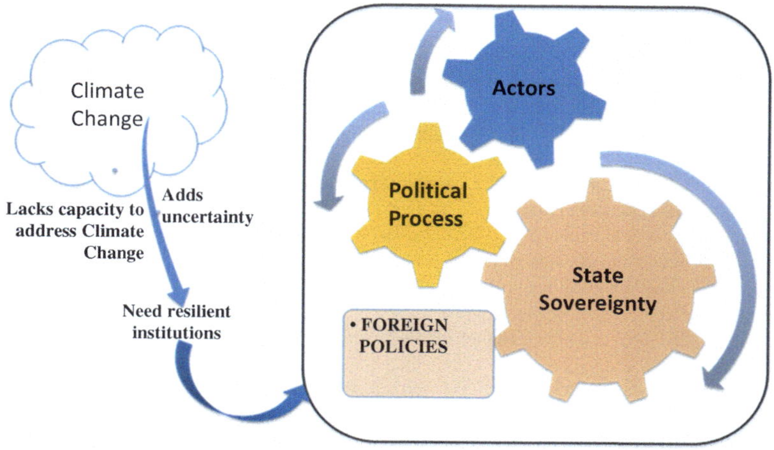

Fig. 5.5 Transboundary water management

5.9 Conclusion

Overall result shows that input of heavy metals is mainly through geogenic sources. Statistical analyses show that metals were mostly released from terrigenous source. As weathering rates increased, concentration of metal release through geogenic source also increased. Hence, metal behaviour is indirectly related with climate change. Significant positive correlation between weathering rate and CO_2 consumption rate signifies increase of CO_2 concentration in the atmosphere which induces weathering rate, thus causing metal release. Therefore, release of metals from natural sources is indirectly affected by the extent of weathering and CO_2 increase due to greenhouse gas emissions. This further indicates that major river systems do respond to the increases of CO_2 concentration in the atmosphere by establishing a new equilibrium at the interface of water-air. This is a result of increased CO_2 dissolution and bicarbonate acid formation inducing weathering and resultant release of metals locked in silicate structures of clay minerals and metal carbonate.

The study of metal quantification with respect to weathering rate will provide understanding of anthropogenic and natural contribution of heavy metal in the sediments. The present study will probably be the first attempt to relate metal behaviour with weathering and carbon dioxide consumption rate in the Brahmaputra River, the fifth largest river in the world. The study of natural and anthropogenic contribution of heavy metal will provide better understanding of metal behaviour under different environmental condition. The study will contribute effective management strategy for freshwater resource with the potential to extrapolate the present case of the Brahmaputra River with major rivers in the world and their possible future scenarios. The integrated results as a whole will also contribute to the development of sustainable freshwater resource management and thus will be helpful to solve one out of the two most vital challenges the world is facing today, i.e. lack of freshwater and need for sustainable energy production.

References

Aris A Z (2009) Genotoxicity assessment of mercuric chloride in the marine fish Therapon jaruba. Environ Asia 2:50–54

Alloway BJ (2010) Heavy metals in soils. Trace metals and metalloids in soil and their bioavailability. Springer, Dordrecht

Bates B, Kundzewicz Z, Wu S (2008) Climate change and water, intergovernmental panel on climate change, Technical Paper VI

Berner RA, Lasaga AC, Garrels RM (1983) The carbonate–silicate geochemical cycle and its effect on atmospheric carbon dioxide over the past 100 million years. Am J Sci 284:641–683

Christensen TH (1989) Cadmium soil sorption at low concentrations. VII. Correlation with soil parameters. Water Air Soil Poll 44:71–82, 1989

Cohen SJ, Waddell MW (2009) Climate change in the 21st century. McGill-Queen's Press, pp 315–316

Das A, Krishnaswami S, Sarin MM (2005) Chemical weathering in the Krishna Basin and Western Ghats of the Deccan Traps, India: rate of basalt weathering and their controls. Geochim Cosmochim Acta 66:3397–3416

Das P, Sarma KP, Jha PK, Herbert JR, Kumar M (2016) Understanding the cyclicity of Chemical weathering and associated CO_2 consumption in the Brahmaputra River Basin (India): the role of major rivers in climate change mitigation perspective. Aquat Geochem 22:1–27. https://doi.org/10.1007/s10498-016-9290-6

Eriksson M, Xu JC, Shrestha AB, Vaidya RA, Santosh N (2009) The changing Himalayas: impact of climate change on water resources and livelihoods in the greater Himalayas. ICIMOD, Kathmandu

Gaillardet J, Dupre B, Allegre CJ (1999) Global silicate weathering and CO_2 consumption rates deduced from the chemistry of large rivers. Chem Geol 159(1–4):3–30

Jha PK, Tiwari J, Singh UK, Kumar M, Subramanian V (2009) Chemical weathering and associated CO_2 consumption in the Godavari river basin, India. Chem Geol 264(1–4):364–374

Kumar M, Furumai H, Kurisu F, Kasuga I (2013a) Tracing source and distribution of heavy metals in road dust, soil and soakaway sediment through speciation and isotopic fingerprinting. Geoderma 211:8–17

Kumar M, Furumai H, Kurisu F, Kasuga I (2013b) Potential mobility of heavy metals through coupled application of sequential extraction and isotopic exchange; comparison of leaching tests applied to soil and soakaway sediment. Chemosphere 90:796–804

Lei D, Yongzhang Z, Jin M, Yong L, Qiuming C (2008) Using multivariate statistical and geostatistical methods to identify spatial variability of trace elements in agricultural soils in Dongguan City, Guangdong, China. J China Univ Geosci 19(4):343–353

Nagarajan N, Devi VJ, Devi CA, Kumaraguru AK (2009) Genotoxicity assessment of mercuric chloride in the marine fish Therapon jaruba. Environ Asia 2:50–54

Rasul G (2015) Water for growth and development in the Ganges, Brahmaputra, and Meghna basins: an economic perspective. Int J River Basin Man 13:387. https://doi.org/10.1080/15715124.2015.10125

Shrestha A, Aryal R (2011) Climate change in Nepal and its impact on Himalayan glaciers. Reg Environ Chang 11(1):65–77. https://doi.org/10.1007/s10113-010-0174-9

Singh SK, Sarin MM, France –LC (2005) Chemical erosion in the eastern Himalaya: major ion composition of the Brahmaputra and δ 13 C of dissolved inorganic carbon. Geochim Cosmochim Acta 69(1):3573–3588

Thorne CR, Russell APG, Alam MK (1993) Planform pattern and channel evolution of the Brahmaputra River, Bangladesh. In: Best JL, Bristow CS (eds) Braided Rivers. The Geological Society, London, pp 257–276

Wu W, Xu S, Yang J (2008) Silicate weathering and CO_2 consumption deduced from seven Chinese rivers originating in the Qinghai–Tibet plateau. Chem Geol 249:307–320, 2008

Chapter 6
Effects of Climate Change on the Use of Wastewater for Aquaculture Practices

Subhendu Adhikari and Rathindra Nath Mandal

Abstract Higher water temperature, sea-level rise and lower precipitation are the manifestations of climate change. Higher temperatures of water and changes in extreme events, like floods and droughts, are responsible for affecting water quality. These extreme events can also pollute water from different sources, like sediments, nutrients, pathogens, pesticides, salt and thermal incidence. This polluted water can have negative impacts on ecosystems, human health, water system reliability, etc. In addition, sea-level rise is responsible for extending the areas of groundwater salinization and estuaries, which can decrease the availability of freshwater for humans and ecosystem in coastal areas.

With anticipated higher temperature, the quality of wastewater could be very poor. Higher water temperature will facilitate the excessive growth of algal bloom in wastewater which will in turn reduce the dissolved oxygen concentration. Higher temperature will also increase the biological activity of wastewater; the growth of microbial population will be more, which will increase the biological oxygen demand (BOD) of wastewater. Because of higher temperature, the organic matter decomposition could increase, and thereby the release of nutrients, viz. nitrogen and phosphorus, will be more in water, which will in turn increase the chance of eutrophication of wastewater. At the same time, the availability of heavy metals and pesticides present in the wastewater could be more to the organisms cultured in the wastewater. If the chance of rainfall is in excess amount, then the wastewater will be diluted, and all the negative qualities of wastewater will be reduced, and the use of the wastewater will be of no problem for aquaculture and agriculture practices. However, if the precipitation reduces particularly in tropics as predicted because of climate change, then the chance of deterioration of wastewater quality will be more. Thus, the operational cost for the wastewater use for aquaculture purposes will be more. In the present paper, these aspects of wastewater use for aquaculture purposes have been discussed.

S. Adhikari (✉) · R. N. Mandal
Regional Research Centre, ICAR-Central Institute of Freshwater Aquaculture, Kolkata, West Bengal, India

R. P. Singh et al. (eds.), *Water Conservation, Recycling and Reuse: Issues and Challenges*, https://doi.org/10.1007/978-981-13-3179-4_6

Keywords Climate change · Wastewater · Use · Aquaculture

6.1 Introduction

Increasing urbanization and industrialization emphasizes the need to explore possibilities of recycling wastes. The aquaculture authorities directed their efforts for obtaining new water resource towards recycling and use of wastewater in line with the country's policy to control pollution. Recycling of sewage wastes through aquaculture assumes importance as it is an ecologically sound practice for handling wastes. For developing countries, a simple system like having stabilization pond is suggested for waste recycling as it is cost-effective. Sewage water is rich in organic matter, heavy metals and essential inorganic nutrients in readily available form for fertilizing fish pond. Sewage can be profitably used for fish culture due to utilization of its smaller detritus particles coated with bacteria, and acceptable by zooplankton and benthos for their sustenance, larger particles of sewage are directly utilized by fish. The content of heavy metals in fish pond water gets reduced in the process of dilution by sedimentation process, but they usually accumulate in bottom sediment or within the weeds grown at the pond.

The quantity of wastewater is increasing with rapid expansion of cities and domestic water supply. As per CPHEEO (Central Public Health and Environmental Engineering Organization, Government of India) estimates, about 70–80% wastewater is generated from the total water supplied for domestic use. The per capita wastewater generation by the class I cities and class II towns represents 72% of urban population in India which has been estimated to be around 98 lpcd (litres per capita daily). The same from the National Capital Territory of Delhi alone, 3663 MLD (million litres per day) of wastewaters (61% of which is treated) is over 220 lpcd (CPCB 1999). As per CPCB estimates, the total wastewater generation from class I cities (498) and class II towns (410) in the country is around 35,558 and 2696 MLD, respectively. A gap of 26,468 MLD in sewage treatment capacity is found as the installed sewage treatment capacity is just 11,553 and 233 MLD, respectively (Kaur et al. 2012). In India, Maharashtra, Delhi, Uttar Pradesh, West Bengal and Gujarat are the major contributors of wastewater (63%; CPCB 2007a). Bhardwaj (2005) estimated that about 48.2 BCM (132 billion litres per day) of wastewaters (with a potential to meet 4.5% of the total irrigation water demand) would be generated by 2050. Thus, overall analysis of water resources indicates that in coming years, there will be twin problems to deal with reduced freshwater availability and increased generation of wastewaters because of increased population and industrialization.

In India, there are 234 sewage water treatment plants (STPs). Most of these were developed under various river action plans (from 1978 to 1979 onwards) and are located in only 5% of the cities/towns along the banks of major rivers (CPCB 2005a). Dissolved air floatation, dual media filter, activated carbon filter, sand filtration and tank stabilization, flash mixer, clariflocculator, secondary clarifiers, sludge drying beds, etc. are the different treatment methods adopted in these plants.

6.2 Causes of Global Water Crisis

The world's total water resource is constant; however, the wastewater production is increasing, and the infrastructure and management systems are not adequate for this increasing volume. Globally, 2 million tonnes of sewage, industrial and agricultural waste excluding the unregulated or illegal discharge of contaminated water is discharged into the world's waterways. This wastewater contaminates freshwater and coastal ecosystems which threaten food security and access to safe drinking and bathing water. This is also creating a major health and environmental management challenge.

Faulty ways of food production lead to the use of 70–90% of the available freshwater, and much of this water returns to the system with additional nutrients and contaminants. Further downstream, agricultural pollution is joined by human and industrial waste. This wastewater flows untreated up to 90% into the densely populated coastal zones. This can contribute to the growth of marine dead zones, which already cover an area of 245,000 km^2, approximately the same area as the entire world's coral reefs. This leads to further losses in biodiversity and ecosystem resilience, which in turn undermines prosperity and efforts towards a more sustain able future (https://www.greenfacts.org).

The per capita average annual availability of freshwater has been reducing from 5177 m^3 in 1951 to 1588 m^3 in 2010 due to population increase and all-round development in India. It can be further reduced to 1341 m^3 in 2015 and 1140 m^3 in 2050 (Kaur et al. 2012). In addition, global warming scenario will also aggravate the water crisis more.

6.3 Role of Agriculture on the Generation of Wastewater

About 70% of total global freshwater has been estimated to be used for the agriculture practices. The daily requirement of drinking water per person is 2–4 l, but it takes 2000–5000 l of water to produce one person's daily food. Proper and appropriate agricultural practices, including irrigation techniques, fertilization practices and reducing water evaporation and crop selection, can save significant amounts of water with subsequent reduction in wastewater generation.

The wastewater produced from rural agriculture, livestock production and inland urban areas represents indeed the first phase in wastewater generation and pollution. Thus, it constitutes a considerable challenge for downstream users. It is characterized by organic and inorganic contaminants, originating from dissolved contents of fertilizers, chemical runoff (such as pesticides), human waste, livestock manure and nutrients.

Agricultural activities take place in both upper and lower catchments. However, upper catchments may be the first cause of contamination in the water basin due to agricultural practices. Agricultural activities take place also at downstream, where the water may be already polluted by other man-made activities which result in domestic and industrial waste. Thus, there is a complex relationship between water quality, agriculture and food quality that is in turn linked to human and ecological health. The excess phosphorus and nitrogen entered in their natural cycles drive algal blooms, including toxic red tides and devastating hypoxic phenomena which can impact fish stocks and/or human health (https://www.greenfacts.org).

6.4 Impact of Industrial Activities of Wastewater Generation

Overall, industry uses 5–20% of total water, which is required for many industrial processes such as heating, cooling, production, cleaning and rinsing, and in doing so, a substantial proportion of total wastewater is being generated. The food and agriculture processing industry is also a major producer of wastewater particularly organic waste with high biochemical oxygen demand, and ultimately this results in low oxygen levels or even anoxic conditions in natural waters. Slaughterhouses may also contribute biological material such as blood-containing pathogens, hormones and antibiotics to the natural waters and thus generate wastewater.

About 13,468 MLD of wastewater is generated by industries, of which 60% is treated apart from domestic sewage (CPCB 2005b; Kaur et al. 2012). Cooling waters are used in the industrial processes of steel manufacture and coke production. These processes at an elevated temperature can have adverse effects on biota and can also become contaminated with a wide range of toxic substances. Mining is a major source of unregulated wastewater discharge in developing countries where more than 70% of industrial wastes are dumped untreated into waterways where they pollute the usable water supply. It percolates into the ground, contaminating aquifers and wells. Both human health and environmental disasters could occur, if the complex organic compounds and heavy metals used in modern industrial processes are released into the environment. The mine waste contaminants may be carcino genic or neurotoxic to people as they contain heavy metals like lead and mercury. The mine waste also contains copper which can be extremely toxic to aquatic organisms. The discharge of toxic mine waste into natural waters could damage the aquatic organisms as it contains many persistent environmental contaminants (https://www.greenfacts.org).

6.5 Climate Change and Wastewater

Large cities are producers of concentrated wastewater (Satterthwaite 2008). The emission of methane (CH_4) during transport, treatment and disposal of wastewater, including sludge, was estimated to be 3–19% of global anthropogenic methane emissions (IPCC 1996). Human sewage and wastewater treatment are the major sources of the greenhouse gas nitrous oxide (N_2O) (IPCC 2007). The emissions of CH_4 and N_2O wastewater are expected to increase by about 50 and 25%, respectively, in the next several decades. Thus, the mitigation of greenhouse gas emissions can be achieved through improvements in collection and management of urban wastewaters, using technologies most appropriate to the economies and settings involved in the process (IPCC 2007). Technologies already exist for reducing, and perhaps reversing, these emission growth rates (Major et al. 2011).

Wastewater treatment facilities in cities of developed nations are sometimes major emitters of greenhouse gas, but those emissions have been identified as important avenues for the reduction of overall greenhouse gas emissions (Rosenzweig et al. 2007). The emission of a large proportion of the greenhouse gas from urban wastewater is expected from the developing countries (Al-Ghazawi and Abdulla 2008) and from informal urban settlements. The rapid population growth and urbanization without concurrent development of sufficient wastewater collection, treatment and disposal infrastructures in many of those developing countries and informal urban settlements results in very large and unmitigated emissions of greenhouse gas. Open sewers along with non-existent sewer systems, ponding and unchecked releases of untreated wastewaters are facts of life in the informal sectors of cities in both developed and developing countries (IPCC 2007; Foster 2008). Improved sanitation facilities, infrastructures, treatment and disposal systems in these settings would not only mitigate emissions but will also offer substantial public health benefits as well (Al-Ghazawi and Abdulla 2008; IPCC 2007).

6.6 Impact of Climate Change on Wastewater Quality

6.6.1 Physico-chemical Properties

Global warming is the most important consequence of climate change. The increase in water temperature could increase in pH of wastewater, which is good for aquaculture. Dissolved oxygen (DO) is the most critical factor in wastewater aquaculture practices. DO concentration of 5 mg/l or more is optimum for most of the aquatic organisms particularly fish (Stickney 2000). When it drops below 2–3 mg/l, hypoxic conditions are started (Doudoroff and Warren 1965; Kalff 2000). The solubility of oxygen in water is inversely related to the temperature of water. For example, water can hold about 14.6 mg/l oxygen at 0 °C, while water holds 8.3 mg/l oxygen at 25 °C (Kalff 2000). Increased incidence of hypoxia and anoxia in freshwater systems can occur because of decreased DO and increased biological oxygen demand (BOD) that are associated with increasing temperatures as a result of climate change. An increase in temperature can reduce the dissolved oxygen and can increase the BOD (Kalff 2000), as the aerobic metabolic rates increase with temperature for most of cold-blooded aquatic organisms.

Fishes could face an oxygen squeeze where decreased oxygen supply cannot meet the increased demand at elevated water temperature (Ficke et al. 2005). Blumberg and Di Toro (1990) reported that DO could be reduced by 1–2 mg/l in Lake Erie under 3–4 °C warming scenario. Warming of this level could lead to DO concentrations under 5 mg/l in the summer months (July–September) and below 2 mg/l in late August to early September. Climate models predict DO levels in the month of July in Lake Suwa, Japan, to decrease from the current value of 6.1–2 mg/l under 2 °C warming scenario (Hassan et al. 1998).

Increase in temperature could warm the bottom water of a turbid/muddy wastewater pond that is beneficial for aquaculture. Increase in temperature will enhance the mineralization of organic matter, which in turn will increase the release of nutrients. Thus, these nutrients will help in the primary productivity of wastewater. Slight increase in temperature could make the wastewater warm, which in turn could increase the metabolic activity of warm-water fishes that is also beneficial for the growth of the fishes.

6.6.2 Primary Productivity and Nutrient Availability

Climate change can profoundly affect primary production and the trophic state of wastewater by changing water temperature and stratification patterns. The trophic status of different wastewater bodies can be greatly influenced by wind, precipitation and stratification particularly in the tropics.

The natural trophic state of any water body is a function of volume, water residence time and nutrient input from the surrounding watershed. However, human activity can also affect the trophic status of water body through anthropogenic enrichment or nutrient depletion and climate change. The input of excess nutrients from urban and agricultural runoff can create eutrophication. The excess sewage discharge from the densely populated area into the nutrient-rich wastewater can also be a causative agent of eutrophication. The increase in temperature due to climate change can also augment the productivity of a body of water by increasing algal growth, bacterial metabolism and nutrient cycling rates. Though increase temperatures will likely result in a general increase in wastewater trophic status, it is difficult to predict the complex relationship between climate change and eutrophication. The input of anthropogenic pollutants and temperature changes (because of climate change) can both accelerate the eutrophication process and delay recovery from anthropogenic eutrophication in wastewater (Ficke et al. 2007).

Enhanced eutrophic conditions can occur through the stimulating explosive macrophyte growth due to increase in temperature. Kankaala et al. (2002) reported that an increase of 300–500% shoot biomass of the aquatic macrophyte like *Elodea canadensis* occurred due to an increase of 2–3 °C in temperature. An increase in biomass of this magnitude would affect the system in various ways. For example, luxuriant growth of macrophytes will take up the available phosphorus of the sediment, and thereby, the amount of phosphorus immediately available for other primary producers would decline.

The macrophytes upon decomposition after death can release nutrients such as nitrogen and phosphorus into the water column. Addition of high concentrations of nitrate and phosphate into the water column could stimulate algae blooms and help in high macrophyte production. The increased oxygen demand during the microbial decomposition of these macrophytes increases the amplitude of the diurnal oxygen cycle of a system. This can lead to decrease in DO levels in the system, raising the probability of anoxia-related fish mortality or of chronically stressful hypoxic conditions in wastewater. The increased growth of macrophytes can push aquatic systems towards a eutrophic state by trapping sediment and preventing flushing of excess nutrients from the system.

6.6.3 Toxicity of Pollutants

It is generally found that toxic effects of common pollutants, for example, organophosphates and heavy metals, to fish increase at higher temperature (Murty 1986). The increased uptake of pollutants may be due to increased gill ventilation rates at higher temperatures (Roch and Maly 1979; Kock et al. 1996). A faster depuration of toxicants is expected with an increase of fish metabolism (MacLeod and Pessah 1973; Huey et al. 1984). Fishes can experience an increased negative effect at higher temperature in spite of their increased ability of metabolizing pollutants at higher temperature. However, these effects are toxicant and species-specific. As no specific metabolic pathway exists in fish to process heavy metals like lead and cadmium, it is difficult for them to depurate these toxicants (Kock et al. 1996). Thus, heavy metals are accumulated in fish more quickly at higher temperature. A fish may physiologically process toxicants present in their food or in the water; however these processes used for the depuration of these compounds are energetically costly. For example, common carp requires ATP for ammonia detoxification from their body (Jeney et al. 1992), and in doing so, less energy becomes available for other important processes such as growth and reproduction (Ficke et al. 2005).

Increasing water temperature could alter the toxicity and uptake of ammonia. Accumulation of ammonia and its different metabolites such as ammonium, nitrite, etc. is a very important issue in any aquaculture systems. At higher temperature, nitrite uptake rates increase in cultured fishes such as channel catfish (*Ictalurus punctatus*) (Huey et al. 1984) and grass carp (*Ctenopharyngodon idella*) (Alcaraz and Espina 1995). Common carp suffers from gill necrosis due to low dissolved oxygen and sublethal ammonia concentrations at higher water temperature (Jeney et al. 1992). Increased uptake of natural toxicants such as ammonia and the synergy existing among the high water temperature, the presence of ammonia and the poor environmental conditions suggest that an increase in global temperature has the potential to lower productivity in intensive wastewater aquaculture systems.

6.7 Role of Climate Change in Wastewater Management

Global climate change may affect water availability, in the timing and intensity of rainfall or the period of time without rain, and affect the quality of water in rivers and lakes through changes in the timing and volume of peak discharge and temperature. Climatic change conditions affect water availability in both time and space, thereby influencing the use of water practices. Changes in climate will also require adaptation in terms of how wastewater is managed. Wastewater and its management also result in the greenhouse gas emission, particularly carbon dioxide (CO_2), methane (CH_4) and nitrous oxide (N_2O). Methane has 21 times greater impact than the same mass of carbon dioxide, while nitrous oxide has 298 times greater impact than the same mass of CO_2 (IPCC 2007).

Extreme rainfall and frequent drought events because of climate change have impacts for non-existent or old and inadequate treatment facilities of wastewater, highlighting the need for better infrastructure development which can cope with extreme surges of wastewater. Floods in the low-lying areas will also spread diseases and cause diarrhoea through the flooding of open sewage or inadequate sewage infrastructure. In some regions, as other water sources decline, irregular rainfall has pushed the exploitation of groundwater resources because of increasing pressure on water resources. The effects of climate change are also aggravated by deforestation grazing of uplands surrounding cities and the rapidly increasing physical expansion of cities along with the heavy build-up of infrastructure. It is also being influenced by the lack of rain-absorbing green vegetation inside the cities' areas (https://www.greenfacts.org).

All these activities/incidences will make complexes in wastewater treatment under climate change scenario. Under such circumstances, the wastewater could be treated biologically through aquaculture practices already mentioned below.

6.8 Opportunities of Use of Wastewater

Better watershed management will become increasingly essential in the future and will find different ways to recycle water. Use of wastewater is an age-old practice for irrigation as well as for fertilization which can continue to expand this role, particularly for peri-urban or urban agriculture and home gardens. However, before water enters the cities, reducing the production of wastewater should be the primary objective throughout the entire management scheme for maximizing water efficiency in the entire water chain.

Wastewater use in agriculture has direct advantages either from the provision of food mainly vegetables to urban populations or to generate biogas, thereby using the nutrients contained therein into resources. Treated wastewater effluent having typical concentration of nutrients from conventional sewage treatment processes would supply all of the nitrogen and most of the phosphorus and potassium normally required for agricultural crops/vegetable production. The effluent having valuable micronutrients in suitable form and the organic matter would also provide benefits in the crop production.

Ten per cent of the world's population depends on food grown with untreated, contaminated wastewater as per estimates (https://www.greenfacts.org). Though wastewater provides affordable food, its use for food production without proper management can pose a serious risk. The resource-poor farmers, who would have little or no access to water for irrigation, get direct benefits from the use of untreated wastewater in the informal way, and this happens mainly in the unregulated sector. Forests, mangrove forests, wetlands and salt marsh could play an important role in wastewater treatment and management by capturing water, filtering out nutrients and other contaminants and releasing water into coastal seas, rivers and lakes.

The conventional wastewater treatment processes are expensive and require complex operations and maintenance. The most neglected areas in the operation of the sewage treatment plants (STPs) in India are the sludge removal, treatment and handling. The facilities constructed to treat wastewater do not function properly and remain closed most of the time due to improper design, poor maintenance, frequent electricity breakdowns and lack of technical manpower (CPCB 2007b). Utilization of biogas generated from UASB (upflow anaerobic sludge blanket) reactors or sludge digesters is also not adequate in most of the cases. In some cases the gas generated is being flared and not being utilized. None of the available technologies has a direct economic return which is one of the major problems with wastewater treatment methods. Local authorities are generally not interested in taking up wastewater treatment due to no economic return. As a result, though the wastewater treatment capacity in India has increased by about 2.5 times since 1978–1979, yet hardly 10% of the sewage generated is treated effectively. The rest finds its way into the natural ecosystems and is responsible for large-scale pollution of rivers and groundwaters (Trivedy and Nakate 2001).

6.9 Biological Treatment of Wastewater Through Aquaculture

Biological treatment of sewage effluents involves systematic use of natural activity of the bacteria for biochemical reactions, resulting in oxidation of organic matter into CO_2, H_2O, N_2 and SO_4. Activated sludge and trickling filter methods, oxidation/ waste stabilization ponds, aerated lagoons and variations of anaerobic treatment systems are widely used for treatment of domestic sewage. The biological processes are the traditional practices of recycling sewage through agriculture, horticulture and aquaculture and have been in vogue in several countries. The sewage-fed fish culture in *bheries* of Kolkata is world-famous. The emphasis in these practices has been on the recovery of nutrients from the wastewater. Taking cue from these practices and deriving from the new databases in different disciplines of wastewater management, aquaculture is being proposed and standardized as a tool for treatment of domestic sewage.

Sewage intake system, duckweed culture complex, sewage-fed fish pond, depuration pond and outlet systems are the components of the treatment of sewage water through aquaculture. The duckweed culture complex comprises a series of duckweed ponds where aquatic macrophytes like *Spirodela*, *Wolffia* and *Lemna* are grown. The wastewater is taken on gravity or pumped through intake system into duckweed culture system, and it is retained for 2 days before use in fish ponds. It may be necessary to incorporate an anaerobic unit where organic load and BOD levels are very high because the optimum intake BOD levels for effective treatment are in the range of 100–150 mg/l.

The duckweed culture unit helps in the removal of heavy metals and other chemical residues that otherwise get into human food chain through cultured fish. These also serve as nutrient pumps, reduce eutrophication effects and provide oxygen from their photosynthetic activity. A total retention time of 5 days is required for the treatment of the wastewater with BOD_5 levels of about 100 mg/l, and in doing so, BOD_5 levels are brought down to 15–20 mg/l, which meet the required standards of different parameters for discharge into natural waters.

Further, taking advantage of high productivity and carrying capacity in sewage-fed system, a production level of 3–4 tonnes of carps per hectare is achieved in fish culture ponds. A high potential of resource recovery in terms of duckweed and fish can be achieved through this system of biological treatment. The main limitation of this biological treatment system, however, is the reduction of efficiency of treatment during the winter period and the regions with temperate climatic condition. This system can be an ideal one for the treatment of wastewater before discharge into natural waters as it requires comparatively lesser amount of land of about 1.0 ha for the treatment of 1 MLD sewage and also resource earnings partially meet working cost.

6.10 Different Sectors in Wastewater Management

For wastewater management, communities should plan not only for current situations but also for future climate change scenarios. The solutions shall be socially and culturally acceptable. Collaboration and dialogue are required among partners, for example, farmers, public health officials, municipal and waste managers, planners and developers, for wastewater management.

Industry has a corporate responsibility so far as industrial sources of wastewater is concerned for taking action to ensure discharged water having an acceptable standard and can also benefit from access to cleaner water resources. The governments and the public sector should play a central role in monitoring, regulating and implementing policy to reduce toxic waste. Wastewater management requires good finance. Insufficient financing does not produce results leading to diminished public and political confidence (https://www.greenfacts.org).

As the volume of wastewater could increase from the population burden particularly in the developing countries under climate change scenario, countries should adopt a multisectoral approach for wastewater management as a matter of urgency, which can incorporate principles of ecosystem-based management from the watersheds down into the sea, connecting sectors that will reap immediate benefits from better wastewater management.

Rural and urban development planning, across all sectors, and where feasible transcending political, administrative and jurisdictional borders are required for successful and sustainable wastewater management.

6.11 Conclusions

It is evident that increase in population will generate more wastewater and climate change will make this worse, if proper treatment and policy are not made to face this challenge. In developing countries like India, the problems associated with wastewater use arise from its lack of treatment. The challenge is thus to find such low-cost, user-friendly methods, which will not only avoid threatening of our substantial wastewater-dependent livelihoods but also will protect the degradation of our valuable natural resources. The use of constructed wetlands has been recognized as an efficient technology for wastewater treatment. The constructed wetlands need lesser material and energy, are easily operated, have no sludge disposal problems and can be maintained by untrained personnel in comparison to the conventional treatments. Further, these systems have lower construction, maintenance and operation costs as these are driven by natural energies of sun, wind, soil, microorganisms, plants and animals. Hence, there seems to be a need for policy decisions and coherent programmes encompassing low-cost decentralized wastewater treatment technologies for planned, strategic, safe and sustainable use of wastewaters, and these technologies should include biofilters, efficient microbial strains, organic/inorganic amendments, appropriate crops/cropping systems, cultivation of remunerative nonedible crops and modern sewage water application methods.

References

Alcaraz G, Espina S (1995) Acute toxicity of nitrite in juvenile grass carp modified by weight and temperature. Bull Environ Contam Toxicol 55:473–478

Al-Ghazawi ZD, Abdulla F (2008) Mitigation of methane emissions from sanitary landfills and sewage treatment plants in Jordan. Clean Technol Environ Policy 10:351–350

Bhardwaj RM (2005) Status of wastewater generation and treatment in India, IWG-Env. Joint Work Session on Water Statistics, Vienna, 20–22 June 2005

Blumberg AF, Di Toro DM (1990) Effects of climate warming on dissolved oxygen concentrations in Lake Erie. Trans Am Fish Soc 119:210–223

CPCB (1999) Status of water supply and wastewater collection treatment & disposal in class I cities-1999. Control of Urban Pollution Series: CUPS/44/1999–2000

CPCB (2005a) Parivesh sewage pollution – news letter. Central Pollution Control Board, Ministry of Environment and Forests, Govt of India, Parivesh Bhawan, East Arjun Nagar, Delhi 110 032. http://cpcbenvis.nic.in/newsletter/sewagepollution/contentsewagepoll-0205.htm

CPCB (2005b) Performance status of common effluent treatment plants in India. Central Pollution Control Board, India

CPCB (2007a) Evaluation of operation and maintenance of sewage treatment plants in India-2007, Control of Urban Pollution Series: CUPS/68/2007. Central Pollution Control Board, India

CPCB (2007b) Advance methods for treatment of textile industry effluents, Resource Recycling Series: RERES/&/2007. Central Pollution Control Board, India

Doudoroff P, Warren CE (1965) Dissolved oxygen requirements of fishes. In: 1965 biological problems in water pollution (Trans. 1962 Seminar). U S Publ Hlth Serv Publ 999-WP-25, pp 145–155

Ficke AA, Myrick CA, Hansen LJ (2005) Potential impacts of global climate change on freshwater fisheries. Colorado State University, Fort Collins, CO, 80523-1474

Ficke AA, Myrick CA, Hansen LJ (2007) Potential impacts of global climate change on freshwater fisheries. Rev Fish Biol Fish 17:581–613. https://doi.org/10.1007/s11160-007-9059-5

Foster V (2008) Overhauling the engine of growth: infrastructure in Africa. Africa infrastructure country diagnostic, World Bank executive summary. Accessed June 28, 2009. http://siteresources.worldbank.org/INTAFRICA/Resources/AICD_exec_summ_9-30-08a.pdf

Hassan H, Hanaki K, Matsuo T (1998) A modelling approach to simulate the impact of climate change on Lake water quality: phytoplankton growth rate assessment. Water Sci Technol 37 (2):177–185

Huey DW, Beitinger TL, Wooten MC (1984) Nitrite-induced methemoglobin formation and recovery in channel catfish (*Ictalurus punctatus*) at three acclimation temperatures. Bull Environ Contam Toxicol 32:674–681 https://www.greenfacts.org/en/wastewater-management/l-2/index.htm

Intergovernmental Panel on Climate Change (IPCC) (1996) IPCC guidelines for national greenhouse gas inventories: reference manual. National Physical Laboratory, New Delhi, pp 6–15

Intergovernmental Panel on Climate Change (IPCC) (2007) Climate change 2007: impacts, adaptation and vulnerability. Contribution of working group II to the fourth assessment report of the IPCC. Cambridge University Press, Cambridge

Jeney G, Nemcsók J, Jeney Z, Oláh J (1992) Acute effect of sublethal ammonia concentrations on common carp (Cyprinus carpio L.). II. Effect of ammonia on blood plasma transaminases (GOT, GPT), G1DH enzyme activity, and ATP value. Aquaculture 104(1–2):149–156. https://doi.org/10.1016/0044-8486(92)90145-B

Kalff J (2000) Limnology. Prentice Hall, Upper Saddle River

Kankaala P, Ojala A, Tulonen T, Arvola L (2002) Changes in nutrient retention capacity of boreal aquatic ecosystems under climate warming: a simulation study. Hydrobiologia 469:67–76

Kaur R, Wani SP, Singh SK, Lal K (2012) Wastewater production, treatment and use in India, pp 1–13. www.ais.unwater.org

Kock G, Triendl M, Hofer R (1996) Seasonal patterns of metal accumulation in Arctic char (Salvelinus alpinus) from an oligotrophic Alpine lake related to temperature. Can J Fish Aquat Sci 53(4):780–786

MacLeod JC, Pessah E (1973) Temperature effects on mercury accumulation, toxicity, and metabolic rate in rainbow trout (*Salmo gairdneri*). J Fish Res Board Can 30:485–492

Major DC, Omojola A, Dettinger M, Hanson RT, Sanchez-Rodriguez R (2011) Climate change, water, and wastewater in cities. In: Rosenzweig C, Solecki WD, Hammer SA, Mehrotra S (eds) Climate change and cities: first assessment report of the urban climate change research network. Cambridge University Press, Cambridge, pp 113–143

Murty AS (1986) Toxicity of pesticides to fish, vol 2. CRC Press, Boca Raton

Roch M, Maly EJ (1979) Relationship of cadmium-induced hypocalcemia with mortality in rainbow trout (*Salmo gairdneri*) and the influence of temperature on toxicity. J Fish Board Can 36(11):1297–1303

Rosenzweig C, Major DC, Demong K et al (2007) Managing climate change risks in New York city's water system: assessment and adaptation planning. Mitig Adapt Strat Glob Change 12:1391–1409. https://doi.org/10.1007/s11027-006-9070-5

Satterthwaite D (2008) Cities' contribution to global warming: notes on the allocation of greenhouse gas emissions. Environ Urban 20:539–549

Stickney RR (2000) Encyclopedia of aquaculture. Wiley, New York

Trivedy RK, Nakate SS (2001) Treatment of hospital waste and sewage in hyacinth ponds. In: Trivedy RK, Kaul S (eds) Low cost wastewater treatment technologies. ABD, Jaipur, pp 132–163

Chapter 7
Bioprocesses for Wastewater Reuse: Closed-Loop System for Energy Options

Rifat Azam, Arya Pandey, Paul N. Black, V. V. Tyagi, and Richa Kothari

Abstract Wastewater discharged from domestic, agricultural, and commercial industries potentially release significant amounts of toxic and pathogenic contaminants into the environment. Wastewater treatment via bioprocess techniques has received a wide attention due to the benefits of end use, i.e., biochemical compounds with sustainable/eco-friendly bioenergy and bioproduct options. Biological treatment of wastewater is beneficial over physical and chemical treatment technology, as it provides biomass in bulk, which can be used for various purposes. The use of reclaimed water (after biological treatment) is significant in terms of environmental, economic, and social aspects, as it does not harm the ecosystem and produces biomass-based products that generate revenue to maintain economic sustainability and provide job opportunities to maintain social integrity. A central goal of these efforts is identifying the means for closing the loop in production systems that take into account legacy management, waste from fossil fuel industries, and integrated design for advanced renewable energy production systems (biogas from biowaste, bioenergy from crassulacean acid metabolism (CAM) plants) that value five aspects of governance: property rights, policy design, financing, and scale as tools to assist enhanced institutional design for integrative management, technology innovation, and public development. The produced biomass on wastewater can further be used to

R. Azam · A. Pandey
Bio-energy and Wastewater Treatment Laboratory, Department of Environmental Science, Babasaheb Bhimrao Ambedkar University, Lucknow, Uttar Pradesh, India

P. N. Black
Department of Biochemistry, University of Nebraska-Lincoln, Lincoln, NE, USA

V. V. Tyagi
Department of Energy Management, Shri Mata Vaishno Devi University, Katra, Jammu & Kashmir, India

R. Kothari (✉)
Bio-energy and Wastewater Treatment Laboratory, Department of Environmental Science, Babasaheb Bhimrao Ambedkar University, Lucknow, Uttar Pradesh, India

WARI Fellow, Robert B. Daugherty Water for Food Institute, University of Nebraska-Lincoln, Lincoln, NE, USA

R. P. Singh et al. (eds.), *Water Conservation, Recycling and Reuse: Issues and Challenges*, https://doi.org/10.1007/978-981-13-3179-4_7

produce bioenergy (biohydrogen, biogas, bioethanol, and biodiesel) and biobased products that have high market value.

Keywords Wastewater · Biomass · Closed-loop system · Bioenergy · Bioproducts

7.1 Introduction

Industrialization and urbanization have accelerated the consumption of freshwater resources and contributed to substantial increases in wastewater generation. Poor water uses and less effective wastewater management practices have a direct impact on the diversity of aquatic organisms. It is important to consider wastewater management as a part of ecosystem with the view of the dimensions of sustainability, social, environmental, and economical. In addition, long-term energy sustainability represents another major issue that must be considered (Elmekawy et al. 2014). Conventional energy sources such as fossil fuel, coal, and LPG are unsustainable in their long-term applications, and their combustion emits greenhouse gases, which contribute to climatic changes, including global warming. In order to maintain long-term environmental sustainability, wastewater management and energy generation must be fully integrated. Treatment of wastewater is a major practice employed by municipalities and industries to achieve the water quality that meets discharge standards. Various methods have been evolved for the treatment of wastewater that includes physical, chemical, and biological process treatments (Guo et al. 2010; Nah et al. 2000; Marañón et al. 2012). Bioremediation is a natural process that decomposes the pollutants through the action of biological catalysts, which occurs at different levels of the ecosystem. It is in this basic process of sustainable development that both wastewater treatment/remediation and energy generation will be merged and outcomes materialized for long-term sustainability. There are two different types of waste substrates, solid and liquid, for bioenergy/bioproducts generation (Pant et al. 2012; Mohan et al. 2011). Various bioprocess routes (e.g., activated sludge process, aerated lagoons, oxidative ponds, anaerobic ponds, and septic tanks) are currently used in standard wastewater treatment facilities. Among the bio-treatment processes, algal-based systems are very effective as they do not require electricity and result in a significant reduction in wastewater pollution loads (e.g., reduction in nitrate and phosphate concentrations). The biologically treated wastewater, i.e., reclaimed water, can be further used for downstream process including agriculture, irrigation, aquaculture, and composting. Bioprocesses of wastewater treatment that include energy production options represent an attractive closed-loop system with decreased expenditures without harming the environment.

7.2 Biological Processes of Wastewater Treatment

Biological methods of wastewater treatment have significant advantages over physical and chemical methods as they are more economically feasible with respect to operational and maintenance costs. The technical man power required for biological treatments is also decreased (Zhu et al. 2013). The biological methods are also highly efficient and do not require large land areas and can be employed by small communities at a reduced scale that is sustainable. In general, biological methods include aerobic and anaerobic processes, which are discussed as detailed below.

7.2.1 Aerobic Processes

Aerobic treatment process consists of high-rate oxidizers for organic and inorganic present in wastewater. Aerobic treatment of wastewater takes advantage of natural process that can be easily engineered into biological processing unit for high rate of wastewater treatment. This encourages the growth of naturally occurring aerobic microorganisms as a means of renovating wastewater. Such microbes are the engines of wastewater treatment plants. Organic compounds are high-energy forms of carbon. Organic compound are high-energy forms of carbon and conversion of these high-energy forms into lowenergy forms is completed by the microbs. In the sufficient amount of dissolved oxygen, they work well, but if it decreases due to heavy organic load, reaeration demanded by them. So, if the rate of reaeration is not equal to the rate of consumption, the dissolved oxygen concentration will fall below the level needed to sustain a viable aquatic system. Oxygen supplied to them will help in consuming the substrate (organic carbon). As a result, organic pollutant converted to inorganic compounds and new microbial cells. So, net production of cells (creation of new cells versus the die-off of old cells) will form an accumulation of biological material. Organic materials present in the form of carbohydrates, fats, proteins, urea, soaps, and detergents. All of these compounds contain carbon, hydrogen, and oxygen. During biochemical processing/degradation, these elements are biologically transformed from organic forms to mineralized forms. Aerobic processes occur in the presence of oxygen and treatments categorized into activated sludge processes, aerated lagoons and oxidative ponds, membrane bioreactors, trickling filters, and rotating biological contractors as illustrated in Fig. 7.1.

7.2.1.1 Activated Sludge Process

This process is used for industrial wastewater treatments that include pesticides, pharmaceuticals, paint, petrochemical, detergents, plastic, and paper. The activated sludge process has two basins. The bottom of first basin contains perforated pipes where air is pumped, allowing small air bubbles to rise within the water (Gao et al.

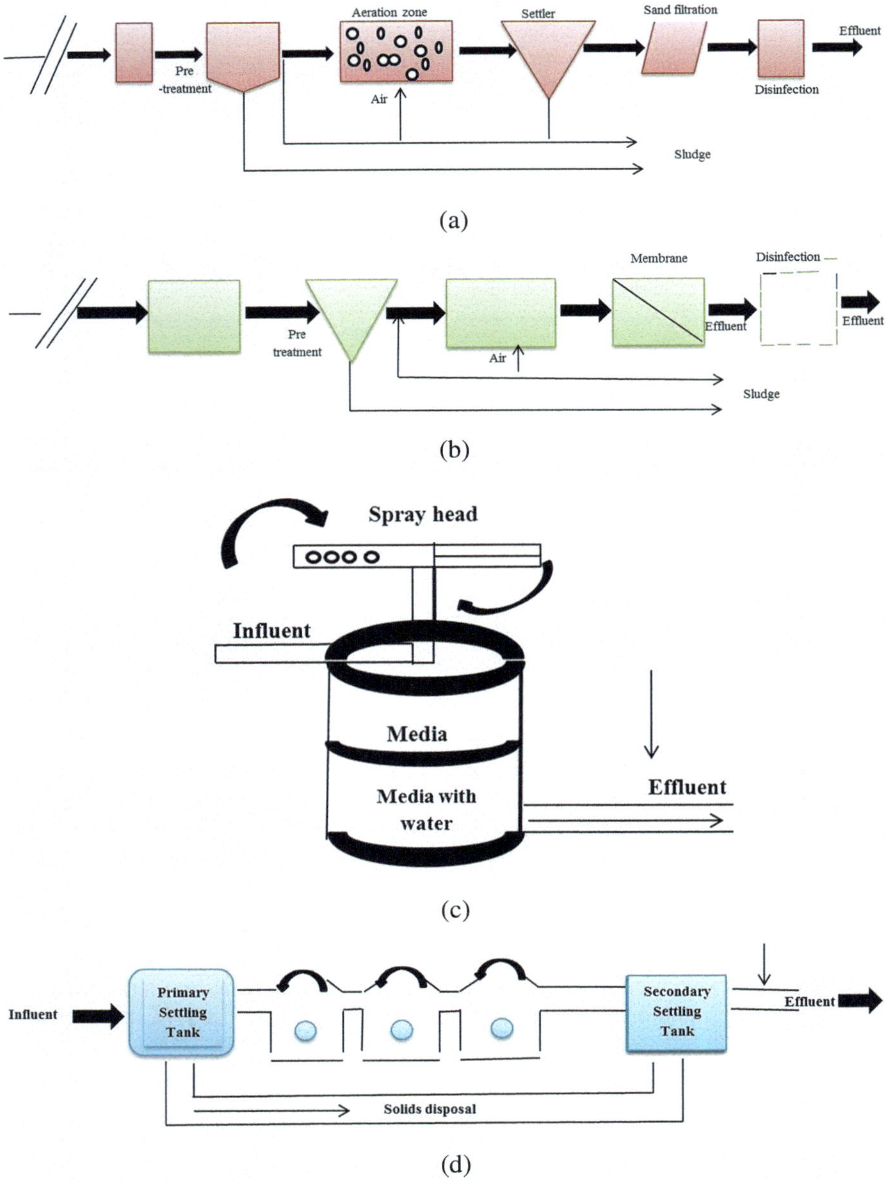

Fig. 7.1 Different aerobic processes of wastewater treatment: (**a**) activated sludge process, (**b**) membrane bioreactors, (**c**) trickling filters, (**d**) rotating biological contractor

2016; Zhang et al. 2012). The second basin is a settling tank, where water flow is greatly reduced allowing large particulates to be removed through gravitational settling. This two-step process is continuous process, and biodegradable wastes are subjected to aerobic microorganism, i.e., activated sludge consists of bacteria (95%), protozoa (4%), and metazoan (1%). The adsorbed organic compounds by the bacteria are being solubilized through hydrolysis. These anaerobic bacteria require pH (7.0–7.5) and 0.1–0.3 mg/L of dissolve oxygen to break down the organic compound (https://www.iowaruralwater.org/tools_tips/toni_glymp/Bacteria-Protozoa.pdf). In industrial wastewater, the impurities including oil and grease are removed by preliminary treatment, and then wastewater is taken for activated sludge process. The BOD and suspended solids are reduced through activated sludge process up to 99%. Figure 7.1a is clearly depicting the structural mechanism for activated sludge process.

7.2.1.2 Aerated Lagoons or Oxidative Ponds

An aerated lagoon or oxidative pond is a treatment pond associated with high intensity of artificial aeration to enhance the biological oxidation of wastewater. Various types of aerated lagoons (such as suspension mixed lagoons and facultative lagoons) are applicable in wastewater treatment. These are shallow ponds (depth of 1 to 2 meters) that allow microorganisms to decompose the primary treated wastewater (Hung et al. 2017). It is also known as high-rate microbial ponds due to the presence of bacteria (*Achromobacter*, *Proteus alcaligenes*, *Chromomonas*, *Zoogloea*, *Pseudomonas*, *Chromatium*, *Thiospirillum*, *Thiopedia*, and *Rhodothecae*), algae (*Chlorella*, *Scenedesmus*, *Euglena*), and protozoans (*Vorticella*, *Macrostoma*, *Paramecium*, and *Podophrya*) (Tharavathy et al. 2014). It has potential to maintain high oxygen rate through the pond due to algal photosynthesis activity. The characteristic feature of these lagoons includes 2–6 days of detention time; BOD loading rate lies between 112 kg/1000 m^3 and 225 kg/1000 m^3 per day and 95% of BOD removal rate (USEPA 2011).

This method of treatment is efficient for small communities as its effluent does not need to disinfection. It handles wide range of wastewater, i.e., municipal, domestic, and kitchen wastewater with low maintenance and operational cost with efficient reduction in pollution load. Aerated lagoons are an efficient and cost-effective system for primary and secondary wastewater treatment in small communities. In this process biological degradation of pollutants is based on attached growth. This growth needs continuous supply with oxygen transfer and effective mixing with organic pollutants. The circulation of wastewater and dissolved oxygen ensures optimal conditions for aerobic growth at the lagoon bottom. As a result, organic pollution is highly reduced and dead zones can be avoided.

7.2.1.3 Membrane Bioreactors

Membrane bioreactor (MBR) technology is being extensively employed to treat municipal and industrial wastewater. Membrane reactor system consists of a combination of membrane units made up of polymeric membranes (polyvinylidene fluoride, polyethersulfone, polyacrylonitrile, polysulfone, polyethylene, polypropylene, etc.) for physical separation, while bioreactor system is basically responsible for biodegradation of organic and inorganic waste compounds. Two main configurations are implemented with this system: (a) external configuration (that involves recirculation of wastewater through membrane) and (b) submerged configuration (membrane modules are usually placed in mixed wastewater) as illustrated in Figure 7.1b. This process uses membrane filtration unit that has potential to replace secondary settler waste with excellent quality of effluent discharge into the river. This technology is widely applicable due to its high rate of disinfection capability, higher volumetric load, less sludge production, technological flexibility for change in influent, and upgraded nitrification (Yoon et al. 2014). These membranes are present outside the aeration zone as it is clearly illustrated in Fig. 7.1b. Garg and Chaudhry (2017) also reported the removal efficiency of MBR with 97.31% of BOD, 98.79% of COD, 97.29% of TDS, and 96.84% of TSS for food industry wastewater.

7.2.1.4 Trickling Filters

Trickling filter technology for wastewater treatment is associated with unnumbered filter, where the influent directly applied at the top of filter media. The trickling filter is an attached growth process that involves a percolating filter with rotational distribution arm that distributes the wastewater above the packing bed of plastic or other coarse materials (Tawfick 2012). Air is able to circulate between the spaces within the packing bed allowing an aerobic environment. The microbes present in wastewater consume organic compounds as a consequence and produced biochemical compound (exocellular lipopolysaccharide) called as slime. The layers of biological slime are rich in gram-negative bacteria such as *Pseudomonas*, *Zoogloea*, *Chromobacter*, *Bdellovibrio bacteriovorus*, and *Flavobacterium*. These bacteria are responsible to break down the organic molecules in the presence of oxygen. Nitrifiers (*Nitrosomonas europaea*, *Nitrobacter hamburgensis*) and denitrifiers (*Thiobacillus denitrificans*) play a significant role to remove nitrogenous compounds present in wastewater. *Desulfotomaculum* and *Desulfovibrio* are responsible to reduce sulfur compound, work in the medium of pH 7, and are inhibited below pH 5.4 and above 9 (Jackson and Mclnerney 2002). Figure 7.1c illustrates the schematics and fundamental mechanism of operation of the trickling filter system.

7.2.1.5 Rotating Biological Contractor

The rotating biological contractor (RBC) consists of a chain of narrowly circular spaced plastic disks that are associated to a rotational hydraulic shaft (Habibi and Vahabzadeh 2013). About 40% of the bottoms of all plates are steeped in discharged wastewater and the film which cultivates the moved disk in and out of the wastewater. Rotating biological contractor consist of a unique and superior alternative to degrade organic and nitrogenous compounds. RBC system is a potential technique to treat wastewater due to its benefits, i.e., less energy consumption, low operation cost, simple and less expensive design of system, low maintenance cost, less energy requirement, and with high treatment efficiency. RBC has potential to eradicate about 99% of fecal coliform bacteria with end number of pathogens present in wastewater. Whereas, Cema et al. (2016) reported 70% of nitrogen removal from coke wastewater by the application of RBC. Figure 7.1d illustrates the mechanism of rotating biological contractors.

7.2.1.6 Microalgae Treatment Systems

Wastewater pollution from different sources including municipal, agricultural, and industrial is also being treated by algal biomass. Industrial wastewater possess mixed composition of complex inorganic compounds (free ammonia, organic nitrogen, nitrites, nitrates, inorganic phosphorous, chloride, sulfate, cadmium, lead, mercury, arsenic, iron, zinc, chromium, etc.) and organic compounds (benzene, xylene, toluene, dichloromethane, trichloroethane, trichloroethylene, organic acids, phenols, starch, sugar, etc.). In order to remove these organic and inorganic pollutant loads, the use of algae alone (*Chlorella pyrenoidosa*, *C. vulgaris*, *Spirulina*, *Chlorella sorokiniana*, *Chlamydomonas*, etc.) or consortium of algae and bacteria (*Chlamydomonas reinhardtii*, *Chlorella vulgaris* and *Scenedesmus rubescens*, *Azospirillum brasilense*, *Azospirillum brasilense*) is potential approach to treat industrial wastewater (Wang et al. 2012; Priyadarshani and Rath 2012). Due to the excess load of pollutants, sometimes industrial wastewater require dilution with media or distilled water and pretreatment process (screening, sieving, autoclaving, etc.) before introduced to algal biomass. A potential reduction in heavy metals (Cr (81.2–96%), Cu (73.2–98%,) Pb (75–98%), and Zn (65–98%)) and nutrients (nitrate (44.3%) and phosphate (95%)) has been reported by Ajayan et al. (2015) to prove the potential of algae for tannery industry wastewater. Hence, use of algae in wastewater remediation represents a significant and sustainable approach with low expenditure to sustain the green therapy for industrial, agricultural, and municipal discharged wastewater.

7.2.2 Anaerobic Processes

In wastewater treatment, anaerobic processes are used for the treatment of organic sludge removed from the wastewater in primary sedimentation and in final sedimentation. This process is worked in the absence of air. Anaerobic process is advantageous over aerobic process as it has high degree of waste stabilization, low production of excess sludge, less requirement of nutrient, no need of oxygen, production of biogas (methane), less requirement of land, etc. But, in terms of treatment efficiency, it is 10–12% less than in activated sludge process. The main backdrops associated with anaerobic process are production of H_2S gas, responsible for corrosion and bad odor. Anaerobic process includes anaerobic ponds, septic tank, and Imhoff tank which are clearly depicted in Fig. 7.2.

7.2.2.1 Anaerobic Ponds

Anaerobic ponds are large stabilization ponds or facultative lagoons, designed to be aerobic at the top, and facultative lagoons are anaerobic at the bottom for treatment of industrial and municipal wastewater. It provides secondary biological treatment of organic compounds by the involvement of natural processes, i.e., sunlight, water, atmospheric oxygen, nutrient (nitrate, phosphate, sulfate, potassium, etc.), and microbes (*Bacillus*, *Bacteroides*, *Bifidobacterium*). Complex organic compounds are broken down by the bacterial activity to produce carbon dioxide, which is used by algal biomass to produce simple sugars through photosynthesis. It is very well-organized technique with numerous inundated inlets and outlets in order to minimize the short circuiting. The efficiency of pollution reduction of urban and industrial wastewater by anaerobic pond has been obtained with BOD (81%), COD (83%), TSS (87%), total nitrogen (33%), and total phosphorous (25%) (Orumieh and Mazaheri 2015). The anaerobic treatment method has also been applied in fisheries wastewater and obtained high rate of pollution removal, i.e., 75–85% of COD (Mendez-Romero et al. 2011). Figure 7.2a provides a schematic of the operational principals for aerobic and anaerobic processes.

7.2.2.2 Septic Tanks

Septic tanks are used for treating municipal wastewater from individual homes or a small group of homes. A well-established septic tank contains a chamber for the settling of solids which form a sludge layer (Bremer and Harter 2012). It works on the principle of anaerobic digestion, i.e., the biodegradable organic compounds of wastewater converted into methane, carbon dioxide, hydrogen sulfide, liquid compounds, etc. Therefore, a continuous and significant reduction obtained in solids sludges accumulated at the bottom of the tank. In this process a thick crust of scum, i.e., anaerobic bacteria are formed on surface of septic tank to maintain the anaerobic condition. Septic tank is basically dominated by non-methanogenic (*Actinomyces*, *Alcaligenes viscolactis*, *A. faecalis*, *Bacillus*, *Bacteroides*, *Bifidobacterium*, *Branhamella catarrhalis*, *Clostridium*, *Corynebacterium*, *Desulfovibrio desulfuricans*, *E. coli*, *Eubacterium*, etc.) and methanogenic (*Methanobacterium*, *Methanobacterium formicicum*, *Methanobacterium ruminantium*, *Methanospirillum* sp., *Methanococcus vannielii*, etc.) bacteria. The efficiency of pollution reduction has been obtained 15–25% in BOD and 40–60% in suspended solid of sewage wastewater (http://www.biologydiscussion.com/waste-management/waste-water treatment/treatment-of-waste-water/24664). The backdrops are associated with its low digestion rate due to low temperature. Septic tanks need further treatment before

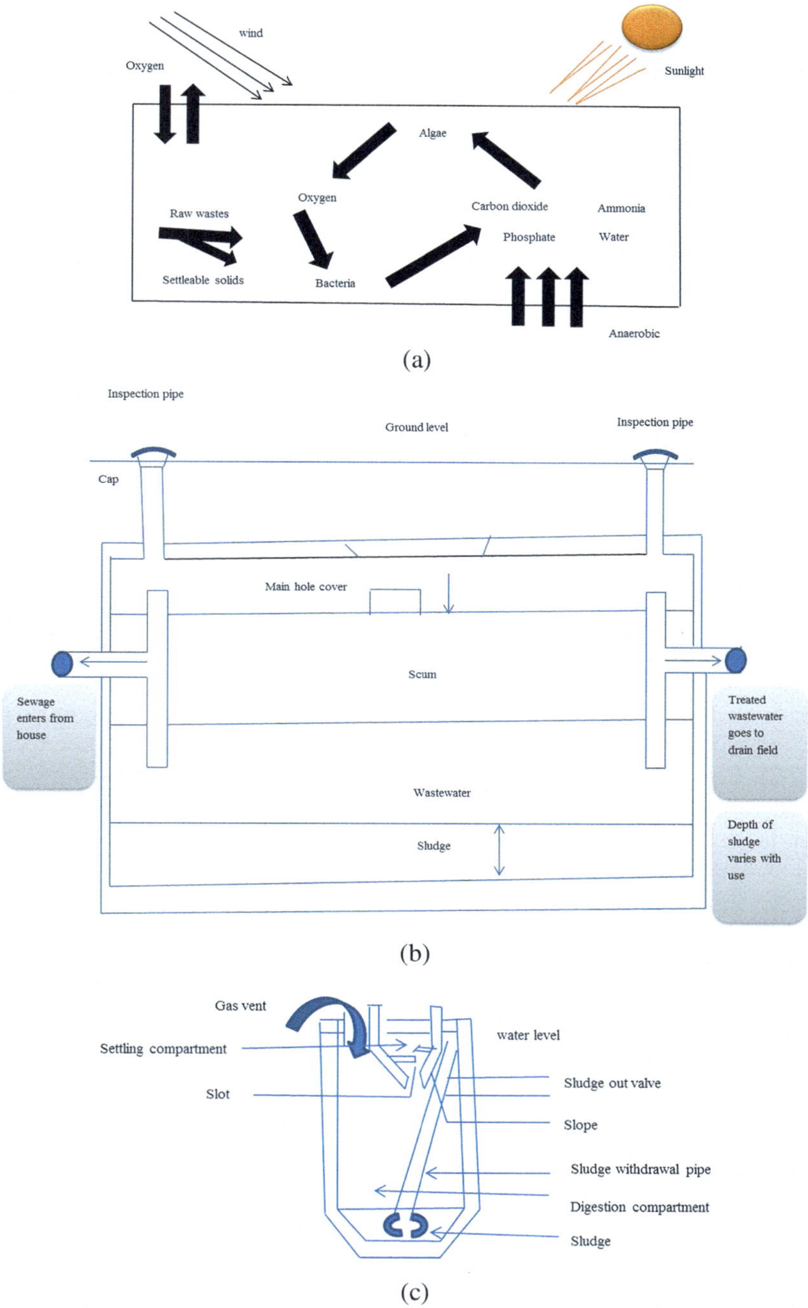

Fig. 7.2 Different schematic structures for anaerobic processes of wastewater treatment: (**a**) anaerobic ponds, (**b**) septic tanks, (**c**) Imhoff tanks

its discharge to the surface waters. Figure 7.2b illustrates the three zones, which operate on the septic tank process.

7.2.2.3 Imhoff Tanks

Imhoff tank is basically used for the treatment of municipal, sewage, and domestic wastewater. It is an improved form of septic tank, which usually consists of a two-story deep tank. The sedimentation or settling usually occurs in the upper story of tank, while the digestion process of settled solid occurs in the lower compartment of tank. Gases (methane, carbon dioxide, etc.) produced by the digestion process escape through the gas vent. The solid sludge present at the bottom can be withdrawn at a regular interval. The process efficiency of Imhoff tank is noted up to 25–35% in BOD and 40–60% in suspended solids (http://www.biologydiscussion.com/waste-management/waste-water-treatment/treatment-of-waste-water/24664). The process of Imhoff tank is clearly illustrated in Fig. 7.2c.

7.3 Use of Reclaimed Water

The fundamental goal as detailed above is the remediation of wastewater from municipal, agricultural, and industrial sources. By developing a closed-loop system, the reclaimed water can be used for downstream applications including municipal water, irrigation, and aquaculture and in industrial cooling systems as described in Table 7.1. Reclaimed water has potential to replace freshwater uses in various fields such as groundwater recharge, aquifer storage, non-portable urban uses, etc. As shown in Fig. 7.3, water reclamation is an essential component of a looped system that maximizes product including clean water while minimizing costs.

7.3.1 For Irrigation

Reclaimed water which arises from dairy and food processing industries have been directly used for irrigation purposes. Polluted (municipal, industrial) wastewater requires treatment process before reuse in irrigation (Mohamed et al. 2014). The viability of using treated wastewater for irrigation is due to its following reasons: (i) the total concentration of dissolved salts such a sodium, potassium, phosphate, sulfate, and nitrates, (ii) crop type (e.g., salt tolerance of particular species), and (iii) process of irrigation (e.g., use of treated wastewater with high salt concentration may cause damages to leaf, stem, buds, etc.). The benefits associated with use of reclaimed water are not restricted with low-cost water source but also enhance in crop yields, minimization of chemical fertilizers, and frost damage protection. Reclaimed water potentially assures the water demands, as it is satisfactorily treated

Table 7.1 Various applications of reclaimed water

Category of use	Specifications of using reclaimed water	Limitations
Landscape irrigation	Park, gardens, cemeteries, sports ground, greenbelts, lawns, etc.	Reclaimed water with high rate of total dissolved solids can adversely affect the flora and fauna
Agricultural irrigation	Edible crops, silviculture, frost protection	Expensive as the source and use are often away from each other
		Health risk may be possible due to the presence of pollutants
		Required extensive treatment before application
Non-portable urban uses	Toilet flushing, commercial laundries, vehicle washing, street cleaning	Dual distribution system cost
Industrial uses	Cooling of towers, boiler feed, washing raw material, cleaning agent of heavy machine, processing water	Treatment required depends on end use
Impoundments	Ornamental, recreational	Dual supply costs of the system
Environmental uses	Stream augmentation, mashes, wetlands	It requires high rate of pollution removal, i.e., ammonia, phosphate, nitrate, BOD, COD, etc.
		Some flora and fauna are sensitive to quality of reclaimed water
Groundwater recharge	Water storage in aquifer, groundwater, etc.	Requires an appropriate hydrogeological condition
Miscellaneous	Aquaculture, snowmaking, soil compaction, dust control	High level of treatment required

to make certain water quality suitable for reuse in irrigation purposes. The appropriate performance of reclaimed wastewater depends on the density of enteric viruses as it is responsible for major health risk. Hence, the cost-effective method of identification and quantification of harmful viruses is important before application of reclaimed water for irrigation to minimize the health risk.

7.3.2 *For Agriculture*

Reclaimed water is applicable in various fields of agriculture such as irrigation of gardens, agriculture and plantation of tree, and silviculture from the off-site treatment plants (Agrafioti and Diamadopoulos 2012); (http://www.ecoefficiency.com.au/Portals/56/factsheets/foodprocess/dairy/ecodairy_fs6.pdf). Therefore, wastewater must be treated and follow the secondary wastewater standards (i.e., < 20 mg/L BOD and < 30 mg/L suspended solid, total coliform <1000 organisms per 100 ml for irrigation). While in case of subsurface irrigation, these requirements are not necessary. The reuse of treated wastewater for vegetable production is directly consumed

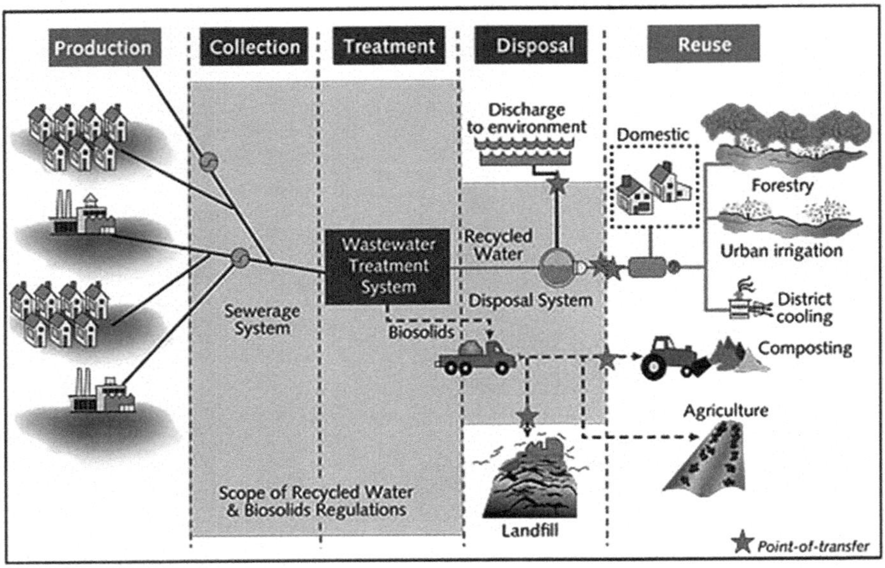

Fig. 7.3 Reuses of treated wastewater as per literature reviewed (Guide to Recycled water and biosolids regulation 2010)

by the human beings much stricter to guidelines of CPCB before application. Therefore, in recent years reuse of treated wastewater has gained significance in water-scarce provinces. The overarching aim of reuse of treated wastewater in agricultural land fulfills the demand of growers to ensure the food safety. Reclaimed water endows with the tremendous environmental benefits in terms of decreasing wastewater discharges, reducing and preventing pollution load in environment. Application of reclaimed water for agriculture is often less than the cost of freshwater used in agricultural fields, because reclaimed water possess valuable nutrients, i.e., nitrate, phosphate, and potassium, responsible for growth and development of plants and simultaneously minimizes the addition of fertilizer cost.

7.3.3 For Aquaculture

The wide application of domestic wastewater, food industry wastewater, dairy industry wastewater, and gray wastewater in aquaculture is increasingly considered a favorable method of integrating reuse of wastewater and nutrient recycling for aquatic cultivation as illustrated in Fig. 7.4. Application of reclaimed water in aquaculture has been accomplished in several nations for a substantial period of time as it serves as a source of essential nutrients required by the plant, i.e., nitrogen, phosphorous, and potassium. It is widely applicable in the tropics. Generally, wastewater is not reused directly in aquaculture system (Bradford et al. 2008). The

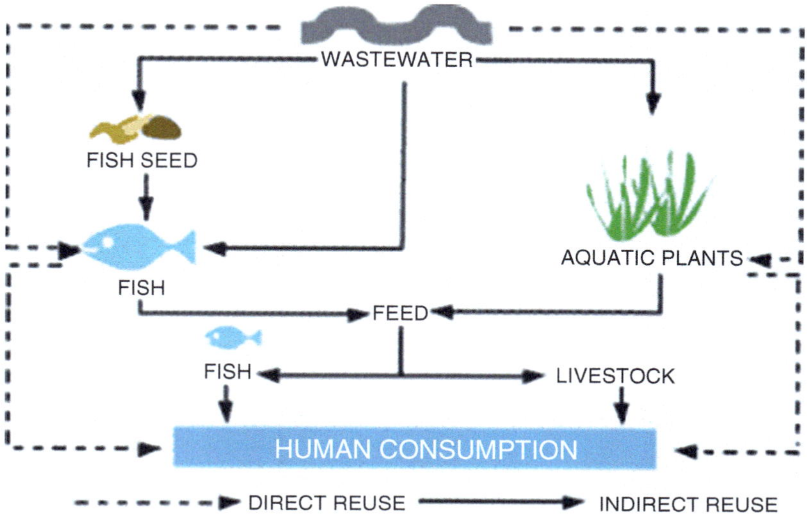

Fig. 7.4 Wastewater reuse in aquaculture (Edwards 2000)

nutrients contained in wastewater are usually preferred for aquaculture as it has potential to provide nutrient for fishes and aquatic weeds, i.e., duckweed which is usually grown for animal feeding. It has also been reported that the microbiological quality for fish culture system in treated wastewater pond is better than freshwater cultured fish (Bansal et al. 2007). In addition to provide an additional, dependable, and locally controlled water supply, reclaimed water can help to find the solutions to diminish the diversion of water from sensitive ecosystems. Using reclaimed water for aquaculture is beneficial as it reduces the crisis of freshwater availability. But it is expensive as it requires high-grade electricity with posttreatment storage and energy intensive process.

7.3.4 For Industrial Applications

Reuse of treated wastewater is also a need of industries, starting from uncontaminated water for boilers to electricity generation and lower water quality for cooling tower. Correas et al. (2016) reported a novel application of nanoparticles, i.e., hydroxyapatite in drug synthesis industries precipitated from wastewater of aquaculture. Complex industrial wastewater such as tannery industry wastewater is found less suitable for biological treatment due to its high acidic and heavy metal concentration. Recently, Yilmaz et al. (2017) explored the use of tannery industry wastewater for electrochemical oxygen generation to reduce the consumption of electricity during the removal of COD (82%). Furthermore, author has shown the feasibility of electrochemical treatment for energy generation using tannery industry wastewater. A power generating plant (620 megawatts production of electricity) uses

Table 7.2 Industrial wastewater, treatment and reuse

Wastewater	Treatment method	Reuse	References
Textile industry wastewater	Membrane filtration	Cleaning of raw materials and various flushing steps during whole processing units	Rossi et al. (1999)
Poultry industry wastewater	Filtration and ozone treatment method	Cleaning agent	Waldroup et al. (1993)
Food and beverages industry wastewater	Two-stage membrane combination of nanofiltration, membrane filtration, and UV disinfection	Boiling and cooling purpose, cleaning agent, transportation, and conditioning of raw material	Rossi et al. (1999)
Paper and cellulose industry wastewater	Biological treatment, membrane filtration (micro-, ultra-, and nanofiltration), membrane bioreactor	Cleaning agent for raw material	Laitinen et al. (2002)

4600 gallons per minute of municipal sewage wastewater by cooling tower (industries.westech-inc.com/blog-commerical-industry/utilizing-unconventional-water-sources-for-industrial-reuse). Table 7.2 describes the multiple applications of treated wastewater.

7.3.5 *Others*

Beside the abovesaid applications of reclaimed water, it can also be utilized in recreational and ornamental impoundments. Recreational impoundments further divided into restricted and nonrestricted impoundments, i.e., non-body (boating and fishing) and body (swimming) contact impoundments, respectively. Reclaimed water can also be used to sustain the wetlands and riparian habitats. Reclaimed water is also applicable in surface water augmentation, aquatic habitat restoration, groundwater recharge, and aquifer storage and recovery. Uses of reclaimed water as portable water also magnify its (reclaimed water) broad-spectrum applicability. Use of reclaimed water for drinking purpose is categorized into two parts, i.e., indirect portable use and direct portable use.

7.4 Bioproducts

A wide range of aquatic flora plays a significant role to remove pollutants prom wastewater. Various industrial wastewater containing hazardous heavy metals (Ag, Pb, Cd, Zn Cu, Hg, and Cr) are being treated by aquatic microorganism and plants, i.e., algae (*C. pyrenoidosa*, *Scenedesmus* sp., and *C. vulgaris*), water hyacinth (*Eichhornia crassipes*), *Fool's Watercress* (*Apium nodiflorum*), *Eurasian watermilfoil* (*Myriophyllum spicatum*), and duckweed (*Lemna trisulca*). After treatment of wastewater, this biomass has various applications like animal feed, fodder, composting, fertilizer, etc.

7.4.1 *Animal Feed*

Biologically treated dairy industry wastewater, i.e., separator de-sludge and product returns, are a good quality source of protein and fat, which can be directly used as animal feed onsite (Bradford et al. 2008). It reduces the transport costs involved in waste disposal from source to dumping area. The dried biomass after removal of extracellular water is applicable in agriculture sector, either as fertilizer (*Laminaria digitata*, *Saccharina latissima*, *Fucus vesiculosus*, *Ascophyllum nodosum*, and *Phymatolithon calcareum*) or animal feed (Delrue et al. 2016). For this purpose, biomass should be free from persistent pollutants, i.e., heavy metals or persisting organic pollutants to minimize the transfer of pollutants into the soil and animals.

7.4.2 *Composting*

The wet biomass produced after treatment of wastewater can be directly used as a feedstock in composting (Mulbry et al. 2008). It can be applied for local/regional area at pilot scale by using micro- and macroalgae (*E. gracilis*, *C. pyrenoidosa*, and *Selenastrum* sp.) as compost.

7.4.3 *Others*

The biomasses produced after treatment of wastewater have various uses in the medicinal and cosmetics industries. These biomasses are widely used to produce high-value chemical compounds such as pigments, carotenes, antioxidants, sun cream, and bioactive compounds.

7.5 Application of Produced Biomass on Wastewater for Bioenergy

To avoid the major economic limitations, integration of biofuel production with wastewater treatment is a best option to provide economics and environmental benefits. Therefore, a comprehensive life cycle assessment study has been performed

Table 7.3 Different ways for energy generation through wastewater (Shoener et al. 2014)

Energy from wastewater	Description	Applications
Biohydrogen/ hydrogen	Biohydrogen can be produced by electrolysis of wastewater using conventional/ nonconventional energy sources	Transportation and electricity generation
Bioelectricity	Chemical energy of wastewater is converted in to electricity through microorganism in microbial fuel cell application (MFC)	Bioelectricity and potable water
Biosolids	Biosolids present in wastewater can be incinerated to produce electricity	Electricity generation
Algae	Wastewater can be used as nutritive medium for algal cultivation, which can be used for various energy products	Produced algae can be used as a source for biofuel application
Heat	Heated water from house and industry is spent to the drain; energy can be recovered as thermal energy	Heat recovery for heating applications
Anaerobic digestion	Sludge can be digested by anaerobic microorganism and produces methane gas	Methane is used for cooking, lighting, and electricity generation

on biomass cultivation and summarized that economic feasibility of biomass-based biofuel would be enhanced significantly by the continuous reuse of treated wastewater (Table 7.3). Nevertheless, 100% recycling of wastewater is not possible even under suitable operational condition. A potential application of treated wastewater for energy saving clearly stated that to replace the 25% of the US transportation fuel by algal-based biofuel, $1.73*10^{12}$ kg/yr of algal biomass is required. To fulfill the demand of such a huge amount of algal biomass, $1100*10^9$ m^3/yr water is required (Farooq et al. 2015). Therefore, use of treated wastewater for biomass (algal) cultivation is a sustainable production pathway significant in terms of bioenergy generation.

Hence, it is very clear from the above points that reuse of treated wastewater can contribute to national development with minimization of environmental damages and health risk caused by the direct discharge of wastewater into the river. In order to promote the reuse of treated wastewater, a positive collaboration between users, authorities, and public is highly needed. Therefore, exchange of technologies and experiences is very important at global level.

7.5.1 Biodiesel

According to USDOE, microalgae have a significance to produce 100 times more oil/acre of land compared to other plants even better than soya beans (Quinn and Davis 2014). There are many microbial species which can be induced to accumulate substantial content of lipid as it is clearly described in Table 7.4. These are food

Table 7.4 Different wastewater and algal biomass for biodiesel production

Algal strain	Wastewater	Biodiesel production	References
Chlorella saccharophila	Dairy industry wastewater	21.82 ± 2. 06	Hena et al. (2015)
Scenedesmus sp.		13.64 ± 0.84	
Consortium		21.34 ± 1.26	
Chlorella pyrenoidosa		6.7 ml/ 6.8 g of dry weight	Kothari et al. (2012)
Chlamydomonas mexicana	Piggery wastewater	33 (% dry weight)	Abou-Shanab et al. (2013)
Scenedesmus obliquus	Piggery wastewater	31 (% dry weight)	Abou-Shanab et al. (2013)
Auxenochlorella protothecoides	Concentrated municipal wastewater	28.9 (% dry weight)	Zhou et al. (2012)
Botryococcus braunii	Carpet mill industry wastewater	9.50 ± 1.24 (%)	Chinnasamy et al. (2010)
Chlorella saccharophila	Carpet mill industry wastewater	17.00 ± 2.89 (%)	Chinnasamy et al. (2010)
Dunaliella tertiolecta	Carpet mill industry wastewater	12.20 ± 1.41 (%)	Chinnasamy et al. (2010)

source for many animals due to presence of polyunsaturated fatty acids, vitamins, minerals, and oil and come in the bottom of food chain being main producer of oxygen on the earth. Algae have the potential to synthesize triacylglycerides (TAGs) and are considered as a resourceful feedstock for biodiesel production (Amaro et al. 2011).

Microalgal biomass has potential to minimize the greenhouse gases through the substitute of fossil fuels, and it is economically feasible as it does not require high operational cost and labor cost. Lipid content of algae plays a crucial role in biodiesel formation. The first step in biodiesel production is (i) microalgal production (open pond system or in closed PBRs which is a state-of-the-art technology, developing rapidly), (ii) harvesting of cultivated algae (including different techniques, i.e., combination of different chemical, mechanical, and thermal method), (iii) oil extraction (disruption of algal cells and intracellular lipid extraction), and (iv) transesterification of extracted oil to produce biofuel in the form of biodiesel. The important benefits of alga-based biodiesel are its biodegradability and nontoxic fuel with less environmental harm over the petrochemical-based fuels.

7.5.2 Biohydrogen

A wide range of wastewater (agricultural wastewater, distillery wastewater, dairy industry wastewater, municipal wastewater, food processing wastewater, beverage industry wastewater, rice slurry wastewater, sugarcane molasses waste, wheat and corn straw) are being used as substrate for biohydrogen production (Senturk and Buyukgungor 2017; Douglas et al. 2017; Alemahdi et al. 2015; Wang et al. 2013). Dark fermentation of these organic wastes with appropriate inoculum (i.e., algal biomass, deoiled algal biomass, bacterial strains) is a promising technology for biohydrogen production. In spite of its significant potential, this technique needs an improved development in utilization of substrate, inoculum, and various operational parameters (i.e., pH, temperature, nutrient, H_2 partial pressure, etc.). Algal and deoiled microalgal biomass (oil-extracted or residual algal biomass) have wide application in biohydrogen production. Oil-extracted algal biomass can be reused as a feedstock for anaerobic digestion to produce biohydrogen. It has been investigated that composition of oil-extracted algal biomass is lignin-free cellulose, associated with starch which is an ideal feedstock for dark fermentation process to produce H_2 gas (Bruno et al. 2009). In dark fermentation process, mixed anaerobic consortia act as biocatalyst which makes it favorable for biohydrogen (H_2) production. Organic compound, wastewater, different organic wastes, and insoluble cellulosic materials are widely used as favorable feedstock for biohydrogen production (Reeta et al. 2013; Mohanakrishna and Venkata 2013). The bioprocess routes of biohydrogen production by using lipid-extracted algae are divided into two steps: (i) the first stage of biohydrogen production involves pretreatment of deoiled algae through hydrolysis as a result of which simple sugar is formed, and (ii) in the second step, the simple sugar formed undergoes in the process of fermentation to form biohydrogen in the occurrence of biocatalyst, i.e., acidogenic bacteria (Venkata 2008). Apart from algal biomass, various bacterial strains *Enterobacter aerogenes*, *Klebsiella* sp., *Halanaerobium hydrogeninformans*, and *T. neapolitana are also able in hydrogen production. Kothari* et al. (2016) *reported a significant amount of biohydrogen production (0.562 L-*H_2L^{-1}*) by using Enterobacter aerogenes,* bacterial strain with dairy industry wastewater. Hence, biohydrogen production by anaerobic digestion and dark fermentation process is significant and economically feasible as algal biomass and hydrogen-producing bacterial strain are easily available and can use wastewater as nutritive supplements for growth and development. Table 7.5 describes the rate of biohydrogen production with selected wastewater and biomass.

7.5.3 Bioethanol

Paper industry, municipal wastewater, and sewage sludge are highly rich in lignin and cellulosic carbohydrate wastes (Ishola et al. 2015; Yan et al. 2015; Thatoi et al. 2014). These wastewater and sludge are suitable feedstocks for hydrolysis to produce ethanol. Fungal (*Saccharomyces cerevisiae*) and bacterial (*Zymomonas mobilis*) culture separately as well as in consortia (*Escherichia coli* and *Saccharomyces* sp.) are widely used in fermentation to produce bioethanol (Bharathiraja et al. 2014; Ivanova et al. 2011). Fermentation of starch-rich algal biomass is used for bioethanol production. The starch and starch-type materials such as polysaccharides are found in alga mainly as a food reserve material among the genera of Chlorophyta,

Table 7.5 Biohydrogen production rate with various types of biomass with wastewater

Inoculum	Wastewater	Biohydrogen production	References
Pond sludge with dense microbes	Cassava wastewater pond sludge	1.91 mol H2/mol glucose	Amorim et al. (2014)
Pretreated sewage sludge	Food waste	5.0 H_2/L/d	Kim et al. (2013)
Rhodobacter sphaeroides O.U.001	Dairy wastewater	3.6 L H2/L dairy wastewater	Seifert et al. (2010)
E. aerogenes	Cheese whey	2.04 mol H2/mol lactose	Rai et al. (2012)
Clostridium acetobutylicum ATCC	Cassava processing wastewater	824 2.41 mol H2/mol glucose	Cappelletti et al. (2011)
C. butyricum	Cornstalk	92.9 mL H2/g dry biomass	Song et al. (2014)
T. thermosaccharolyticum W 16	Cornstalk	89.3 mL H2/g dry biomass	Zhao et al. (2013)
Microalgae	Enriched functional consortia	25.1 mL H2/g dry biomass powder	Ho et al. (2012)

Rhodophyta, Cryptophyta, etc. (John et al. 2011). The biotechnological process for bioethanol production follows the starch production and its conversion through anaerobic pathway using an anaerobic microorganism (Branyikova et al. 2011). These types of conversion pathways involve various intermediate steps which increases the complexity of the process (Ho et al. 2013a, 2013b).

7.5.4 Biogas

The foremost feedstock involved in anaerobic digestion process includes biomass from crop residues, forest residues, lignocelluloses, and dedicated energy crops; however recent advancement involves alga (freshwater, salt water, and cyanobacteria) as a biomass feedstock for biogas production process. Wastewater generated from different industrial plants (e.g., alcoholic beverage plants, rubber plants, ethanol plants, edible oil refineries, cassava plants, paper industries, and oil palm mills) are widely used to generate electricity from biogas (DEDE 2015). In Thailand, around 1459 industrial plants produce 1311 million m^3 of wastewater which has potential to generate 158 MW of electricity from biogas per year (Deeswasmongkol and Paoprasert 2016). Experimental research has been shown that *Chlamydomonas reinhardtii* has potential to produce 587 ml of biogas per volatile solids, while fermentation of *Scenedesmus obliquus* was incompetent with only 287 ml of biogas per volatile solid being produced, and it has been observed that methane content of biogas was 7 to 13% higher than methane content present in biogas produced from maize. The residual algal biomass can be used for methane generation. This conversion process does not require the drying process as the fresh algal biomass is directly used inside the biogas plant (Parmar et al. 2011). The generated biogas typically contains energy carrier in the form of methane and carbon dioxide. It is technique trouble-free to produce biogas in contrast to biodiesel production. The residual product obtained from the biogas plant remains as a nutrient-rich source that can be used in agriculture (Rodolfi et al. 2009). It has been reported that upflow anaerobic sludge bank has potential to convert organic compound of wastewater into methane in small high-rate reactors. Around 60% of the anaerobic full-scale treatment facilities are treating a wide range of industrial wastewater for methane production at global level (Angenent et al. 2016). Bioenergy concept must be implemented to sustain the environment as well as green economy.

Usher et al. (2014) reported energy content of different biomass-based biofuel, viz., significant biomass-based bioenergy options (biodiesel from algae 35–41 MJ/kg, bioethanol 23.4 MJ/kg, biogas 37.2 MJ/kg, bio-oil 33–39 MJ/kg, hydrogen 144 MJ/kg, biodiesel from soya 37.2 MJ/kg) that have significant potential to replace the petrochemical-based fossil fuels (gasoline 45 MJ/kg, diesel 48 MJ/kg) in terms of energy content. Therefore, it is clear from the data that biomass-based biohydrogen is significant in terms of energy content to fulfill the demand of bioenergy. But the main shortcomings of biohydrogen are related to its packaging, storage, and transportation which make this fuel economically unfeasible. Zhang et al. (2014) reported significant production of lipid (162.5 mg L^{-1}) and triacylglycerol (147.5 mg $L^{-}1$) using *Scenedesmus* sp. ZTY1 cultivated in domestic wastewater. The author has also reported a significant reduction in dissolved organic carbon, total nitrogen, and phosphorus, i.e., 60, 92.9, and 99.2%, respectively. Deepanraj et al. (2015) reported significant yield of biogas (9926 ml) with volatile solid removal (46.52 g/L) from domestic waste. The author has obtained good results by the use of pretreated domestic wastewater slurry in comparison to untreated wastewater slurry. Kandasamy et al. (2017) have achieved 911 mg/L of bioethanol by the use of lignocellulosic and bagasse waste pretreated by acid (H_2SO_4) and alkaline (NaOH).

Hence, a significant potential of biofuel can be obtained by using different wastewater, but biogas and bioethanol require a pretreatment process which is economically unviable and harmful in environmental viewpoint. However, biodiesel production does not require these pretreatment processes of wastewater. Therefore, biomass-based biodiesel production integrated with various wastewater treatments is a way forward to sustain the environment and green fuel economy.

7.6 Conclusion

This present chapter is focused and offered a snapshot on different biological techniques of wastewater treatment enabling to recover bioenergy by the application of treated wastewater. Various aerobic and anaerobic methods of wastewater treatment have been clearly described for further application of treated wastewater in various fields (irrigation, aquaculture, industries, etc.). Reuse of treated wastewater with the help of aquatic biomass is significant approach for energy generation as it minimizes the overload on freshwater sources. Therefore, closed-loop systems of bioenergy generation integrated with biological method of wastewater treatment to minimize the environmental risk provide a new insight for clean environment and energy sector and concluded with two aspects, i.e., challenges and ways to move forward in this sector:

7.6.1 Challenges Ahead

Conventional biological techniques for wastewater treatment are not effectively implemented on the part of clean and green society. Hence, in order to achieve widespread acceptance and adoption of techniques with respect to closed-loop system for energy production and wastewater treatment, a transformation of perspectives must be needed on planning and policy making on wastewater management issues.

7.6.2 A Way Forward

A closed-loop system for wastewater treatment with energy production can be a reliable, economical, and regenerative option for problems of energy and environmental crisis in broader sense. Biomass-based energy options in integration with wastewater treatment also helpful in reducing the carbon footprinting from fossil fuel-based transportation sector and utilization of wastewater as a growth media manage to help in reduction of freshwater footprinting, in a sustainable way for clean environment.

References

Abou-Shanab RA, Ji MK, Kim HC, Paeng KJ, Jeon BJ (2013) Microalgal species growing on piggery wastewater as a valuable candidate for nutrient removal and biodiesel production. J Environ Manag 115:257–264

Agrafioti E, Diamadopoulos E (2012) A strategic plan for reuse of treated municipal wastewater for crop irrigation on the Island of Crete. Agric Water Manag 105:57–64

Ajayan KV, Selvaraju M, Unnikannan P, Sruthi P (2015) Phycoremediation of tannery wastewater using microalgae Scenedesmus species. Int J Phytoremediation 17:907–916

Alemahdi N, Che Man H, Abd Rahman N, Nasirian N, Yang Y (2015) Enhanced mesophilic bio-hydrogen production of raw Rice straw and activated sewage sludge by co-digestion. Int J Hydrog Energy 40:16033–16044

Amaro HM, Guedes A, Malcata FX (2011) Advances and perspectives in using microalgae to produce biodiesel. Appl Energy 88:3402–3410

Amorim NCS, Alves I, Martins JS, Amorim ELC (2014) Biohydrogen production from cassava wastewater in an anaerobic fluidized bed reactor. Braz J Chem Eng 31:603–612

Angenent LT, Richter H, Buckel W, Spirito CM, Steinbusch KJJ, Plugge CM, Strik DPBTB, Grootcholten TIM, Buisman CJN, Hamelers HVM (2016) Chain elongation with reactor microbiomes: open culture biotechnology to produce biochemical. Environ Sci Technol 50:2796–2810

Bansal AK, Mitra A, Arora RP, Trilop Gupta, Singvii BSM (2007) Biological treatment of domestic wastewater for aquaculture. JABS. ISSN 19906-6145

Bharathiraja B, Jayamuthunagai J, Praveenkumar R, Vinoth Arulraj J, Vinoshmuthukumar P, Saravanaraj A (2014) Bioethanol production from lignocellulosic materials – an overview. SciTech J 7:20–29

Bradford SA, Segal E, Zheng W, Wang Q, Hutchins SR (2008) Reuse of concentrated animal feeding operation wastewater on agricultural lands. J Environ Qual 37:97–115

Branyikova, Marsalkova B, Doucha J, Branyik T, Bisova K, Zachleder V, Vitova M (2011) Microalage –novel highly efficient starch producers. Biotechnol Bioeng 108:766–776

Bremer JH, Harter T (2012) Domestic wells have high probability of pumping septic tank leachate. Hydrol Earth SystSci 16:2453–2467

Bruno J, Ortega-Lopez V, Coronas A (2009) Integration of absorption cooling systems into micro gas turbine trigeberation systems using biogas: case study of a sewage treatment plant. Appl Energy. 86:837–847

Cappelletti BM, Reginatto V, Amante ER, Antônio RV (2011) Fermentative production of hydrogen from cassava processing wastewater by *Clostridium acetobutylicum*. Renew Energy 36:3367–3372

Cema G, Zabczynski S, Ziembinska-Buczynska A (2016) The assessment of the coke wastewater treatment efficacy in rotating biological contractor. Water Sci Technol 73:1202–1210

Chinnasamy S, Bhatnagar A, Hunt RW, Das KC (2010) Microalgal cultivation in wastewater dominated by carpet mill effluents for biofuel productions. Bioresour Technol 101:3097–3105

Correas C, Gerardo ML, Lord AM, Ward MB, Andreoli E, Barron AR (2016) Nanostructured fusiform hydroxyapatite particles precipitated from aquaculture wastewater. Chemosphere 168:1–7

DEDE, Department of Alternative Energy Development and Efficiency (2015) Alternative Energy Development Plan 2015. Available from: http://www.dede.go.th/download/files/AEDP2015_Final_version.pdf

Deepanraj B, Sivasubramanian V, Jayaraj S (2015) Experimental and kinetic study on anaerobic digestion of food waste: the effect of total solids and Ph. J Renew Sustainable Energy 7:063–104

Deeswasmongkol K, Paoprasert NH (2016) Design of Biogas Production Using Wastewater from industrial plants: financial analysis of various feed-in-tariff remunerations in Thailand. Int J Energy Econ Policy 6:318–334

Delrue F, Álvarez-Diaz PD, Fon-Sing S, Fleury G, Sassi JF (2016) The environmental biorefinery. Using microalgae to remediate wastewater, a winwin paradigm Energy 9–132

Douglas GBT, Dal S, Lucas M, Andreani CL, Carvalho KQD, Coelho CRM, Gomes SD (2017) Production and performance of anaerobic fixed-bed reactors using three support arrangements from cassava starch wastewater. J Braz Assoc Agric Eng 37:160–172

Edwards P (2000) Wastewater-fed aquaculture: state-of-the-art, waste recycling and reuse management in the world. Int Ecol Eng Soc 37–49

Elmekawy A, Hegab H, Dominguez-Benetton X, Pant D (2014) Internal resistance of microfluidic microbial fuel cell: challenges and potential opportunities. Bioresour Technol 142:672–682

Farooq W, Suh WI, Park MS, Yang JW (2015) Water use and recycling in microalgae cultivation for biofuel application. Bioresour Technol 184:73–81

Gao P, Xu W, Sontag P, Li X, Xue G, Liu T, Sun W (2016) Correlating microbial community compositions with environmental factors in activated sludge from four full-scale municipal wastewater treatment plants in Shanghai China. Appl Microbiol Biotechnol 100:4663–4673

Garg S, Chaudhry S (2017) Treatment of wastewater of food industry by membrane bioreactor. IJRASET 4:153. ISSN 2393-8021

Guide to Recycled water and biosolids regulation (2010)

Guo XM, Trably E, Latrille E, Carrère H, Steyer JP (2010) Hydrogen production from agricultural waste by dark fermentation. Int J Hydrog Energy 35:10660–10673

Habibi A, Vahabzadeh F (2013) Degradation of formaldehyde in packed-bed bioreactor by kissiris-immobilized Ralstonia eutropha. Biotechnol Bioprocess Eng 18:455–464

Hena S, Fatimah S, Tabassim S (2015) Cultivation of algae consortium in a dairy farm wastewater for biodiesel production. Water Res Ind 10:1–14

Ho SH, Chen CY, Chang JS (2012) Effect of light intensity and nitrogen starvation on CO2 fixation and lipid/carbohydrate production of an indigenous microalga Scenedesmus obliquus CNW-N. Bioresour Technol 113:244–252

Ho KL, Lee DJ, Su A, Chang JS (2013a) Biohydrogen from lignocellulosic feedstock via one-step process. Int J Hydrog Energy 37:15569–15574

Ho SH, Huang SW, Chen CY Hasunuma T, Kondo A, Chang JS (2013b) Bioethanol production using carbohydrate-rich microalgae biomass as feedstock. Bioresour Technol 1351:91–198

http://www.ecoefficiency.com.au/Portals/56/factsheets/foodprocess/dairy/ecodairy_fs6.pdf

http://www.farmersjournal.ie/septic-tank-inspection-system-to-be-reviewed-in-2016-185709/. Accessed 13 Aug 2015

http://www.waterworld.com/articles/wwi/print/volume-28/issue-1/regional-spotlight/middle-east-africa/stepping-up-monitoring-for-the-gcc-s.html

Hung YT, Ahmad MSA, Yeh RYL, Liu RLH, Fu YP (2017) Oxidation pond for municipal wastewater treatment. Appl Water Sci 7:31–51

industries.westech-inc.com/blog-commerical-industry/utilizing-unconventional-water-sources-for-industrial-reuse

Ishola M, Brandberg T, Taherzadeh MJ (2015) Simultaneous glucose and xylose utilization for improved ethanol production from lignocellulosic biomass through SSFF with encapsulated yeas. Biomass Bioenergy 77:192–199

Ivanova V, Petrova P, Hristov J (2011) Application in the ethanol fermentation of immobilized yeast cells in matrix of alginate/magnetic nanoparticles, on chitosan-magnetite microparticles and cellulose coated magnetic nanoparticles. Int Rev Chem Eng 3:289–299

Jackson BE, Mclnerney MJ (2002) Thiosulfate disproportion by desulfotomaculum thermobenzoicum. Appl Environ Microbial 66:3650–3653

John RP, Anisha GS, Nampoothiri KM, Pandey A (2011) Micro and macro algal biomass: a renewable source for bioethanol. Bioresour Technol 102:186–193

Kandasamy M, Hamawand I, Bowtell L, Seneweera S, Chakrabarty S, Yusaf T, Shakoor Z, Algayyim S, Eberhard F (2017) Investigation of ethanol production potential from lignocellulosic material without Enzymatic hydrolysis using the ultrasound technique. Energies 10:2–12

Kim S, Choi K, Kim JO, Chung J (2013) Biological hydrogen production by anaerobic digestion of food wastes and sewage sludge treated using various pretreatment technologies. Biodegradation 24:753–764

Kothari R, Pathak VV, Kumar V, Singh DP (2012) Experimental study for growth potential of unicellular alga *Chlorella pyrenoidosa* on dairy waste water: an integrated approach for treatment and biofuel production. Bioresour Technol 116:466–470

Kothari R, Kumar V, Pathak VV, Tyagi V (2016) Sequential hydrogen and methane production with simultaneous treatment of dairy industry wastewater: bioenergy profit approach. Int J Hydrog Energy 42:4870–4879

Laitinen N, Kulovaara M, Levane E (2002) Ultrafiltration of stone cutting mine waste water with ceramic membranesa case study. Desalination 149:121–125

Marañón E, Castrillón L, Quiroga G, Fernández-Nava Y, Gómez L, García MM (2012) Co-digestion of cattle manure with food waste and sludge to increase biogas production. Waste Manag 32:1821–1825

Mendez-Romero DC, Lopez-Lopez A, Vallenjo-Rodreguiz R, Leonbecerril E (2011) Hydrodynamic and kinetic assessment of an anaerobic fixed-bed reactor for slaughterhouse wastewater treatment. Chem Eng Process 50:273–280

Mohamed F, Cherki G, Abdelaziz B (2014) Water deficit effect on yield and forage quality of Medicago sativa populations under field conditions in the Marrakesh area of Morocco. AWUT-Ser Biol XVII:1–8

Mohan SV, Devi MP, Mohanakrishna G, Amarnath N, Babu ML, Sarma PNN (2011) Potential of mixed microalgae to harness biodiesel from ecological water-bodies with simultaneous treatment. Bioresour Technol 102:1109–1117

Mohanakrishna G, Venkata MS (2013) Multiple process integrations for broad perspective analysis of fermentative H_2 production from wastewater treatment: technical and environmental considerations. Appl Energy 107:244–254
Mulbry W, Kondrad S, Buyer J (2008) Treatment of diary and swine manure effluents using freshwater algae fatty acid content and composition of algal biomass at different manure loading rates. J Appl Phycol 20:1079–1085
Nah YM, Ahn KH, Yeom IT (2000) Nitrogen removal in household wastewater treatment using an intermittently aerated membrane bioreactor. Environ Technol 21:107–114
Orumieh H, Mazaheri R (2015) Efficiency of stabilization ponds under different climate condition in Iran. J Life Sci 5(1):794–805
Pant D, Singh A, Van BSG, Irving OS, Singh NP, Diels L (2012) Bio electrochemical systems (BES) for sustainable energy production and product recovery from organic wastes and industrial wastewaters. RSC Adv 2:1248
Parmar A, Singh NK, Pandey A, Gnansounou E, Madamwar D (2011) Cyanobacteria and microalgae: a positive prospect for biofuels. Bioresour Technol 102:10163–10172
Priyadarshani I, Rath B (2012) Commercial and industrial applications of micro algae – a review. J Algal Biomass 3:89–100
Quinn JC, Davis R (2014) The potentials and challenges of algal based biofuels. Bioresour Technol 184:444–452
Rai PK, Singh SP, Asthana RK (2012) Biohydrogen production from cheese whey wastewater in a two-step anaerobic process. Appl Biochem Biotechnol 167:1540–1549
Reeta RS, Anil KP, Gwendoline C, Pierre F, Larroche C (2013) Biological upgrading of volatile fatty acids, key intermediates for the valorization of bio waste through dark anaerobic fermentation. Bioresour Technol 145:166–174
Rodolfi L, Zittelli GC, Bassi N, Padovani G, Biondi N, Bonini TMR (2009) Microalgae for oil: strain selection, induction of lipid synthesis and outdoor mass cultivation in a low-cost photobioreactor. Biotechnol Bioeng 102:100–112
Rossi A, Malpei F, Bonomo L, Bianchi R (1999) Textile wastewater reuse in northern Italy. Milan
Sarma SJ, Pachapur V, Brar SK, Le Bihan Y, Buelna G (2015) Hydrogen biorefinery: Potential utilization of the liquid waste from fermentative hydrogen production. Renew Sust Energ Rev 50:942–951
Seifert K, Waligorska M, Laniecki M (2010) Hydrogen generation in photo biological process from dairy waste water. Int J Hydrog Energy 35:9624–9629
Senturk I, Buyukgungor H (2017) Bio hydrogen production by anaerobic fermentation of sewage sludge – effect of initial pH. Res Environ Ecol 5:107–111
Shoener BD, Bradley IM, Crusick RD, Guest JS (2014) Energy positive domestic wastewater treatment. Environ Sci Processes Impacts 16:1204–1222
Slavov AK (2017) General characteristics and treatment possibilities of dairy wastewater–a review. Food Technol Biotechnol 55:14–28
Song ZX, Li XH, Li WW, Bai YX, Fan YT, Hou HW (2014) Direct bioconversion of raw corn stalk to hydrogen by a new strain clostridium sp. FS3. Bioresour Technol 157:91–97
Tawfick A (2012) Polyurethane trickling filter in combination with anaerobic hybrid reactor for treatment of tomato industry wastewater. INTECH Open Science Open Minds. ISBN 978-953-51-0726-2
Tharavathy NC, Krishnamoorthy M, Hosetti BB (2014) Oxidation pond. A tool for wastewater treatment. J Ecol Environ Sci 2:1–4
Thatoi BH, Dash PK, Mohapatra S, Swain MR (2014) Bioethanol production from tuber crops using fermentation technology: a review. Int J Sustainable Energy 35:443–468
US EPA Report (2011) U.S. Environmental protection agency (EPA) decontamination research and development conference. U.S. Environmental protection agency, Washington, DC, EPA/600/R/12/557, 2012
Usher PK, Ross AB, Camargo –VMA, Tomlin AS, Gale WF (2014) *An* Overview of the potential environmental impacts of large scale microalgae cultivation. Biofuels 5:331–349
Venkata M (2008) Fermentative hydrogen production with simultaneous wastewater treatment: influence of pretreatment and system operating conditions. J Sci Ind Res 67:950–961

Waldroup AL, Hierholzer RE, Forsythe RH (1993) Recycling of poultry chill water using ozone. Fayetteville

Wang Y, Sheng G, Shi B, Li W, Yu H (2012) A novel electrochemical membrane bioreactor as a potential net energy producer for sustainable wastewater treatment. Sci Rep 3:1864

Wang H, Paul DR, Chung TS (2013) Surface modification of polyimide membranes by diethylenetriamine (DETA) vapor for H_2 purification and moisture effect on gas permeation. J Membr Sci. 430:223–233

Yan J, Wei Z, Wang Q (2015) Bioethanol production from sodium hydroxide/hydrogen peroxide-pretreated water hyacinth via simultaneous saccharification and fermentation with a newly isolated Thermotolerant *Kluyveromyces marxianus* strain. Bioresour Technol 93:103–109

Yilmaz JL, Lim ZL, Beganovic M, Breazeale S, Andre C, Stymne S, Vrinten P, Senger T (2017) Determination of substrate preferences for desaturases and elongases for production of docosahexaenoic acid from oleic acid in engineered canola. Lipids 52:207–222

Yoon Y, Hwang Y, Kwon M, Jung Y, Hwang TM, Kang JW (2014) Application of O_3 and O_3/H_2O_2 as post-treatment processes for color removal in swine wastewater from a membrane filtration system. J Ind Eng Chem 20:2801–2805

Zhang T, Shao MF, Ye L (2012) Pyro sequencing reveals bacterial diversity of activated sludge 743 from 14 sewage treatment plants. ISME J 6:1137–1147

Zhang TY, Wu Y, H Hu HY (2014) Domestic wastewater treatment and biofuel production by using microalga *Scenedesmus*sp. ZTY1. Water Sci Technol 69:2496–2496

Zhao L, Cao GL, Wang AJ, Guo WQ, Ren HY, Ren NQ (2013) Simultaneous saccharification and fermentation of fungal pre-treated cornstalk for hydrogen production using *Thermoanaerobacterium thermosaccharolyticum* W16. Bioresour Technol 145:103–107

Zhou W, Li Y, Min M, Hu B, Zhang H, Ma X, Li L, Cheng Y, Chen P, Ruan R (2012) Growing wastewater-born microalga Auxenochlorella protothecoides UMN280 on concentrated municipal wastewater for simultaneous nutrient removal and energy feedstock production. Appl Energy 98:433–440

Zhu X, Tian J, Liu C, Chen L (2013) Composition and dynamics of microbial community in a zeolite bio filter-membrane bioreactor treating coking wastewater. Appl Microbiol 97:8767–8877

Chapter 8
Subsurface Processes Controlling Reuse Potential of Treated Wastewater Under Climate Change Conditions

Pankaj Kumar Gupta and Brijesh Kumar Yadav

Abstract In the last few decades, investigation on ecohydrological interaction including biogeochemical characteristics has been an important research topic in hydrology due to its role in natural resource management. Despite the research focus on this subject, quantifying the geochemical processes controlling the moisture flow and solute transport remains awaited especially under climate change conditions. At the same time, the wastewater treated in conventional wastewater treatment systems are reported with the inefficient removal of many emerging contaminants particularly in rural areas and remote communities in low socioeconomic conditions. As a result, in many instances, the wastewater may get disposed without appropriate advance treatment that further contaminate the soil-water system. Thus, there are urgent needs of the research to assess the risk posed to groundwater and to develop improved knowledge frame of hydrologic and biogeochemical processes and of geologic features controlling contaminant migration in the subsurface. Therefore, the main focus of this chapter is to present the different biogeochemical processes controlling reuse potential of treated wastewater in the subsurface under climate change conditions. The different geochemical process involved during the fate and transport in the subsurface are clearly elaborated and exemplified. Further, the role of varying climatic conditions on biogeochemical makeup and transport is discussed thoroughly. A state of the art of the different aspects of modeling and practical approaches to quantify the governing biogeochemical processes in the subsurface is reviewed comprehensively. Finally, the methodological framework is charted on the basis of the technical and socioeconomic aspect to implement the potential reuse of wastewater and remedial measures in the field. The outcomes of this chapter are of direct use in applying remediation technique in the field and for the decision-making related to the planning of (waste) water under varying environmental conditions.

P. K. Gupta (✉) · B. K. Yadav
Department of Hydrology, Indian Institute of Technology Roorkee, Roorkee, Uttarakhand, India

R. P. Singh et al. (eds.), *Water Conservation, Recycling and Reuse: Issues and Challenges*, https://doi.org/10.1007/978-981-13-3179-4_8

Keywords Treated wastewater · Contaminant transport · Subsurface Environment · Bioremediation · Groundwater resources · Climate change

8.1 Introduction

Subsurface pollution has become a global concern due to the rapid growth of industrialization, urbanization, and modern development under climate change conditions (Gupta and Joshi 2017; Gupta et al. 2013, 2018a, b). Industries are key performers in the economy of developing countries but are also considered as one of the major polluters due to disposal of huge amount of untreated/partially treated wastewater into the environment, which creates serious soil and water pollution and causes severe toxic effects in living beings (Mustapha et al. 2018). The wastewater discharged from various industries carry various organic and inorganic pollutants, which are used in the processing of raw materials to obtain a good quality of products within a short period of time and in an economic way; however, their toxicity is usually ignored that poses a serious challenge for the safety of the environment and human health. Further, technological innovations in industries have given rise to new products and new pollutants in abundant level, which is above the self-cleaning capacity in the environment (Gupta and Sharma 2018; Ranjan et al. 2008).

A large number of impoverished communities around the world face challenges with respect to treatment of sewage and domestic wastewater. In particular, the rural areas and remote communities in low socioeconomic conditions may lack conventional centralized wastewater treatment systems. As a result, in many instances, the wastewater may get disposed without appropriate treatment that contaminates drinking water resources. Even if the communities choose a low-cost system for primary treatment, such as waste stabilization ponds that do not require much operation cost and skilled supervision, the existing wastewater does not undergo traditional secondary treatment for considerable reduction in biodegradable organic material, PPCPs, pathogens, nutrients, etc. from the wastewater. Treatment of wastewater is expensive, and in developing countries, the infrastructure for domestic wastewater treatment prior to disposal is poor in small cities and may be nonexistent in the rural and other remote communities. In developing nations, the discharge of untreated wastewater is the main cause for widespread pollution of surface and groundwater resources since there is often a large gap between generation and treatment of domestic wastewater. For example, according to a recent report (CPCB, India 2009), out of ~38 billion L/day of sewage/wastewater generated, treatment capacity exists for only ~12 billion L/day in India. The problem has reached a crisis proportion as both rural and urban communities may be suffering from pathogen-related health issues (Suthar et al. 2009) due to contamination of precious drinking water sources.

The land irrigated with (un)-treated wastewater containing emerging pollutants including PPCPs, biosolids, and petroleum hydrocarbons, during fertigation, has deteriorated large volumes of soil and groundwater (Clement et al. 2000). These emerging pollutants infiltrate into the subsurface along with soil moisture; it will move into the partially saturated zone. The advective-dispersive mechanisms cause spreading of these pollutants in the partially saturated zone and creates a redox environment in and around the polluted subsurface zones (Christensen et al. 2000). The redox environment controls the microbial numbers, diversity, and activity, which may decrease the natural attenuation of several emerging pollutants. Similarly, adsorption-desorption and or (de)attachment causes the alternation in biogeochemical behaviors of these pollutants in the subsurface.

Impacts of climate change on subsurface water quality are not investigated thoroughly, and very few studies were found relevant. Green et al. (2011a, b) showed that the indirect impacts of climate change are more likely to affect subsurface water quality issues. The geochemical makeup of subsurface environment is under direct influences of the surface or atmospheric conditions; thus, the variations in ambient temperatures, precipitations, stream flow, and other dominating variables affect the subsurface water quality (Klein and Nicholls 1999; Sharif and Singh 1999; Pierson et al. 2001; Ranjan et al. 2006; IPCC 2007a, b). At the same time, the pollution due to the release of several pollutants is a major concern to subsurface water quality. Natural attenuation of these pollutants is likely to be affected by environmental variability as microorganisms present in subsurface work differently under various environmental conditions. Further, rapid groundwater table fluctuations along with high pore water velocities are expected in shallow aquifers under climate change conditions (Dobson et al. 2007a, b) which ultimately affect subsurface water resources. Changing groundwater flow velocity and groundwater table dynamics causes stronger advective transport and enhances the dissolution rate of immiscible pollutants.

Thus, the objective of this book chapter is to present the subsurface processes controlling reuse potential of treated wastewater under climate change conditions. The information of this chapter will help in management and remediation of polluted sites under site prevailing conditions.

8.2 Treated Wastewater: As a Source of Subsurface Pollutants

The application of treated wastewater gains more attention due to water scarcity in most of the parts of the world and water conservation being so important at the present time (Kinney et al. 2008).The largest source of marginal water for agriculture is treated wastewater in many areas. Likewise, in urban areas, most of the treated wastewater is used for garden irrigation. Wastewater does not undergo higher treatment for considerable reduction in biodegradable organic material, TSS, salts (mainly Na, Cl, and bicarbonates), nutrients, microelements, emerging pollutants like pharmaceuticals and personal care products (PPCPs), nanoparticles, pathogens, etc. from the wastewater. Thus, higher quantities of these pollutants in the treated water than in fresh water are reported in most of the literature (Bernstein et al. 2006). To understand the suitability of treated wastewater in fertigation, a comparative account of physicochemical characteristics is presented in this section.

8.2.1 Electrical Conductivity (EC)

The US EPA recommended EC as a significant parameter for reclaimed water quality in different fields such as agriculture and industrial reuse (US EPA 2004). Treated wastewater from municipal and industrial treatment plants, especially pulp and paper industries, has been shown to have high concentrations of ions, which subsequently increases the EC. The high value of EC in treated wastewater than freshwater was reported in the literature. Bernstein et al. 2006 show that the EC value of treated wastewater was 2.0–2.5 dS m^{-1}, which was higher than the potable water. The higher EC values in treated wastewater due to insufficient equalization of wastewater during primary/secondary treatment lead to less removal of total dissolved solids.

8.2.2 pH, Salt Composition, and Chemical Oxygen Demand (COD)

The low performances of conventional treatment resulted in highly dissolved organic matters and salt contents in treated wastewater, which altered the salt composition and pH values too. Generally, the high values of salt concentration and pH were found in previous studies. The highly dissolved organic matters containing treated wastewater resulted in high chemical oxygen demand. Bernstein et al. (2006) reported 86 mg/l of COD in treated wastewater. Similarly, Kushwah et al. (2012) show that the COD varied from 346.4–792.4 mg/l in the influent wastewater and 182.6–86.4 mg/l in the final treated wastewater at Kotra wastewater treatment plant. The pH, salt concentration, and COD of treated wastewater vary with sources of influent wastewater which originated from different industries.

8.2.3 Hydrocarbon Contaminants

Wastewater from petroleum refining and petrochemical industries contain BTEX compounds, polyaromatic hydrocarbons (PAHs), oil and grease, large amounts of suspended particulate matter, sulfides, ammonia, and phenol (Tobiszewski et al. 2012; Mustapha et al. 2015). These discharges are one of the major environmental hazards to human and animals (Seeger et al. 2011). Furthermore, oily wastewater can lead to the loss of biodiversity, destruction of breeding habitats of aquatic organisms, and hazard to the biota including humans (Mustapha et al. 2011). Traditional physicochemical, mechanical, and biological technologies cannot effectively treat the effluents from petroleum industries to meet strict effluent discharge regulations (Saien and Shahrezaei 2012; Wu et al. 2015).

8.2.4 Trace Elements

Research on the fate of heavy metals in treated wastewater shows significant residual occurrence due to low removal efficiencies of many trace elements by secondary as well as advance treatments (Karvelas et al. 2003; Sharma et al. 2007; Chary et al. 2008). Karvelas et al. (2003) investigated occurrence and fate of heavy metals in the wastewater treatment processes and found 47–63% of Cd, Cr, Pb, Fe, Ni, and Zn remains in treated wastewater effluent. Similarly, Sharma et al. (2007) investigated heavy metal contamination of soil resulting from wastewater irrigation in suburban areas of Varanasi, India. The results indicate that concentration of Cd was higher than the permissible limits of the Indian standard during summer, whereas Pb and Ni concentrations were higher in both summer and winter seasons. Furthermore, bioaccumulation/biomagnification enhanced further concentration in biomass. Thus, the reuse potential of wastewater depends on the concentration of these trace metals as well as their bioaccumulation/biomagnification rates (Anitha et al. 2012).

8.2.5 Emerging Pollutants

Emerging pollutants includes pharmaceuticals, and personal care products (PPCPs) designed to have biological effects even at low concentration and concern to cause ecological adverse effects. The main source of these pollutants is (un)-treated wastewater discharged from wastewater treatment plants/sewage treatment plants (WTPs/STPs) and land application of biosolids (Chefetz et al. 2008; Caliman and Gavirilescu 2009). Okuda et al. (2008) evaluated the removal efficiency of 66 pharmaceuticals during wastewater treatment process including conventional activated sludge (CAS) and biological nutrient removal (BNR) processes (Ebele et al. 2017). They show that the 30–80% removal efficiency was achieved, in which removal efficiencies of carbamazepine and crotamiton were less than 30% during wastewater treatment processes. Arye et al. (2010) reported about 50% removal of carbamazepine and other pharmaceuticals in soil irrigated with treated wastewater. These studies show that a lot of emerging pollutants remain in treated wastewater even after complete advance treatment processes. Thus, the land irrigated with (un)-treatment wastewater having a high risk of further contamination by emerging pollutants.

8.2.6 Microbial and Colloids/Nanoscale Pollutants

Treatment of wastewater is expensive, and in developing countries, the infrastructure for domestic wastewater treatment prior to disposal is poor in small cities and may be nonexistent in the rural and other remote communities. In developing nations, the discharge of untreated wastewater is the main cause for widespread pollution of surface and groundwater resources since there is often a large gap between generation and treatment of domestic wastewater. For example, according to a recent report (CPCB, India 2009), out of ~38 billion L/day of sewage/wastewater generated, treatment capacity exists for only ~12 billion L/day in India. The problem has reached a crisis proportion as both rural and urban communities may be suffering from pathogen-related health issues (Suthar et al. 2009). The pathogens of the major threat to the ecological health are viruses and the protozoa. The *cryptosporidium* and *Giardia* are well-known protozoa contaminates. Viruses are the most critical for the groundwater among the microbiological contamination; they are much smaller in size than others and are not filtrated out to the same extent in the porous soil matrix. Rose et al. (1996) reported that water from land irrigated with wastewater contains a significant number of pathogens including viruses and colloids. Likewise, Levantesi et al. (2010) quantify the pathogenic microorganisms at two wastewater reclamation sites and found a high concentration of *Salmonella* gene copies, *Clostridium* spores, and *Giardia* cysts in treated wastewater. Thus, the reuse potential of treated wastewater depends upon the degree of occurrences of these pollutants and their biochemical activation and deactivation characteristics.

Other than nZVI, the nanoparticles tested for remediation purpose are metal and their oxides (Ag, TiO_2, ZrO_2) and carbon-based nanomaterials (EPA 2008; Karn et al. 2009; Watlington 2005). TiO_2 nanoparticles can serve both as reductive and oxidative catalysts and have been explored for the photo degradation of various pollutants that includes dyes, inorganic pollutants, and organic pollutants (Savage and Diallo 2005). After the discovery of carbon nanotubes (CNT) in 1991, its application in water purification has attracted great attention. Carbon nanotubes are basically the graphene sheets in tube form and that can be single-walled or double-walled depending upon the manufacturing conditions (Li et al. 2003).

8.3 Fate and Transport of Emerging Pollutants in Subsurface

The release of (un)-treated wastewater containing emerging pollutants including PPCPs, biosolids, and petroleum hydrocarbons, during fertigation, has deteriorated large volumes of soil and groundwater (Clement et al. 2000). If agriculture land irrigated with treated wastewater, emerging pollutants infiltrate into the subsurface along with soil moisture, it will move into the partially saturated zone (Gharaibeh et al. 2007). Because water and air already make a two-fluid system in the partially saturated soil-water system, the introduction of second liquid results in a three-fluid porous-medium system. In two-fluid systems, two fluids and one solid, and between these three substances, form three possible interface combinations: interfaces between the two fluids, interfaces between one fluid and the solid, and interfaces between a second fluid and the solid. When three fluids are present, there are six possible interface types (interface pairs), so the system becomes quite complex. An additional complication to the overall multi-fluid problem is the observation that immiscible organic fluid is often composed of a number of components. The multi-fluid porous-media systems are characterized by a solid phase within which interconnected pore space allows fluids to flow along with nonaqueous liquids, like petrochemical products. For example, gasoline has a number of components in it, some of which are lighter fractions like benzene, toluene, ethylbenzene, and xylene (BTEX).

In partially saturated zones, pollutants fractionalized and cause partitioning into air phase, aqueous phase, solid phase, and pure phase itself. The remaining mass of pollutants moves downward through the partially saturated zone, in which lighter pollutants like LNAPLs are generally retained by the water table due to their lighter density to water, while dense pollutants like DNAPLs penetrate the water table and move downward till they are retained by an impermeable layer (USEPA 1995). At the water table, LNAPL starts dissolving with groundwater and subsequently moves to surrounding downgradient locations due to advection, diffusion, and dispersion mechanisms of mass transport (Dobson et al. 2007a, b; Powers et al. 1991).

Advection, dispersion, and liquid diffusion contribute to contaminant spreading in the water-saturated area, with the last being the slowest process. Advective transport driven by groundwater flow depends on groundwater velocity. Mechanical dispersion causes migration in both a parallel as well as in a perpendicular direction with respect to the main groundwater flow direction. Dispersion dominates over diffusion at groundwater velocities higher than 0.1 m day-1 (McCarthy and Johnson 1993). Diffusion, driven by solute concentration gradients over space, may dominate at low groundwater velocities.

Let us consider the subsurface contaminant mass balance in an elementary root zone layer of Δz thickness shown in Fig. 8.1. The contaminant influx into the layer is denoted by $J(z)$ (mg per unit area per unit time) $[ML^{-2}S^{-1}]$, and the outflux from the layer is $J(z + \Delta z)$. S^C is the sink term denoting the amount of solute taken up by plant

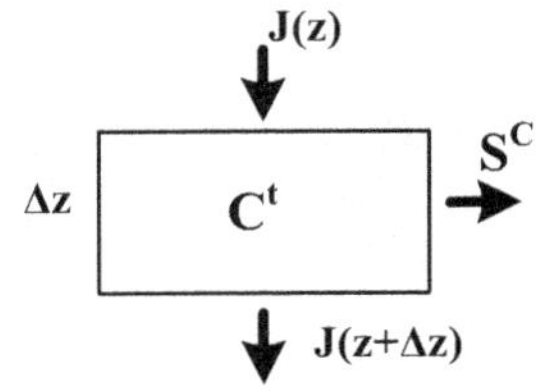

Fig. 8.1 Diagram showing a soil-water system having subsurface contaminant influx (J_z) and outflux ($J_{z+\Delta z}$) with sink term (S^C)

roots and degradation of the contaminant by microbial biomass [$ML^{-3}S^{-1}$]. Then, the change in total contaminant concentration C^t over time t can be written as:

$$\frac{\partial C^t}{\partial t} = \frac{J(z) - J(z + \Delta z)}{\Delta z} - S^C \tag{8.1}$$

Diffusive Flux Diffusion is mathematically described by Fick's law as the net rate of contaminant transport proportional to the negative gradient of its concentration and can be modified for the unsaturated porous medium as:

$$J_{\text{diff}} = -\tau D_o \theta \frac{\partial c}{\partial z} = -D_m \theta \frac{\partial c}{\partial z} \tag{8.2}$$

where τ is the tortuosity factor (dimensionless) which accounts for the increased distance of transport due to tortuous path of the solute particle in a porous media. D_o and D_m are the free-water diffusivity and molecular diffusion coefficients, respectively [$L^2 T^{-1}$].

Dispersive Flux The dispersion at a microscopic scale occurs due to the variation of velocity within the pores and due to the tortuous movement of the fluid around the soil grains. Macroscopic dispersion refers to the dispersion resulting from the interfingering of materials of different permeability. Mechanical dispersion is mathematically described in the same way as molecular diffusion by using the Fick's law as:

$$J_{\text{dis}} = -\alpha_L v \frac{\partial C}{\partial z} = -\alpha_L \frac{q}{\theta} \frac{\partial C}{\partial z} \tag{8.3}$$

where α_L is the longitudinal dispersivity of the porous media in the direction of flow [L] and v is the pore velocity. Now by adding all the abovementioned fluxes, we get the resultant flux as in Eq. 8.4.

$$J = J_{\text{adv}} + J_{\text{dis}} + J_{\text{diff}} = qC - D\theta \frac{\partial C}{\partial z} \tag{8.4}$$

where $D = \tau D_O + \alpha_L \frac{q}{\theta} = \tau D_O + \alpha_L \nu$ here; D is the diffusion-dispersion or hydrodynamic dispersion coefficient, which is the pore water velocity-dependent function [$L^2\ T^{-1}$]. The modified form of advection-dispersion as:

$$\frac{\partial(\theta c + \rho_b s)}{\partial t} = -\frac{\partial}{\partial z}\left(qC - \theta D\frac{\partial C}{\partial z}\right) - S^C \tag{8.5}$$

Biodegradation/Natural Attenuation A general expression of organic contaminants like PPCPs, hydrocarbons depletion in soil, in which microbial densities and contaminant concentration determine the degradation kinetics, can be written as (Yadav and Hassanizadeh 2011):

$$-\frac{\partial C}{\partial t} = \mu_{\max} C \frac{(C_O + X_O - C)}{(K_S + C)} \tag{8.6}$$

where $\mu_{\max}$ is the maximum growth rate, C is the contaminant concentration at time t, C_0 is the initial contaminant concentration, X_0 corresponds to the contaminant required to produce initial microbial density, and K_s is the half saturation constant also known as a growth-limiting concentration. The above equation reflects a linear relationship for changes in microbial density and the nonlinear relationship of changes in contaminant concentration on the rate of contaminant degradation. Furthermore, different simplified degradation kinetic models can be approximated considering extreme ratios of initial contaminant concentration (C_0) to K_s or initial microbial densities (X_0) to C_0 in Eq. 8.6.

Adsorption/Desorption Adsorption/desorption is an important geochemical process controlling the fate and transport of pollutants in the subsurface. The solid phase mass partition, i.e., adsorption is governed by:

$$S = K_d C \tag{8.7}$$

where S is the mass of adsorbed contaminants, C is the mass of aqueous phase, and K_d is the distribution coefficient product of organic carbon partitioning coefficient ($K_{oc)}$ and organic carbon content of the soil f_{oc} as:

$$K_d = K_{oc} f_{oc} \tag{8.8}$$

The retardation factor (R) is mainly used to estimate the adsorption in subsurface, and it is calculated using following equation:

$$R = 1 + \frac{\rho_b}{n} K_d \tag{8.9}$$

where R, retardation factor; ρ_b , soil bulk density (g/cm^3); n, porosity; and K_d, soil distribution coefficient.

Volatilization Mass transfer between the air, water, or pure phases, commonly known as volatilization, is an important geochemical process controlling the fate and transport of emerging pollutants in the subsurface. Between air and water phases, Henry's law is used to estimate the volatilization rate, given by:

$$p = k_H C_w \tag{8.10}$$

where p is the partial pressure of the contaminant in the gas phase, k_H is Henry's law constant (a function of temperature), and C_w is the aqueous concentration of the contaminant.

8.4 Subsurface Processes Controlling Reuse Potential of Treated Wastewater

8.4.1 Redox Environment and Redox Buffering

Sufficient organic matter and other reduced pollutants introduced in subsurface due to heavy irrigation using wastewater create a redox environment in and around the polluted subsurface zones (Christensen et al. 2000). The redox system of contaminant plumes which originated from the application of (un)-treated wastewater in subsurface may involve gases (O_2, N_2, CH_4, CO_2), and dissolved components (NO_3^-, NH_4^+, CH_2O, Fe^{2+}, Mn^{2+}, SO_4^{2-}, HS^-, H^+) as well as solids (FeOOH, MnO_2) and components Fe^{2+}, Mn^{2+}, NH_4^+ associated with the solids by ion exchange. The redox conditions of a contaminant plume constitute an important part of the chemical framework controlling the behavior of the contaminants in the plume. The redox conditions developed in plume zone is dominating by O_2 reduction by aerobic respiration/denitrification; iron (III), Mn (IV), and sulfate reduction; and methanogenic production as shown in Fig. 8.2. Within this zone and downgradient from it, sulfate reduction may take place. Iron reduction takes place further downgradient, where the conditions are less reducing (Albrechtsen and Christensen 1994). Zones of manganese and nitrate reduction have been observed, sometimes overlapping the iron-reducing zone. Finally, aerobic conditions may exist in the outskirts of the reduced plume, if the pristine aquifer is oxidized and contains significant amounts of dissolved oxygen (Barcelona and Holm 1991). In this process, the hydrolysis of the organic pollutant takes places, which produces the sufficient amount of CO_2 (Borden et al. 1995).

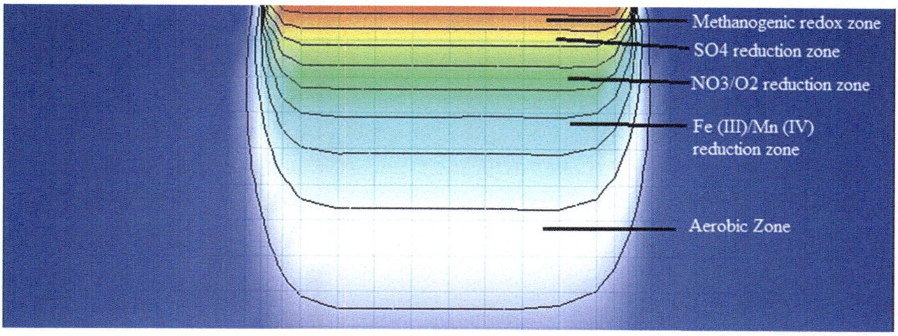

Fig. 8.2 Development of redox condition in the subsurface due to the release of pollutants

The set of reactions that create the complex redox environment of contaminants plumes originated from the application of wastewater consists of combinations of two half-reactions: oxidation half-reaction and reduction half-reaction. During organic matter oxidation, it is evident that when all electron acceptors are present, oxygen will be used first, followed by nitrate, manganese, iron, and sulfate. Finally, methanogenesis and fermentation reactions dominate, when the most favorable electron acceptors are depleted. The numbers, diversity, and ability of microorganisms to attenuate different organic substrates are reduced under these conditions. Therefore, the natural attenuation of several emerging pollutants is reported to be quite slow under redox conditions. Thus, redox environment which originated from the application of (un)-treated wastewater may lead high risk and vulnerability of subsurface natural resource pollution.

8.4.2 *Adsorption/Desorption Processes*

Sorption and desorption are the major processes influencing the fate (uptake, biodegradation, chemical degradation, photodegradation, and mobility) of organic contaminants in the environment (Petersen et al. 1996; Jeribi et al. 2002; Sparks 2003). Generally, adsorption of the pollutants depends on the K_d values, which correlated with organic carbon content (f_{or}) of soil. The organic carbon content (f_{oc}) of soil increases as it is irrigated with treated wastewaters. Furthermore, the changing pH values cause the ionization of complex organic pollutants, PPCPs, which reduce their lipophilicity (K_{ow}). Ternes et al. (2007) investigated the adsorption of pharmaceuticals and musk fragrances in soil irrigated with treated wastewater in Braunschweig, Germany. They estimated the K_d values of different pharmaceuticals and musk fragrances compounds and show that adsorption is not more effective for pharmaceutical compounds due to its corresponding low K_d values. Whereas, comparatively high adsorption was found for Musk fragrances having high K_d values. Chefetz et al. (2008) investigated sorption and mobility of pharmaceutical compounds in soil irrigated with treated wastewater. The results of the adsorption-desorption batch experiment show that the adsorption isotherms were more linear for soil irrigated with treated wastewater than freshwater. They suggested that the adsorption isotherm of the pollutants depends upon the soil organic matter (SOM). They also suggest that the higher mobility of PPCPs in soil column systems indicates that their residues in soils irrigated with treated wastewater can leach from the root zone and be transported to the groundwater after rain events. Thus, the reuse potential of the treated wastewater depends on the degree of SOM present in wastewater as well as its mobility in the subsurface. It is generally observed that the humification of the SOM as well as the level of clay and organo-clay complexes affect significantly the adsorption of several pollutants originated from the application of treated wastewater (Senesi and Plaza 2007).

8.4.3 Attachment and Detachment Processes

Land irrigated with treated wastewater contains a significant amount of nanoscale emerging pollutants like viruses/colloids. The behavior of these nanoscale pollutants in subsurface depends upon their interaction with solids to be the result of differences in the electrical charge and the hydrophobicity of pollutant surface. Solids having high isoelectric points are better virus adsorbents than those with low isoelectric points (Bales et al. 1991). Thus, the electrostatic conditions of soil matrix control the attachment/detachment of nanoscale pollutants, especially viruses/colloids (Yates et al. 1987). Attachment of viruses is more in soil having higher cation exchange capacity (Burge and Enkiri 1978), exchangeable iron, and iron oxides (Lipson and Stotzky 1984; Chu et al. 2003). Moore et al. (1982) reported that the attachment shows more in media having higher specific surface areas. Generally, granular soils are weaker adsorbents than clays and minerals (Sobsey et al. 1980; Moore et al. 1981). Clays may have surfaces that have a very heterogeneous charge distribution. Vilker et al. (1983) suggested that poliovirus 1 be attached to the edges of montmorillonite particles in regions of positive charges due to the presence of aluminum ions. Farrah and Preston (1993) show that viruses/colloids are detached more in soil modified with precipitation of metallic salts than unmodified sand. Furthermore, the higher pH than 3.9 creates a negative charge, and less pH than 3.9 creates a positive charge, which may control the attachments and de-attachment behaviors of these nanoscale pollutants (Schijven and Hassanizadeh 2000). Similarly, high electrical conductivity due to the application of treated wastewater may increase the attachment rate, which acts as long-term residual pollutant in the subsurface. An increasing organic contents by regular application of treated wastewater may reduce the attachment sites on soil solids, which may result in additional pollutants load on underlying groundwater resources. Thus, attachment and de-attachment behavior of nanoscale pollutants control the reuse potential of treated wastewater (Sinton et al. 1997).

8.4.4 Mobile–Immobile Regions

The nonuniform flow of water which takes place leads to the non-equilibrium conditions in the variably saturated zone. Water mostly flows from macropores by passing the rest of the soil matrix resulting in a nonuniform wetting of the soil as flowing water has very less time to equilibrate with the soil matrix. These non-equilibrium conditions were termed to be most frustrating processes in terms of hampering accurate predictions of contaminant transport in soils and fractured rocks by Simunek et al. (2003). Preferential flow and transport hasten the movement of contaminants like fertilizers, pesticides, pathogens, and trace elements originating from the application of treated wastewater to the underlying vadose zone (Gardenas et al. 2006, Wang 2008).

Studies for predicting the water flow in subsurface rely on the assumption of continuity, i.e., hydraulic properties over the domain are uniform. Since subsurface is composed of various soil materials like sand, clay, loam etc., hydraulic properties of the soil vary with space (Botros et al. 2009). Many researchers had studied the effects of preferential flow paths in soil on fluid and solute transport (Harter et al. 2005) and tried to model this phenomenon using various approaches (Simunek and Genuchten 2009). Because of preferential flow of water and solutes through the macropores, there is very less time for water and solutes to equilibrate with the rest of the soil matrix creating non-equilibrium conditions in the soil matrix.

Dual-porosity and dual-permeability models have been extensively used for modeling preferential flow phenomenon, where flow region is divided into two: mobile and immobile regions. Simunek et al. (2003) extensively reviewed various conceptual models which can be used for modeling non-equilibrium flow and transport of water and solute in the heterogeneous vadose zone. To investigate the spatial and temporal distribution of contaminants originated from the application of treated wastewater, Richard's equation and advection-dispersion equations are mostly used for simultaneous water and solute transport considering mobile-immobile regions. Total moisture content (θ) in the soil domain is considered to be the sum of moisture content in fractures (θ_{m}) and matrix (θ_{im}). The moisture flow equation in its mixed form coupled by a nonuniform sink function for water uptake by plants can be written in 3D form as:

$$\frac{\partial \theta_{\mathrm{m}}}{\partial t} = \frac{\partial}{\partial x}\left[K_{\mathrm{x}}(h)\frac{\partial h}{\partial x}\right] + \frac{\partial}{\partial y}\left[K_{\mathrm{y}}(h)\frac{\partial h}{\partial y}\right] + \frac{\partial}{\partial z}\left[K_{\mathrm{y}}(h)\frac{\partial h}{\partial z} + K(h)\right] - S_{\mathrm{m}} - \Gamma_{\mathrm{w}} \tag{8.11}$$

$$\frac{\partial \theta_{\mathrm{im}}}{\partial t} = -S_{\mathrm{im}} + \Gamma_{\mathrm{w}} \tag{8.12}$$

where θ is the volumetric water content defined as the volume of water per unit volume of soil (dimensionless), h is the pressure head of mobile region [L], S is a sink function that represents the water extraction by surface vegetation [T^{-1}], z is the depth of vadose zone measured positive upward [L], Γ_w is the water transfer rate between mobile and immobile region [T^{-1}], K is the hydraulic conductivity of the soil [LT^{-1}], and t is the time [T]. The water transfer Γ_w rate between mobile and the immobile region is assumed to be proportional to the difference in effective saturation between mobile and immobile regions.

$$\Gamma_w = \omega (S_{e,m} - S_{e,im}) \tag{8.13}$$

where ω is a first-order water transfer coefficient and $S_{e,m}$ and $S_{e,im}$ are the effective saturation of the mobile and immobile regions defined as:

$$S_{e,m} = \frac{\theta_m - \theta_{m,r}}{\theta_{m,s} - \theta_{m,r}}$$

$$S_{e,im} = \frac{\theta_{im} - \theta_{im,r}}{\theta_{im,s} - \theta_{im,r}} \tag{8.14}$$

Similarly, the solute transfer function Γ_s between mobile and the immobile region is considered to be proportional to the difference in solute concentration between mobile and immobile regions.

$$\Gamma_s = \omega_s (C_m - C_{im}) + \begin{cases} \Gamma_w C_m & \Gamma_w > 0 \\ \Gamma_w C_{im} & \Gamma_w < 0 \end{cases} \tag{8.15}$$

The following texts describe factors affecting mobility-immobility of contaminants originated due to land application of treated wastewater in the subsurface.

- Subsurface heterogeneity creates mobile-immobile zones, which may cause the complexity in water and solute flow regions. Thus, the subsurface zone having a high degree of formations of heterogeneity holds more pollutants in pore spaces (Simunek et al. 2003).
- Similarly, the pollutant heterogeneity due to the application of treated wastewater having trace metals, hydrocarbons, etc. causes more vulnerability to soil and underlying groundwater resources (Yadav and Junaid 2014).
- The bulk density, particle size distribution, pH, redox conditions, ionic exchange capacity, organic matter contents, types and the amount of metallic oxides, and types and the amount of clay minerals significantly affect mobile-immobile regions of water flow and solute transport in the subsurface.

8.5 Role of Climate Change In Biogeochemical Processes

Observational records and climate projections provide abundant evidence that like other ecosystems, subsurface processes have dynamic interactions with the ground surface and its prevailing environmental conditions. Climate variability affects subsurface processes both directly by altering surface water and heat flux and indirectly via changes in groundwater extraction patterns. Subsurface water resources are vulnerable and have the potential to be strongly affected by climate change with wide-ranging consequences (Green et al. 2011a, b). Impacts of climate change to the subsurface natural resources are further compounded due to accelerating demands arising from increasing population, urbanization, deforestation, and intensification of agriculture. Subsurface recharge and increased groundwater discharge due to climate change significantly disturb the moisture flow pattern, which further affects the biogeochemical characteristics of the subsurface system (Green et al. 2011a, b). In this situation, land irrigated with treated wastewater causes additional pollution load on subsurface natural resources. The following text describes the effects of climate change on biogeochemical processes controlling reuse potential of treated wastewater in the subsurface.

- Temperature is a key factor in reaction kinetics and dissolved oxygen concentrations, and hence, a small change in subsurface water temperatures could have a significant impact on subsurface water quality (Gunawardhana and Kazama 2012).The increased subsurface temperature may alter the geochemical processes (particularly redox reactions) that can exert control on the dissolved concentration and mobility of a wide variety of chemical contaminants (e.g., nutrients, trace metals, iron, and manganese) released by application of treated wastewater (Destouni and Darracq 2009). This may be of particular concern for supply wells that derive their water from riverbank infiltration. Rising temperatures could potentially increase soil mineralization rates of organic nitrogen to nitrate, leading to an increase in the potential for nitrate leaching to the water table in agricultural areas that irrigated with treated wastewater (Stuart et al. 2011).
- Modifications in biogeochemical interactions take place due to subsurface water warming by increasing temperature-propagated heat flux to sufficient depth, which can affect (co)-metabolic actions of soil microbiotas responsible for degrading subsurface pollutants (Gupta and Yadav 2017).
- Likewise, temperature plays a significant role in controlling the nature and extent of microbial metabolisms that are responsible for degradation of several pollutants (Yadav and Hassanizadeh 2011). Bioavailability and solubility of pollutants are also temperature-dependent. Low-temperature conditions usually result in increased viscosity, reduced volatilization, and decreased water solubility of pollutants, and thus delayed the onset of biodegradation process.

- Alternation in subsurface physical and functional interactions is mostly caused by the increasing evapotranspiration (ET) resulting from rising temperature. The increasing ET accelerates the subsurface moisture losses, which directly affects microbial cellular water loss in the subsurface. In particular, biodegradation process is strongly affected by the soil moisture contents. Thus, lands irrigated with treated wastewater in the semiarid and arid region are more vulnerable to underlying natural resources (Dettinger et al. 2004).
- Furthermore, in semiarid and arid regions, climate variability and changes accelerate evapotranspiration rates and cause the significant subsurface water losses, which ultimately lead to the high salinity problems (Earman and Dettinger 2011).
- The ratio of soil moisture and air in unsaturated zone has a direct effect on pollutants fate and transport under varying climatic conditions (Yadav and Hassanizadeh 2011). The low soil moisture content results in greater air-filled porosity, which should improve oxygen mass transfer to the pollutant-degrading microbial assemblage. However, there is likely to be a trade-off between improved oxygen availability and soil moisture content (Arora et al. 1982; Alvarez and Illman 2006).
- However, pollutant movement is getting retarded by adsorption on organic and/or mineral components of soil solids when air-filled porosity increases at low water contents. Thus, irrigation with treated wastewater reduces air-filled porosity and causes more downward movement of pollutants (Petersen et al. 1994).
- Extreme rainstorm intensity related to climate change may increase the downward flux of pollutants (e.g., nitrate, chloride) introduced by application of treated wastewater in vadose zone (Taylor et al. 2013). Alternatively, intense storms could produce precipitation rates that quickly exceed soil infiltration capacities and may flush additional pollutants mass to underlying groundwater table.
- The increase in rainfall intensity and surface flooding expected to accompany under climate change may drive the expansion of irrigated wastewater with runoff to the subsurface. This could potentially increase the amount of dissolved toxic chemicals and nutrients in runoff ultimately increasing the vulnerability of shallow aquifers to contaminations (Clifton et al. 2010; Green et al. 2011a, b).
- The continuous irrigation by treated wastewater causes accelerated advective and dispersive flux, which increases pore water velocities in partially saturated zones. Rapid groundwater table fluctuations along with high pore water velocities can enhance the mobilization of pollutants considerably (Dobson et al. 2007a, b).
- Large subsurface zones are polluted by advective and dispersive flux under high groundwater flow velocities caused by climatic variations. The fast groundwater velocity also enhances dissolution of pollutants, which may increase the pollution load in downgradient locations (Gupta and Yadav 2017).
- The dynamics of groundwater table accelerate (up)-downward movement of the pollutants causing their entrapment in pore space, which ultimately increases their coverage area. Pollutants entrapped in the form of isolated blobs or ganglia in

pore spaces increase water interfacial area responsible for their enhanced dissolution rates in the aqueous phase (Soga et al. 2004). Thus, pollutants trapped in the porous media act as long-lasting sources of groundwater pollution (Yadav and Hassanizadeh 2011).
- The wide coverage of pollutants due to advective and dispersive flux resulted by rapid groundwater table and pore water velocities creates large redox environment in the subsurface. Underlying redox environment may reduce the degradation of upcoming pollutants from treated wastewater applied at the land surface.
- To sum up, the climatic variation response to dynamic soil moisture flow, nature of underlying groundwater flow, and ambient temperature profile and pollutant regimes not only affect the microbial degradation rate but also the transport of substrate throughout the variably saturated zone. Thus, the climate change significantly affects the biogeochemical processes controlling reuse potential of treated wastewater in the subsurface.

8.6 Groundwater Pollution Load Resulted from Land Application of Treated Wastewater: A Case Study

Commonly, groundwater aquifer vulnerability has been assessed by data-based models like DRASTIC, modified DRASTICA, etc. incorporating the major geohydrological factors that affect and control the groundwater contamination movements. But these models are not effectively able to implement the contamination characteristics and biogeochemical behaviors to evaluate the major risk to underlying resources. Thus, an effort has been made to map the vulnerability of shallow groundwater to (sub)-surface pollutants resulting from land irrigated with wastewater of study area, using soil moisture flow and contaminant transport modeling.

8.6.1 Study Area

Kishanghrah Tehsil of Ajmer district, Rajasthan, is selected for this study (Fig. 8.3). The study area is a semiarid climatic zone in the map of India. The area is dry throughout the years except for monsoon or rainy season. The mean annual rainfall (1957–2012) of the district is 323 mm. Almost 95% of the total annual rainfall is received during the southwest monsoon, which enters the district in the last week of June and withdraws in the middle of September. The annual potential evapotranspiration in the district is 1565.6 mm and is the highest (243 mm) in the month of May (CGWB 2008).

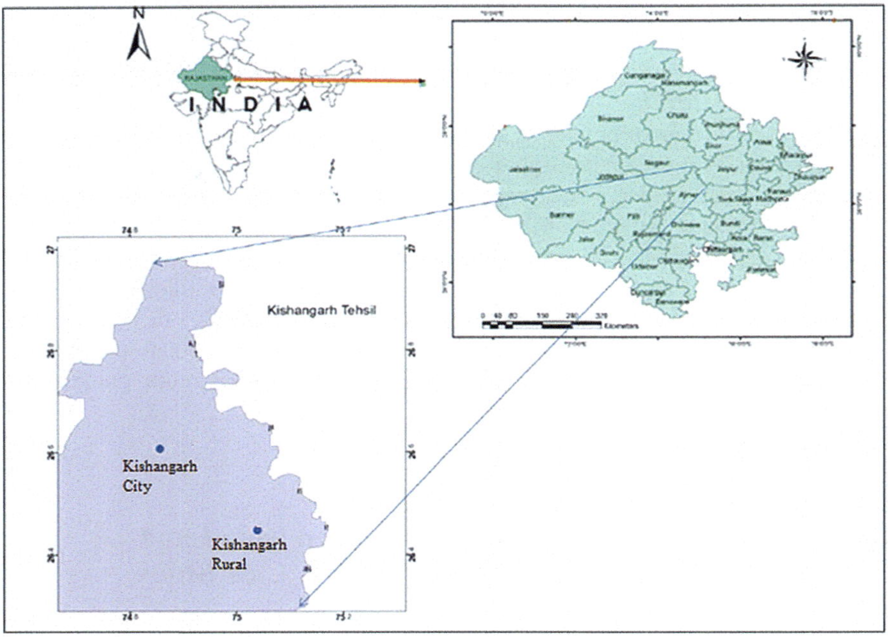

Fig. 8.3 Location map of the study area showing the two point locations (Kishangarh city and Rural area) for which groundwater contamination resulted from the land application of treated wastewater has been assisted using moisture flow and solute transport modeling

8.6.2 Methodology

HYDRUS-1D is used to model the solute transport to the underlying groundwater resource. The classical advection-dispersion equation coupled with Richard's equation is numerically simulated at different point locations for assessing the intrinsic vulnerability. The governing flow and transport equations (refer Eq. 8.9) are solved by Galerk in finite element scheme. Simulations are carried for two approaches: equilibrium approach and non-equilibrium approach using dual porosity approach. The methodology follows the physics of the soil moisture and contaminant movement in variably saturated porous media. This work demonstrates the potential vulnerability of shallow aquifer regions irrigated with treated wastewater in Kishangarh, Rajasthan. The soil moisture flow and solute transport regimes of the vadose zone associated with specific hydrogeological conditions play a crucial role in pollution risk assessment of the underlying groundwater resources. The role of soil type, slope, and the land-use cover is considered for estimating the transient flux at the top boundary from daily precipitation and evapotranspiration data of the study area. A constant flux of 150 mL/hr. of BTEX, a mixture of hydrocarbons, generally found in treated wastewater is applied as solute flux at top boundary.

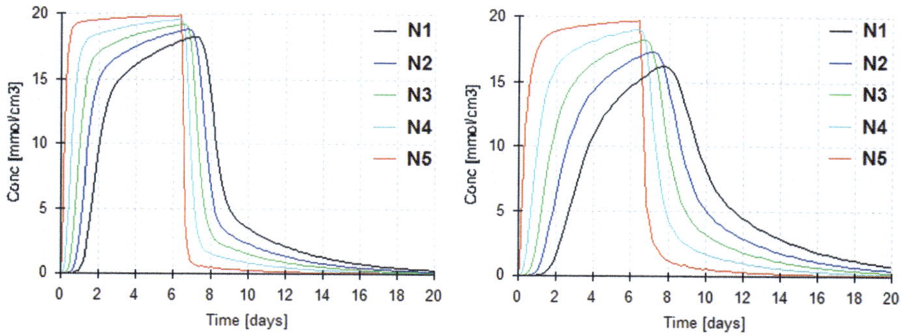

Fig. 8.4 Simulated BTCs for hydrocarbon pollutants resulted from land irrigation of wastewater in the subsurface. N1 represents observation node at top boundary and N5 at water table location

8.6.3 Results and Discussion

The computed solute transport profiles of varying hydrogeological conditions associated with different locations in the study area are shown in Fig. 8.4. The intrinsic vulnerability is assessed by comparing the transit time for the solute peak to reach the saturated zone. Figure 8.3 shows the variation in hydrocarbon concentrations with time for a constant flux boundary condition for all locations. The N5 line in each graph represents the concentration of solute at the water table, and the N1 lines represent the same at the observation point located at the top boundary and remaining (N2-N4) in the middle of the vadose zone. Results show that the high vulnerability of groundwater in the area irrigated with wastewater in the Kishangarh city, whereas low vulnerability is observed in the rural area of Kishangarh. The study may assist in decision-making related to the planning of application of treated wastewater and the sustainable water resource development of the selected semiarid area.

8.7 Conclusion and Future Prospects

Land application of treated wastewater results in more vulnerability of underlying subsurface resources. This is due to insufficient treatment capacities of wastewater treatment plants as well as the presence of PPCPs, BTEX, nanoscale particles, viruses, and other low-degradable emerging pollutants. The natural attenuation of these pollutants is quite slow in the subsurface and acts as long-term source of pollution. The sufficient amount of these pollutants is introduced in the subsurface when land is irrigated with treated wastewater. Thereafter, these pollutants start moving downward due to gravity and reach to groundwater table. The wide area coverage by the pollutants creates a redox environment in and around the polluted subsurface zones. Similarly, the mass partitioning takes place in the partially saturated zone by adsorption/desorption and or attachment/attachments to solid particles. Further, the mobile-mobile regions of subsurface control the reuse potential of treated wastewater. On the other hand, the climatic variables affect the solubility of different pollutants present in treated wastewater. In this book chapter, the physicochemical properties of a typical treated wastewater are presented. Then, the fate and transport processes of emerging pollutants are described in depth. Thereafter, the subsurface process that controls the reuse potential of treated wastewater is elaborated. Similarly, the role of climate change on underlying geochemical processes is highlighted. A case study on this topic is presented here to demonstrate the potential vulnerability of shallow aquifer regions irrigated with treated wastewater in Kishangarh, Rajasthan. In the future, there is a need to improve the performance of the wastewater treatment plants. Furthermore, developing some appropriate in situ bioremediation techniques is required to decontaminate the polluted soil-water resources under climate change conditions.

Acknowledgment The University Grant Commission (UGC) Fellowship received by Mr. Pankaj K Gupta is duly acknowledged.

References

Albrechtsen H-J, Christensen TH (1994) Evidence for microbial iron reduction in a landfill leachate-polluted aquifer Vejen, Denmark. Appl Environ Microbiol 60:3920–3925

Alvarez PJJ, Illman WA (2006) Bioremediation and natural attenuation, process fundamentals and mathematical models. Wiley-Interscience. ISBN-10 0-471-65043-9

Anitha K, Bolan NS, Müller K, Laurenson S, Naidu R, Kim WI (2012) The influence of wastewater irrigation on the transformation and bioavailability of heavy metal(loid)s in soil. Adv Agron 115:215–297

Arora HS, Cantor RR, Nemeth JC (1982) Land treatment: a viable and successful method of treating petroleum industry wastes. Environ Int 7:285–291

Arye G, Dror I, Berkowitz B (2010) Fate and transport of carbamazepine in soil aquifer treatment (SAT) infiltration basin soils. Chemosphere 82(2011):244–252

Bales RC, Hinkle SR, Kroeger TW, Stocking K, Gerba CP (1991) Bacteriophage adsorption during transport through porous media: chemical perturbations and reversibility. Environ Sci Technol 25:2088–2095

Barcelona MJ, Holm TR (1991) Oxidation–reduction capacities of aquifer solids. Environ Sci Technol 25:1565–1572

Bernstein N, Bartal A, Friedman H, Rot I, Snir P, Chazan A, Ioffe M (2006) Application of treated wastewater for cultivation of roses (Rosa hybrida) in soil-less culture. Sci Hortic 108:185–193

Borden RC, Gomez CA, Becker MT (1995) Geochemical indicators of intrinsic bioremediation. Groundwater 33(2):180–189

Botros FE, Harter T, Onsoy YS, Tuli A, Hopmans JW (2009) Spatial variability of hydraulic properties and sediment characteristics in a deep alluvial unsaturated zone. Vadose Zone J 8 (2):276–289

Burge WD, Enkiri MK (1978) Virus adsorption by five soil. J Environ Qual 7:73

Caliman FA, Gavirilescu M (2009) Pharmaceuticals, personal care products and endocrine disrupting agents in the environment – a review. Clean 37:277–303

Central Ground Water Board, Ministry of Water Resources, Government of India (2008) Groundwater Scenario Ajmer District Rajasthan. District Groundwater Brochure, Western Region Jaipur

Chary NS, Kamala CT, Raj DSS (2008) Assessing risk of heavy metals from consuming food grown on sewage irrigated soils and food chain transfer. Ecotoxicol Environ Safe 69:513–524

Cheftez B, Mualem T, Ben-Ari J (2008) Sorption and mobility of pharmaceutical compounds in soil irrigated with reclaimed wastewater. Chemosphere 73:1335

Christensen TH et al (2000) Characterization of redox conditions in groundwater contaminant plumes. J Contam Hydrol 45(3–4):165–241

Chu Y, Jin Y, Baumann T, Yates MV (2003) Effect of soil properties on saturated and unsaturated virus transport through columns. J Environ Qual 32:2017–2025

Clement TP, Johnson CD, Sun Y, Klecka GM, Bartlett C (2000) Natural attenuation of chlorinated ethene compounds: model development and field-scale application at the Dover site. J Contam Hydrol 42(2–4):113–140

Clifton C et al. (2010) Water and climate change: impacts on groundwater resources and adaptation options. Water Working Notes, World Bank Group Water Sector Board Note No 25, 76 p

CPCB (2009) Status of water supply, wastewater generation and treatment in Class I cities and Class II towns of India. Central Pollution Control Board

Destouni G, Darracq A (2009) Nutrient cycling and N_2O emissions in a changing climate: the subsurface water system role. Environ Res Lett 4(3):035008

Dettinger MD, Cayan DR, Meyer MK, Jeton AE (2004) Simulated hydrologic responses to climate variations and change in the Merced, Carson, and American River Basins, Sierra Nevada, California, 1900–2099. Clim Chang 62:283–317

Dobson R, Schroth MH, Zeyer J (2007a) Effect of water-table fluctuation on dissolution and biodegradation of a multi-component, light nonaqueous-phase liquid. J Contam Hydrol 94:235–248

Dobson R, Schroth MH, Zeyer J (2007b) Effect of water-table fluctuation on dissolution and biodegradation of a multi-component, light nonaqueous-phase liquid. J Contam Hydrol 94:235–248

Earman S, Dettinger M (2011) Potential impacts of climate change on groundwater resources – a global review. J Water Climate Change 02(4):213–229 http://www.iwaponline.com/jwc/002/jwc0020213.htm

Ebele AJ, Abou-Elwafa Abdallah M, Harrad S (2017) Pharmaceuticals and personal care products (PPCPs) in the freshwater aquatic environment. Emerg Contam 3(2017):1–16

EPA (2008) Nanotechnology for site remediation fact sheet. U.S. Environmental Protection Agency, (October), pp 1–17

Farrah SR, Preston DR (1993) Adsorption of viruses to sand modified by in situ precipitation of metallic salts. Wiener Mitteilungen Wien 12:25–29

Gardenas A, Simunek J, Jarvis NJ, van Genuchten MT (2006) Two-dimensional modelling of preferential water flow and pesticide transport from a tile-drained field. J Hydrol 329:647–660

Gharaibeh MA, Eltaif NI, Al-Abdullah B (2007) Impact of field application of treated wastewater on hydraulic properties of vertisols. Water Air Soil Pollut 184:347–353

Green TR et al (2011a) Beneath the surface of global change: impacts of climate change on groundwater. J Hydrol 405:532–560

Green TR et al (2011b) Beneath the surface of global change: impacts of climate change on groundwater. J Hydrol 405:532–560

Gunawardhana LN, Kazama S (2012) Statistical and numerical analysis of the influence of climate variability on aquifer water levels and groundwater temperatures: the impacts of climate change on aquifer thermal regimes. Global Planet Chang 86–87:66–78

Gupta PK, Shashi R, Yadav BK (2013) BTEX Biodegradation in soil-water system having different substrate concentrations. Int J Eng 2(12)

Gupta PK, Yadav BK (2017) Chapter 8: Bioremediation of Non-Aqueous Phase Liquids (NAPLS) polluted soil and water resources. In: Environmental pollutants and their bioremediation approaches. CRC Press, Taylor and Francis Group, Boca Raton. ISBN 9781138628892

Gupta PK, Joshi P (2017) Assessing groundwater resource vulnerability by coupling GIS based DRASTIC and solute transport model in Ajmer District, Rajasthan. Journal of Geological Society of India (Springer), DOI: https://doi.org/10.1007/s12594-018-0958-y

Gupta PK, Sharma D (2018) Assessments of hydrological and hydro-chemical vulnerability of groundwater in semi-arid regions of Rajasthan, India. *Sustainable Water Resources Management*, 1-15. https://doi.org/10.1007/s40899-018-0260-6.

Gupta PK, Ranjan S, Kumar D (2018a) Groundwater pollution by emerging industrial pollutants and its remediation techniques. Chapter 2, In Recent advances in environmental management, CRC Press Taylor & Francis Group, ISBN 9780815383147, Vol 1

Gupta PK, Abhishek YBK (2018b) Impact of hydrocarbon pollutants on partially saturated soil media in batch system: Morphological analysis using SEM techniques. Chapter 5, water quality management; Water Sci Technol, ISBN: 978-981-10-5794-6, Vol. 79, Springer

Harter T, Ginn TR, Onsoy YS, Horwath WR (2005) Spatial variability and transport of nitrate in a deep alluvial vadose zone. Vadose Zone J 4:41–54

IPCC (2007a) Climate change (2007). The physical science basis. In: Solomon S et al (eds) Contribution of working group I to the fourth assessment report of the intergovernmental panel on climate change. Cambridge University Press, Cambridge/New York, p 996

IPCC (2007b) Climate change (2007). Impacts, adaptation and vulnerability. In: Parry ML, Canziani OF, Palutikof JP, Linden PJVD, Hanson CE (eds) Contribution of working group II to the fourth assessment report of the intergovernmental panel on climate change. Cambridge University Press, Cambridge/New York

Jeribi M, Almir-Assad B, Langevin D, Henaut I, Argillier JF (2002) Adsorption kinetics of asphaltenes at liquid interfaces. J Colloid Interface Sci 256(2):268–272

Karn B, Kuiken T, Otto M (2009) Nanotechnology and in situ remediation: a review of the benefits and potential risks. Environ Health Perspect 117(12):1823–1831

Karvelas M, Katsoyiannis A, Samara C (2003) Occurrence and fate of heavy metals in the wastewater treatment process. Chemosphere 53(10):1201–1210

Kinney CA, Furlong ET, Kolpin DW, Burkhardt MR, Zaugg SD, Werner SL, Bossio JP, Benotti MJ (2008) Bioaccumulation of pharmaceuticals and other anthropogenic waste indicators in earthworms from agricultural soil amended with biosolid or swine manure. Environ Sci Technol 42:1863–1870

Klein RJT, Nicholls RJ (1999) Assessment of coastal vulnerability to climate change. Ambio 28 (2):182–187

Kushwah SK, Malik S, Singh A (2012) Water quality assessment of raw sewage and final treated water with special reference to waste water treatment plant Bhopal, MP, India. Res J Recent Sci 1(ISC-2011):185–190

Levantesi C, La Mantia R, Masciopinto C, Böckelmann U, Ayuso-Gabella MN, Salgot M et al (2010) Quantification of pathogenic microorganisms and microbial indicators in three wastewater reclamation and managed aquifer recharge facilities in Europe. Sci Total Environ 408 (21):4923–4930

Li YH, Wang S, Luan Z, Ding J, Xu C, Wu D (2003) Adsorption of cadmium(II) from aqueous solution by surface oxidized carbon nanotubes. Carbon 41(5):1057–1062

Lipson SM, Stotzky G (1984) Effect of proteins on Reovirus adsorption to clay minerals. Appl Environ Microbiol 8:525–530

McCarthy KA, Johnson RL (1993) Transport of volatile organic compounds across the capillary fringe. Water Resour Res 29(6):1675–1683

Moore RS, Taylor DH, Sturman LS, Reddy MM (1981) Poliovirus adsorption by 34 minerals and soils. Appl Environ Microbiol 42:963–975

Moore RS, Taylor DH, Reddy MM, Sturman LS (1982) Adsorption of reovirus by minerals and soils. Appl Environ Microbiol 44:852–859

Mustapha HI, Rousseau D, van Bruggen J, Lens P (2011) Treatment performance of horizontal subsurface flow constructed wetlands treating inorganic pollutants in simulated refinery effluent. In: 2nd Biennial Engineering Conference. School of Engineering and Engineering Technology, Federal University of Technology, Minna

Mustapha HI, van Bruggen JJ, Lens PN (2015) Vertical subsurface flow constructed wetlands for polishing secondary Kaduna refinery wastewater in Nigeria. Ecol Eng 84:588–559. https://doi.org/10.1016/j.ecoleng.2015.09.060
Mustapha IH, Gupta PK, Yadav BK, van Bruggen JJA, Lens PNL (2018) Performance evaluation of duplex constructed wetlands for the treatment of diesel contaminated wastewater. In: Chemosphere. https://doi.org/10.1016/j.chemosphere.2018.04.036
Okuda T, Kobayashi Y, Nagao R, Yamashita N, Tanaka H, Tanaka S, Fujii S, Konishi C, Houwa I (2008) Removal efficiency of 66 pharmaceuticals during wastewater treatment process in Japan. Water Sci Technol 57(1):65–71
Petersen LW, Rolston DE, Moldrup P, Yamaguchi T (1994) Volatile organic vapor diffusion and adsorption in soils. J Environ Qual 23:799–805
Petersen LW, El-Farhan YH, Moldrup P, Rolston DE, Yamaguchi T (1996) Transient diffusion, adsorption, and emission of volatile organic vapors in soils with fluctuating low water contents. J Environ Qual 25:1054–1063
Pierson WL, Nittim R, Chadwick MJ, Bishop KA, Horton PR (2001) Assessment of changes to saltwater/freshwater habitat from reductions in flow to the Richmond river estuary, Australia. Water Sci Technol 43(9):89–97
Powers SE, Loureiro CO, Abriola LM, Weber WJ (1991) Theoretical study of the significance of non-equilibrium dissolution of nonaqueous-phase liquids in subsurface systems. Water Resour Res 27(4):463–477
Ranjan SP, Kazama S, Sawamoto M (2006) Effects of climate and land use changes on groundwater resources in coastal aquifers. J Environ Manag 80(1):25–35
Ranjan S, Gupta PK, Yadav BK (2018) Application of nano-materials in subsurface remediation techniques – challenges and future prospects. Chapter 6, In Recent advances in environmental management, CRC Press Taylor & Francis Group, ISBN 9780815383147, Vol 1
Rose JB, Dickson LJ, Farrah SR, Carnahan RP (1996) Removal of pathogenic and indicator microorganisms by a full-scale water reclamation facility. Water Res 30:2785–2797
Saien J, Shahrezaei F (2012) Organic pollutants removal from petroleum refinery wastewater with nanotitania photocatalyst and UV Light Emission. Int J Photoenergy 2012:1–5
Savage N, Diallo MS (2005) Nanomaterials and water purification: opportunities and challenges. J Nanopart Res 7(4–5):331–342
Schijven JF, Hassanizadeh SM (2000) Removal of viruses by soil passage: overview of modeling, processes, and parameters. CRC Crit Rev Environ Sci Technol 30:49–127
Seeger EM, Kuschk P, Fazekas H, Grathwohl P, Kaestner M (2011) Bioremediation of benzene-, MTBE- and ammonia-contaminated groundwater with pilot-scale constructed wetlands. Environ Pollut 199:3769–3776
Senesi N, Plaza C (2007) Role of humification processes in recycling organic wastes of various nature and sources as soil amendments. Clean-Soil Air Water 35:26–41. https://doi.org/10.1002/clen.200600018
Sharma RK, Agrawal M, Marshall F (2007) Heavy metal contamination of soil and vegetables in suburban areas of Varanasi, India. Ecotoxicol Environ Saf 66(2):258–266
Sherif MM, Singh VP (1999) Effect of climate change on sea water intrusion in coastal aquifers. Hydrol Process 13(8):1277–1287
Simunek J, van Genuchten MT (2009) Modelling nonequilibrium flow and transport processes using HYDRUS. Vadose Zone J 7:782–797
Simunek J, Jarvis NJ, van Genuchten MT, Gardenas A (2003) Review and comparison of models for describing non-equilibrium and preferential flow and transport in the vadose zone. J Hydrol 272:14–35
Sinton LW, Finlay RK, Pang L, Scott DM (1997) Transport of bacteria and bacteriophages in irrigated effluent into and through an alluvial gravel aquifer. Water Air Soil Pollut 98:17–42
Sobsey MD, Dean CH, Knuckles ME, Wagner RA (1980) Interactions and survival of enteric viruses in soil materials. Appl Environ Microbiol 40:92–101
Soga K, Page JWE, Illangasekare TH (2004) A review of NAPL source zone remediation efficiency and the mass flux approach. J Hazard Mater 110(1–3):13–27

Sparks DL (2003) Environmental soil chemistry. Academic/China Translation & Printing Services Ltd., Amsterdam/Hong Kong
Stuart M et al (2011) A review of the impact of climate change on future nitrate concentrations in groundwater of the UK. Sci Total Environ 409(15):2859–2873
Suthar S, Chhimpa V, Singh S (2009) Bacterial contamination in drinking water: a case study in rural areas of northern Rajasthan, India. Environ Monit Assess 159:43–50
Taylor RG et al (2013) Ground water and climate change. Nat Clim Chang 3:322–329
Ternes TA, Bonerz M, Herrmann N, Teiser B, Andersen HR (2007) Irrigation of treated wastewater in Braunscheug, Germany: an option to remove pharmaceuticals and musk fragrances. Chemosphere 66:894–904
Tobiszewski M, Tsakovski S, Simeonov V, Namieśnik J (2012) Chlorinated solvents in a petrochemical wastewater treatment plant: an assessment of their removal using self-organising maps. Chemosphere 87:87. https://doi.org/10.1016/j.chemosphere.2012.01.057
US EPA (2004) Guidelines for water reuse, EPA/625/R-04/108 September
USEPA (U.S. Environmental Protection Agency) (1995) Light nonaqueous phase liquids. Office of solid waste and emergency response, Washington, DC. EPA/540/S-95/500
Vilker VL, Meronek GC, Bulter PC (1983) Interaction of poliovirus with montmorillonite clay in phosphate-buffered saline. Eviron Sci Technol 17:631–634
Wang F (2008) Modeling the spatial distribution of nitrogen leaching from dairy farmland. Vadose Zone J 7(2):439–452
Watlington K (2005) Emerging nanotechnologies for site remediation and wastewater treatment. Environmental Protection Agency, Washington DC
Wu S, Wallace S, Brix H, Kuschk P, Kirui WK, Masi FD (2015) Treatment of industrial effluents in constructed wetlands: Challenges, operational strategies and overall performance. Environ Pollut 201:107–120
Yadav BK, Hassanizadeh SM (2011) An overview of biodegradation of LNAPLs in coastal (semi)-arid environment. Water Air Soil Pollut 220:225. https://doi.org/10.1007/s11270-011-0749-1
Yadav BK, Junaid S (2014) Groundwater vulnerability assessment to contamination using soil moisture flow and solute transport modelling. J Irrig Drain Eng 141:04014077-1. https://doi.org/10.1061/(ASCE)IR.1943-4774.0000841
Yates MV, Yates SR, Wagner J, Gerba CP (1987) Modeling virus survival and transport in the subsurface. J Contam Hydrol 1:329–345

Chapter 9
Removal of Organic Pollutants from Industrial Wastewaters Treated by Membrane Techniques

M. Soniya and G. Muthuraman

Abstract Water is the most commonly used resource in the world. The world's supply of sanitary water is limited and exposed to contamination. Increasing demands for water are required for cultivation, manufacturing, and urban development, but these are more important to control than the distribution of limited freshwater resources. Various industries use water for diverse processes and then discharge it back into the surroundings. In recent times, researchers have focused on effluent treatment by a variety of processes with low costs and high removal efficiency. This study examines the removal of contamination from industrial effluents by a liquid membrane method. Water is recovered and reused through the liquid membrane, which has significant ecological benefits and decreases the effects of wastewater discharge on environmental water quality. Financial and environmental benefits have been recognized in this recently developed liquid membrane method.

Keywords Recovery · Effluent · Water · Environment · Demand · Liquid membrane

Abbreviations

BAHLM	Bulk aqueous liquid membrane
BOHLM	Bulk organic liquid membrane
DBL	Diffusion boundary layer
FLM	Flowing liquid membrane
HF	Hollow fiber
HFCLM	Hollow fiber liquid membrane
HFLM	Hollow fiber liquid membrane
HLM	Hybrid liquid membrane
L/L	Liquid-liquid extraction
LM	Liquid membrane

M. Soniya · G. Muthuraman (✉)
Department of Chemistry, Presidency College, University of Madras, Chennai, India

R. P. Singh et al. (eds.), *Water Conservation, Recycling and Reuse: Issues and Challenges*, https://doi.org/10.1007/978-981-13-3179-4_9

MBSE	Membrane based solvent extraction
MBSS	Membrane based solvent stripping
MHS	Multi membrane hybrid system
O/W/O	Oil-in-water-in-oil

9.1 Introduction

Water is the major resource for all living beings. However, water pollution has become one of the most dangerous issues in the world. Water resources have been repeatedly contaminated by coloring agents discharged from many industries, including textiles, manufacturing and other industries. This contamination has increased significantly in recent years and is leading to increased toxicity, chemical oxygen demand, and interference with the transmission of sunlight in waterways, which decreases photosynthetic activity (Ravindra and Sunil 2009). In particular, the textile industry uses large amounts of water during the dyeing process and discharges wastewater into the environment that can contain 5000 tons of dyeing materials. These poisonous materials absorb oxygen in the water (Pirkaramia et al. 2013).

The complex structure of dyes are resistant to fading on exposure to light and water; therefore, it is difficult to decolorize them (Poots et al. 1976). Most of the dyes are harmful and may affect aquatic life, even the food chain, and some of them cause an allergic reaction (Banat et al. 1996). Dyes are difficult to treat because their synthetic complex aromatic molecular structures are quite stable. Therefore, it is extremely important to remove dyes from industrial effluents before they are discharged into water bodies (Khan and Husain 2007).

In recent decades, methods used to improve conventional water treatment processes for the removal of dyes from wastewater include the following (Hassan et al. 2013): photocatalytic degradation, electrochemical degradation, photo-Fenton oxidation, flocculation, ozonation, coagulation, ion exchange, and adsorption methods (Stephenson and Sheldon 1996; Chiou and Chuang 2006; Salem and El-maazawi 2000). However, these physical and chemical methods cannot completely remove dyes due to several limitations, including their high cost, production of concentrated sludge, and lack of capability to treat large amounts of pollutants. Recently, liquid membrane technology has become an alternative and safe separation process for the removal of pollutants from wastewater.

9.2 Treatment of Wastewaters from the Textile Industry

Wastewater treatment methods can be classified into physical, chemical, and biological processes. However, these three methods are insufficient to remove dyes from wastewater. An alternative treatment method is required to remove dyes from wastewater. Li (1968) first proposed the use of a liquid membrane (LM), which has attracted increasing attention in recent years. An LM has many other applications in separation, macromolecular chemistry, and membrane technology. It can be used to simultaneously extract and strip dye from water. The LM serves as a membrane barrier between two phases of aqueous solutions (dye and strip phase). The advantages of an LM include its high selectivity, high efficiency, and ability to achieve particular molecular recognition.

9.3 Types and Transport Methods of Liquid Membranes

A LM is relatively high in efficiency. There are three types of LM processes: emulsion liquid membranes, bulk liquid membranes, and supported liquid membranes. Liquid membranes have two liquid phases, the donor and stripping phases, which are separated by an immiscible membrane phase. Based on the mechanism of dye transport from the donor phase through the liquid membrane into the acceptor phase, the LM technique can be divided into three major types of transport:

(a) Simple transport: The dye moves through the membrane phase (Nath 2008).
(b) Facilitated transport: A carrier-mediated membrane phase accelerates the dye diffusion through the liquid membrane (Teng et al. 2014).
(c) Active transport: Oxidation-reduction and biochemical conversions take place in the donor-membrane and membrane-acceptor phase interfaces (Paugam and Smith 1993). The transport mechanisms are shown in Figs. 9.1, 9.2, and 9.3.

Detailed discussions of these membranes used for the transport of dyes from water and wastewater are described in the following chapters.

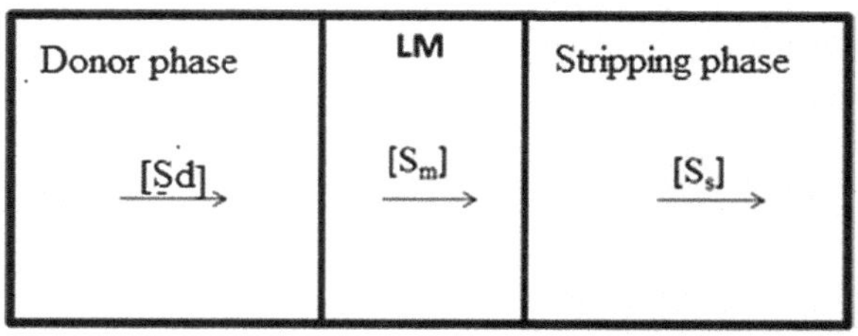

Fig. 9.1 Simple transports of species (S) from the donor phase (F), through the liquid membrane (M) to the stripping phase (S)

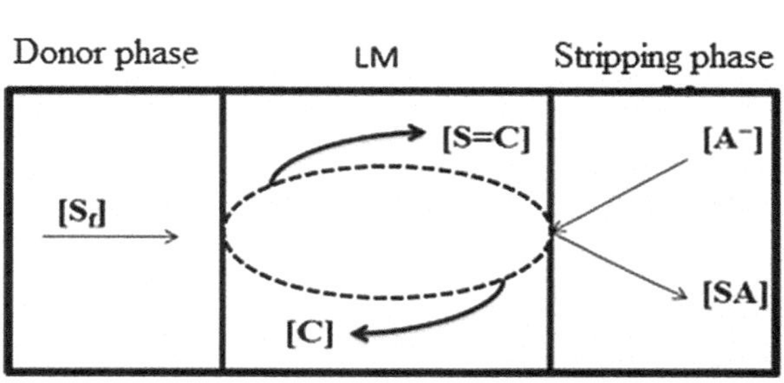

Fig. 9.2 Facilitated transport of species (S) through the liquid membrane

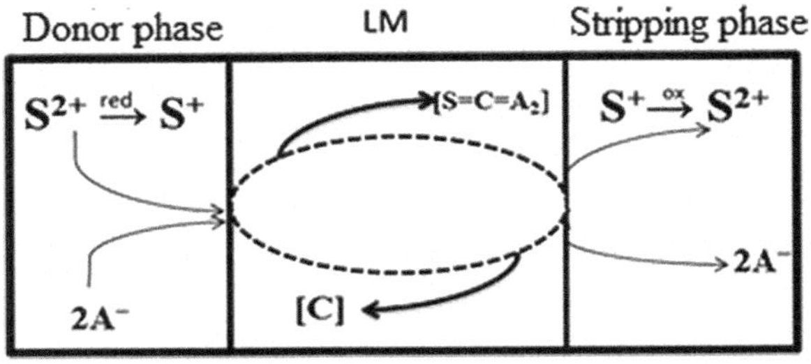

Fig. 9.3 Active transport of species (S) through the liquid membrane

9.4 Wastewater Treatment by the Bulk Liquid Membrane Method

9.4.1 Introduction

Liquid membrane systems have been the subject of numerous studies. They are used in separation processes to recover metals in hydrometallurgical processes, to separate radioactive elements, and to treat wastewater. The effective application of liquid membranes is determined by their stability (Wells 1993). The main process that occurs in a liquid membrane system is extraction. The liquid membrane methods can be used for ecological treatment as well as efficient and economical leading of the process.

A liquid membrane is a thin liquid layer that divides two homogenous liquids or gases (donor and stripping phase). The liquid membrane should be immiscible and insoluble in the external solutions (Wódzki 1997). The transport of substances between phases is caused by a driving force, which is a concentration gradient. The selectivity and rate of transport through the membrane depend on the variation in the solubility and diffusion coefficients of components in the membrane and the concentration of components in the donor- and acceptor-phase solutions. To improve the selectivity and influence the process capacity, the carrier is introduced into the membrane phase. The carrier has an affinity for the donor solution, and a complex soluble in the membrane is formed by a reversible reaction. As mentioned, there are three types of liquid membranes with different structure and transport mechanisms: emulsion liquid membranes, bulk liquid membranes, and supported liquid membranes. The bulk liquid membrane is an easy and efficient method (Ma et al. 2002), as well as the cheapest transport method because of its low capital cost. Bulk liquid membranes are described in this section.

9.4.2 Definition

A bulk liquid membrane contains a relatively thick layer of immiscible organic solution is used to separate the donor and strip phase. In this case the immiscible organic phase was used only one support material in bulk liquid membrane.

9.4.3 Types and Design of Bulk Liquid Membrane

A bulk liquid membrane consists of three phases: two aqueous phases (donor and stripping) and one membrane phase. The membrane phase divides the aqueous phases into two parts, which are separated by the barrier. The barrier is designed to be cylindrical in shape and is placed between the donor and strip phases. The former consists of rectangular and cylindrical vessels, whereas the latter includes H- and U-tube vessels. An experimental setup of H- and U-type bulk liquid membranes is shown in Fig. 9.4.

9.4.4 Theory of Bulk Liquid Membranes

Two methods can be used for the removal of industrial waste in wastewater and separation of the reaction mixtures (Schlosser 2000a, b; Malinowski 2001): membrane-based solvent extraction (MBSE) and membrane based solvent stripping (MBSS), in which the solvent can be reused. The two methods can remove organic compounds and metal ions through the bulk liquid membrane.

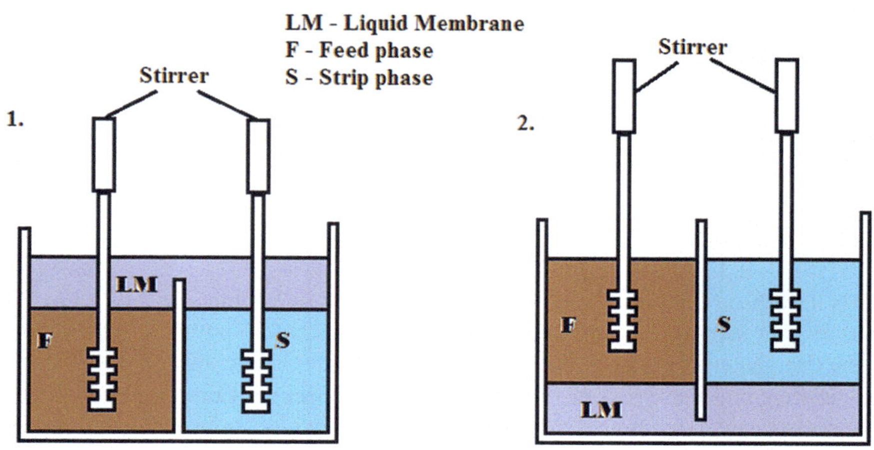

Fig. 9.4 Schematic of an experimental setup for a bulk liquid membrane: 1) the density of the liquid membrane is less than the aqueous phase; 2) the density of the liquid membrane is greater than the aqueous phase

9.4.5 Applications and Disadvantages of Bulk Liquid Membranes

The bulk liquid membrane method is mostly used in metal separation, biotechnologies, organic compounds, gas separation, and wastewater treatment, as described in the following sections.

9.4.5.1 Industrial Applications

Mercury is extensive in cultivation (Katzung 1987) and is very dangerous to human health. For that reason, the removal of mercury is very important. Liquid membranes have potential for eliminating mercury from wastewater containing calixarenes as carriers (Alpoguz et al. 2004). Ersoz (2007) reported a three-phase method was used for the removal of mercury from industrial wastewater.

9.4.5.2 Disadvantages

The most important disadvantage of bulk liquid membrane is the creation of emulsions, which can remove protein from microbial cells and thus lead to cell breakdown (Van Sonsbeek et al. 1993).

9.5 Supported Liquid Membranes and Their Applications in Wastewater Treatment

9.5.1 Introduction

LMs have been used in various industries to separate the components of a mixture and recover some metal ions and other contaminants before polluting the environment (Benjjar et al. 2012). These LM methods combines extraction and stripping in a single step, and the liquid membranes are highly selective (Loiacono et al. 1986). Supported liquid membranes are often used for their wider applications in industry. These systems can be developed in the presence of an inert polymer support and contains small amounts of organic solvents.

The supported liquid membrane is a three-phase membrane system. The organic liquid, which is held in the pores of a microporous polymeric support, separates two aqueous phases. The transport through supported liquid membranes is a result of permeate dissolution in the membrane phase and its diffusion to the acceptor phase. Generally, in supported liquid membranes, the combination of three simultaneous processes occurs: 1) extraction of analyte from an aqueous donor phase to the organic membrane phase; 2) its diffusion through the hydrophobic, organic liquid membrane; and finally, 3) re-extraction to an aqueous acceptor phase. In general, this

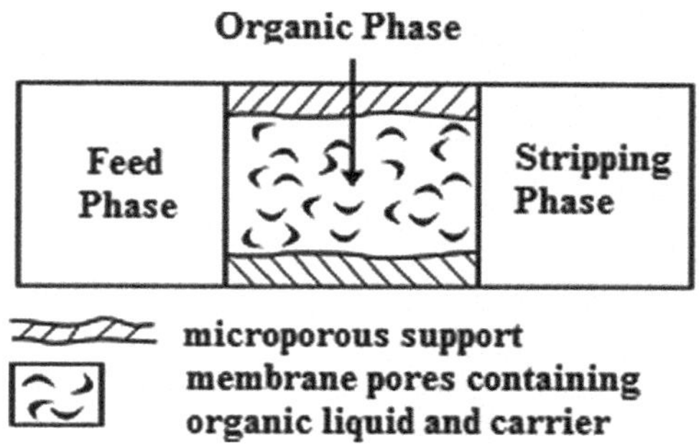

Fig. 9.5 Schematic diagram of a supported liquid membrane

three-phase system has been proven to provide efficient selectivity and clean extracts.

9.5.2 *Working Theory of the Supported Liquid Membrane*

A supported liquid membrane consists of an organic solvent immobilized in the pores of a polymer supported by capillary forces. This membrane separates two compartments: the donor compartment containing the substrate solution and the receiving compartment containing pure water. The support of these membranes is usually an inert, hydrophobic microporous polymer.

The passage of chemical species through these membranes is an interfacial phenomenon. Therefore, the use of a support with large porosity is very important and necessary to increase the contact area and ensure the best transport, separation, and selectivity conditions for the transported species across the supported liquid membrane. Figure 9.5 shows a schematic diagram of a supported liquid membrane.

9.5.3 *Classification*

Based on the size, shape, surface area, and application, supported liquid membranes can be classified into two types: flat-sheet supported liquid membranes and hollow-fiber supported liquid membranes.

9.5.3.1 Flat-Sheet Supported Liquid Membranes

The flat-sheet supported liquid membrane acts as a microporous solid support in supported liquid membranes. The solid support is impregnated with the extractant and is clamped between two half cells using gaskets, thus forming two compartments (Fig. 9.6). One compartment is for the donor solution and the other

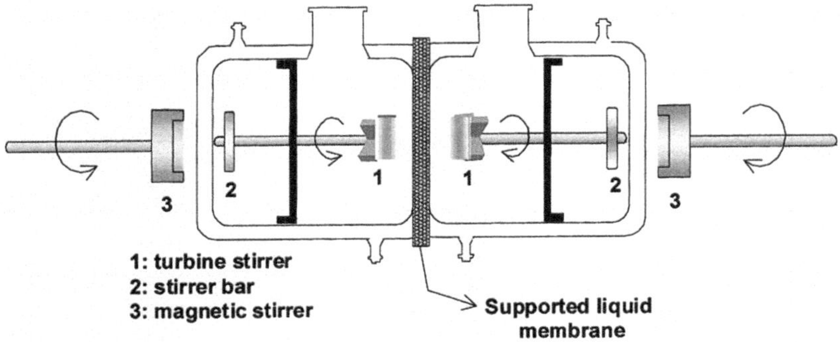

Fig. 9.6 Schematic figure of the flat-sheet supported liquid membrane transport cell

compartment isfor the strip solution. Both the phases are stirred by mechanical stirrers (Kemperman et al. 1996).

9.5.3.2 Hollow-Fiber Supported Liquid Membranes

Hollow-fiber supported liquid membranes are used for the removal of metal ions. The outer cell of the module is a single nonporous material through which the solution present inside cannot be transported. Inside the shell, many thin fibers are packed in neat rows. The source phase passes through the fibers and the receiving phase passes through the shell side with the help of pumps.

9.5.4 Transport Mechanism

Several models have been proposed to explain the method of facilitated transport. The theory is useful to understand the current transport systems and to design more selective and effective systems in the future (Eljaddi et al. 2014).

Three mechanisms for the facilitated transport phenomenon are known. The first is a solution-diffusion mechanism, in which the carrier and the complex are mobile in the membrane phase (Fig. 9.7a). The second mechanism occurs by jumping on fixed sites: The carrier is fixed in the pores of the membrane and the substrate moves successively by association with several carriers. This mechanism is well known in polymer inclusion membranes (Fig. 9.7b). The third mechanism occurs by jumping on mobile sites, in which the substrate is moving in series by linking to multiple mobile carriers (Fig. 9.7c).

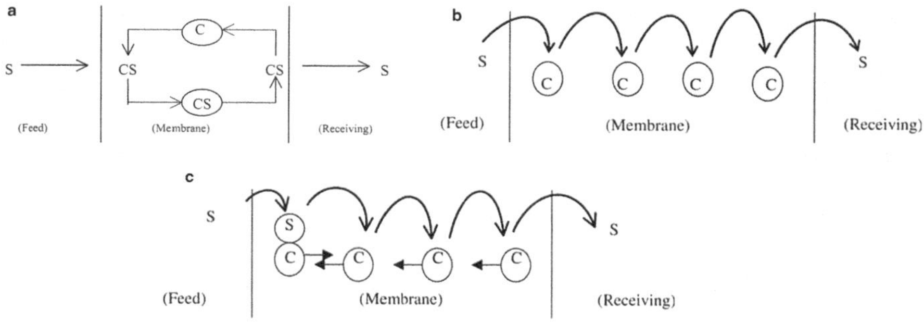

Fig. 9.7 (**a**) Solution-diffusion Mechanism- carrier and the complex are mobile in the membrane phase; (**b**) Mechanism by jumping on fixed sites, in which the carrier is fixed in the pores of the membrane and the substrate (S) moves successively by an association with several carriers (C), this mechanism is well known in polymer inclusion membranes (PIM); (**c**) Mechanism by jumping on mobile sites in which the substrate (S) is moving in series by linking to multiple mobile carriers (C)

9.5.5 Advantages and Disadvantages of Supported Liquid Membranes

9.5.5.1 Advantages of Supported Liquid Membranes

Supported liquid membranes have many advantages, such as low energy requirements, low working cost, simple scale-up, and nonstop operation. Another advantage is that the hollow fiber modules have low investment and working costs (Mulder 1991).

9.5.5.2 Disadvantages of Supported Liquid Membranes

The most important disadvantage of supported liquid membranes is membrane instability due to the partition of the organic solvent/carrier to the aqueous phases. Many studies have tried to overcome this problem, with several successes (Kemperman et al. 1996; Kemperman 1995; Kemperman et al. 1998).

9.6 Removal of Pollutants from Industrial Wastewater by Emulsion Liquid Membranes

9.6.1 Introduction

Liquid membranes are broadly used in fields including wastewater treatment (Muthuraman et al. 2009; Talebi et al. 2012). These methods are used for various

Fig. 9.8 Schematic of the emulsion liquid membrane setup

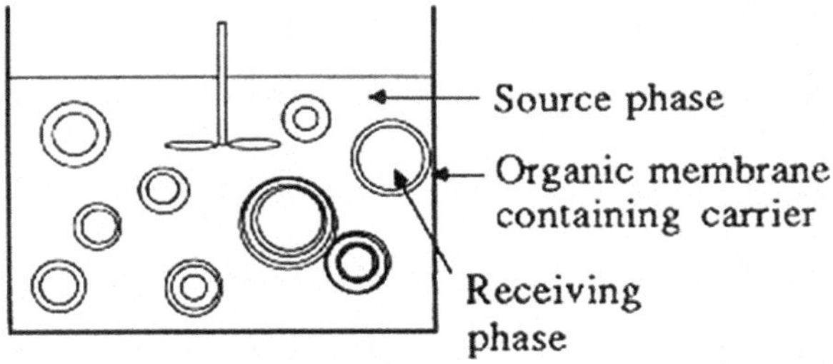

purposes, such as the transport and recovery of metal ions, removal of aromatic hydrocarbons, and transport of dyes from textile wastewaters (Chang et al. 2011). The emulsion liquid membrane is a separation method that requires an organic solvent, extractant, and surfactant for its formulation. An emulsion liquid membrane can be visualized as a bubble within a bubble, where the inner bubble is the receiving phase and the outer bubble contains the carriers. Anything outside the bubble is the source phase. The emulsion liquid membrane setup is shown schematically in Fig. 9.8. Emulsion liquid membranes were developed as a form of solvent extraction, with extraction and stripping occurring simultaneously in the same stage. Emulsion liquid membranes have good stability. Those that provide rapid extraction have micro-droplet sizes in the range of 0.3–10 μm (Li et al. 1983; Chanukya and Rastogi 2013).

9.6.2 Definition

The emulsion liquid membrane is a system in the form of double emulsions. An emulsion liquid membrane may be one of two types: a water-in-oil emulsion dispersed in an external aqueous phase or an oil-in-water emulsion dispersed in an outer organic phase. The membrane phase in the the water-in-oil-in-water type is an immiscible oil phase separating the aqueous phases. The emulsion liquid membrane consists of three steps: 1) preparation and diffusion in fine droplets in oil, 2) permeation of the solute through the membrane, and 3) stabilization of the solution.

9.6.3 Transport Mechanism

Emulsion liquid membranes have been widely used to investigate ion transport against the concentration gradient by a coupled transport mechanism. Ion transport through an emulsion liquid membrane plays a significant role in partition technologies. The principal rate-determining step in solute penetration through a liquid membrane is solute dispersion through the membrane. However, separation can be enhanced by the use of additives, specific carriers, and chemical reagents. The different partition mechanisms are largely classified into two types: simple and facilitated mechanism.

9.6.4 Emulsion Liquid Membrane Design Considerations

9.6.4.1 Emulsification

Emulsification is the method of diffusing one liquid into another one if the two liquids are immiscible (Matos et al. 2013). It is a two-step method, as follows: 1) droplets are formed in one liquid, and 2) the newly-formed boundary between the two liquids is stabilized by an emulsifier (Floury et al. 2004; Perrier-Cornet et al. 2005; Tesch and Schubert 2002). Issues occur with the thermodynamic instability of the newly produced droplets and the turbulence associated with emulsification (Jafari et al. 2008). Droplet coalescence can be prevented by increasing the viscosity of the emulsion liquid membrane (Tesch and Schubert 2002; Behrend et al. 2000).

9.6.4.2 Demulsification

Demulsification depends on the emulsion liquid membrane's surfactant concentration, microdroplet size, and the solvent viscosity (Sun et al. 1998). The emulsion liquid membrane's removal efficiency and metal recovery are highly dependent on the demulsification (Sun et al. 1998). The two main approaches for the demulsification are chemical and physical methods. Chemical methods involve the accumulation of a demulsifier in the emulsion. Physical methods include heating, centrifugation, solvent dissolution, high shear, and use of high-voltage electrostatic fields.

9.6.4.3 Stability of Emulsion Liquid Membranes

The stability of emulsion liquid membranes is one of the most important considerations for this method. An emulsion liquid membrane process that is used to remove industrial pollutants has a number of advantages. However, several factors decrease the emulsion stability, including membrane formulation, preparation, and pH (Hou and Papadopoulos 1996).

9.6.5 Applications and Disadvantages of Emulsion Liquid Membranes

9.6.5.1 Applications

The most important applications of the emulsion liquid membrane method are as follows:

1. Large interfacial vicinity for mass transfer
2. High dispersal speed of metal ions through the membrane
3. Simultaneous concert of transport and stripping in a similar method

9.6.5.2 Disadvantages

The emulsion liquid membrane method has several disadvantages, including the following:

1. If, for any reason, the membrane does not remain intact during the process, the separation attained to that point is destroyed.
2. Emulsion instability is one of the most important disadvantages to this method.

References

Alpoguz HK, Memon S, Ersoz M, Yy'lmaz M (2004) Transport kinetics of Hg2þ through bulk liquid membrane using calix[4]arene ketone derivative as carrier. Sep Sci Technol 39:799–810

Banat IM, Nigam P, Singh D, Marchant R (1996) Microbial decolourization of textile-dye-containing effluents. Bioresour Technol 58:217–227

Behrend O, Ax K, Schubert H (2000) Influence of continuous phase viscosity on emulsification by ultrasound. Ultrason Sonochem 7(2):77–85

Benjjar A, Hor M, Riri M, Eljaddi T, Kamal O, Lebrun L, Hlaibi M (2012) A new supported liquid membrane (SLM) with methyl cholate for facilitated transport of dichromate ions from mineral acids: parameters and mechanism relating to the transport. J Mater Environ Sci 3(5):826–839

Chang SH, Teng TT, Norli I (2011) Optimization of Cu (II) Extraction from Aqueous Solutions by Soybean-Oil-Based Organic Solvent Using Response Surface Methodology. Water Air Soil Pollut 217:567–576

Chanukya BS, Rastogi NK (2013) Extraction of alcohol from wine and color extracts using liquid emulsion membrane. Sep Purif Technol 105(0):41–47

Chiou MS, Chuang GS (2006) Competitive adsorption of dye metanil yellow and RB15 in acid solutions on chemically cross-linked chitosan beads. Chemosphere 62:731–740

Eljaddi T, Kamal O, Benjjar A, Riri M, Elatmani MELH, Lebrun L, Hlaibi M (2014) New supported liquid membranes containing TBP and MC as carriers for the facilitated transport of cadmium ions from acidic mediums: Parameters and mechanism. J Mater Environ Sci 5 (6):1994–1999

Ersoz M (2007) Transport of mercury through liquid membranes containing calixarene carriers. Adv Colloid Interface Sci 134–135:96–104

Floury J, Legrand J, Desrumaux A (2004) Analysis of a new type of high pressure homogeniser. Part B. Study of droplet break-up and recoalescence phenomena. Chem Eng Sci 59 (6):1285–1294

Hassan AA, Victor K, Nidal H (2013) Hybrid ion exchange – Pressure driven membrane processes in water treatment: A review. Sep Purif Technol 116:253–264

Hou W, Papadopoulos KD (1996) Stability of waterin- oil-in-water type globules. Chem Eng Sci 51 (22):5043–5051

Jafari SM, Assadpoor E, He Y, Bhandari B (2008) Re-coalescence of emulsion droplets during high-energy emulsification. Food Hydrocoll 22(7):1191–1202

Katzung BG (1987) Basic and clinical pharmacology, 3rd edn. Appleton and Lange, New York, pp 734–735

Kemperman AJB (1995) Stabilization of supported liquid membranes, Ph.D. Thesis, University of Twente, The Netherlands

Kemperman AJB, Bargeman D, Van den Boomgaard TH, Strathmann H (1996) Stability of supported liquid membranes: state of the art. Sep Sci Technol 31(20):2733–2762

Kemperman AJB, Rolevink HHM, Bargeman D, Van den Boomgaard TH, Strathmann H (1998) Stabilization of supported liquid membranes by interfacial polymerization top layers. J Membr Sci 138:43–55

Khan AA, Husain Q (2007) Decolorization and removal of textile and non-textile dyes from polluted wastewater and dyeing effluent by using potato (Solanum tuberosum) soluble and immobilized polyphenol oxidase. Bioresour Technol 98:1012–1019

Li NN (1968) Separating hydrocarbons with liquid membranes. US patent 3:410–794

Li N, Cahn R, Naden D, Lai R (1983) Liquid membrane processes for copper extraction. Hydrometallurgy 9(3):277–305

Loiacono O, Drioli E, Molinari R (1986) Metal ion separation and concentration with supported liquid membranes. J Membr Sci 28:123–138

Ma M, He DS, Liao SH, Zeng Y, Xie QJ, Yao SZ (2002) Kinetic study of L-isoleucine transport through a liquid membrane containing di(2-ethylhexyl) phosphoric acid in kerosene. Anal Chim Acta 456:157

Malinowski JJ (2001) Two-phase partitioning bioreactors in fermentation technology. Biotechnol Adv 19:525

Matos M, Suarez MA, Gutierrez G, Coca J, Pazos C (2013) Emulsification with microfiltration ceramic membranes: A different approach to droplet formation mechanism. J Membr Sci 444 (1):345–358

Mulder M (1991) Basic principles of membrane technology. Kluwer Academic Publishers, Dordrecht

Muthuraman G, Teng TT, Leh CP, Norli I (2009) Use of bulk liquid membrane for the removal of chromium (VI) from aqueous acidic solution with tri-n-butyl phosphate as a carrier. Desalination 249:884–890

Nath K (2008) Membrane separation processes. PHI Learning Pvt. Ltd, New Delhi

Paugam MF, Smith BD (1993) Active transport of uridine through a liquid organic membrane mediated by phenylboronic acid and driven by a fluoride ion gradient. Tetrahedron Lett 34 (23):3723–3726

Perrier-Cornet J, Marie P, Gervais P (2005) Comparison of emulsification efficiency of protein-stabilized oil-in-water emulsions using jet, high pressure and colloid mill homogenization. J Food Eng 66(2):211–217

Pirkaramia A, Olya ME, Limaee Y (2013) Decolorization of azo dyes by photo electro adsorption process using polyaniline coated electrode. Prog Org Coat 76:682–688

Poots VJP, McKay G, Healy JJ (1976) The removal of acid dye from effluent using natural adsorbents-I peat. Water Res 10:1061–1066

Ravindra WG, Sunil AMK (2009) Studies on auramine dye adsorption on psidium guava petals. Korean J Chem Eng 26(1):102–107

Salem IA, El-maazawi M (2000) Kinetics and mechanism of color removal of methylene blue with hydrogen peroxide catalyzed by some supported alumina surfaces. Chemosphere 41:1173–1180

Schlosser S (2000a) Membrane based processes with immobilized inter-face. In: Bako K, Gubicza L, Mulder M (eds) Integration of membrane processes into bioconversions. Kluwer Academic, New York, p 55

Schlosser S (2000b) Pertraction through liquid and polymeric membranes. In: Bako K, Gubicza L, Mulder M (eds) Integration of membrane processes into bioconversions. Kluwer Academic Publishers, New York

Stephenson RJ, Sheldon JB (1996) Coagulation and precipitation of a mechanical pulping effluent-I. Removal of carbon, colour and turbidity. Water Res 30:781–792

Sun D, Duan X, Li W, Zhou D (1998) Demulsification of water-in-oil emulsion by using porous glass membrane. J Membr Sci 146(1):65–72

Talebi A, Teng TT, Abbas FMA, Norli I (2012) Optimization of nickel removal using liquid–liquid extraction and response surface methodology. Desalin Water Treat 47(1–3):334–340

Teng TT, Talebi A, Muthuraman G (2014) In: Hamidi A, Mojiri A (eds) Wastewater Engineering: Advanced Wastewater Treatment Systems. IJSR Books

Tesch S, Schubert H (2002) Influence of increasing viscosity of the aqueous phase on the short-term stability of protein stabilized emulsions. J Food Eng 52(3):305–312

Van Sonsbeek HM, Beeftink HH, Tramper J (1993) Two-liquid-phase bioreactors. Enzym Microb Technol 15:722–729

Wells AF (1993) Structual inorganic chemistry. Wydawnictwo Naukowo Techniczne, Warszawa

Wódzki R (1997) Liquid membrane. Structure and mechanism of action, [in]: Narębska A. (red.): Membranes and membrane separation technology. Wydawnictwo Uniwersytetu Mikołaja Kopernika, Toruń

Chapter 10
Assessing the Impacts of Temperature, Precipitation and Land Use Change on Open Water Bodies of Middle Ghaghara River Basin

Rajesh Kumar, Prem Chandra Pandey, Pyarimohan Maharana, Hemant Gautam, and Vijendra Kumar Pandey

Abstract Wetlands are one of the most productive and fragile ecosystems on the Earth's surface. They have been providing essential services (irrigation, groundwater recharge, flood control and drinking water) to the society since the beginning of civilisation. The middle Ghaghara River basin is located in middle Ganga alluvial plain and is mainly drained by Ghaghara, Sarju, Rapti, Burhi Rapti and Kuwano rivers. In this study, Landsat 5 and 8 images of the post-monsoon months of 1989 and 2015 were used to extract the open water bodies of wetlands. Density slicing of the shortwave infrared band and the modified normalised water difference index (MNDWI) were utilised for the extraction of open water bodies. The total area of open water bodies was 472 km^2 in 1989, while it reduced to 317 km^2 in 2015. Thus, the reduction in open water bodies of wetlands is about 33%.

Precipitation and temperature collectively play a major role in nurturing the wetlands. Therefore, in this study, trends in these climate parameters were analysed using non-parametric Mann–Kendall (MK) and Sen's slope methods. A significant increasing trend in annual mean temperature and a decreasing trend in total annual precipitation caused to shrink in open water bodies of the wetlands. Besides this, encroachment of cropland and built-up areas on wetlands denotes anthropogenic reason for the reduction in their areal extent. This study is beneficial to the planners

---

R. Kumar (✉) · H. Gautam · V. K. Pandey
Centre for the Study of Regional Development, School of Social Sciences, Jawaharlal Nehru University, New Delhi, India

P. C. Pandey
Institute of Environment and Sustainable Development, Banaras Hindu University, Varanasi, India

Department of Geography, School of Geosciences, Faculty of Exact Sciences, Tel Aviv University, Tel Aviv, Israel

P. Maharana
Delhi School of Arts and Commerce, University of Delhi, New Delhi, India

R. P. Singh et al. (eds.), *Water Conservation, Recycling and Reuse: Issues and Challenges*, https://doi.org/10.1007/978-981-13-3179-4_10

and environmentalists for the restoration of shrinking open water bodies of wetlands in the middle Ghaghara River basin.

Keywords Middle Ghaghara River basin · Landsat 8 OLI-TIRS · MNDWI · Density slicing · Open water body · Mann–Kendall test · Sen's slope

10.1 Introduction

Inland freshwater has become a subject of scientific investigation for several factors and, thus, has gained tremendous attention all around the globe for its conservation (Woodward et al. 2010; Woodward 2009). The reason for this particular concern is freshwater biodiversity and its coverage (0.8 percent of the globe) around the world (Dudgeon et al. 2006). Moreover, freshwater ecosystems have distinctive properties with wide variation in diversity (Olden et al. 2010; Whittaker et al. 2005). Among inland freshwater, wetlands are one of them and considered as natural reserves, inhabiting various aquatic flora and fauna (Mitsch and Gosselink 2000).

Freshwater bodies are particularly vulnerable to climate change and change in land use patterns, affecting their areal coverage and limits (Woodward et al. 2010). Foremost reasons for these changes in the freshwater body are due to its relative isolation and physical fragmentation within a terrestrial landscape and its exploitation by humans for their benefits and development (Woodward 2009). These exploitations are mainly for obtaining 'goods and services' from the existing freshwater and surrounding landscape (Woodward et al. 2010; Woodward 2009). Since these freshwater bodies are hotspot for human activities, therefore it is imperative for the existence of human. The uncontrolled activities such as development, flow regulation, water extraction, fisheries over-exploitation and pollution led to widespread habitat degradation, shrinkage and conversion from one land use to another (Varghese et al. 2008; Strayer and Dudgeon 2010). In the absence of human activities in the freshwater/wetlands, it may expand up to a certain extent, resulting from rare events such as river capturing and flooding (Banarescu, 1990; Burridge et al. 2006; Olden et al. 2010). The Intergovernmental Panel on Climate Change (IPCC) (2007) predicted temperature increase in tropics, impacting biodiversity of the freshwater bodies (Heino et al. 2009; Poff et al. 2009), while model from Deutsch et al. (2008) predicted a severe impact on the tropical biodiversity by rising temperature and falling precipitation. Earlier, few conservationists can see the impact of human-induced climate change as a threat to freshwater biodiversity (Firth and Fisher 1992). Evidence for the beginning of human-induced climate change impact on freshwater bodies was in research (Woodward et al. 2010). Many authors have already warned of possible rising water temperature and changes in the geographic ranges or phenology of freshwater species (Ashizawa and Cole 1994, Magnuson et al. 2000, Parmesan 2006, Heino et al. 2009). Brinson and Malvárez (2002) identified six geographical regions, namely North America, South America,

Northern Europe, northern Mediterranean, temperate Russia and South–South East Asia where the loss of biodiversity caused reduction in wetland areas. Although the geographical determination for changes in wetland ecosystem is quite variable for species belonging to native regions (Brinson and Malvárez 2002), the consequences of climatic change have immense potential to affect the species and their control either directly or indirectly, such as establishing new invasive species and their alteration (switchgrass, *Panicum virgatum*; prairie species), impact on existing species (*Styela clava* and *Molgula manhattensis*: both are exotic ascidians), their geographical distribution (*Salvelinus fontinalis* also known as brook trout and *Salmo trutta* as brown trout) and management's strategy to control them (water hyacinth, *Eichhornia* spp., and salt cedar leaf beetle, *Diorhabda elongata*) (Carlton 2000; Rahel and Olden 2008; U.S. EPA 2008).

An increasing temperature and less precipitation will directly affect the freshwater conditions either by drying up or limiting water level. Thus, these factors could impact either its decrease in the areal coverage or reduction in biodiversity of species that have narrow thermal tolerances (Nakano et al. 1996; Allan et al. 2005; Sinha 2011). Different models predicted an increase in the frequency and severity of floods and droughts (IPCC 2007; Hirsch and Archfield 2015), resulting from significant variation in hydrological status, caused by changes in the amount, timing and duration of precipitation and large difference in minimum and maximum temperature (Allan et al. 2005; Bassi et al. 2014). As a result, the size and permanency of running and standing waters will change, along with modifications in biodiversity. Later on, these conditions prompted several meetings and discussions for conservation and planning of freshwater, marine and terrestrial ecosystems (Heino et al. 2009, Poff et al. 2009). In combination with land use pattern changes, these conditions affect the freshwater areal coverage and cause shrinkage or drying of some parts thereof. A brief overview of climatic variability and land use change impacts on open water bodies seems warranted, considering the topic of the research in the present study.

In the recent years, wetlands are being converted into urban and cropland due to the rapid growth of population and growing global economy (Pandit and Qadri 1990; Sinha 2011). In addition, an increase in demand of freshwater for agriculture results in negative ecological impacts (Jackson et al. 2001; Verma 2001; Dudgeon et al. 2006; Pekel et al. 2016). Even human beings trap about 25% of freshwater in the reservoirs before it met the oceans (Vörösmarty and Sahagian 2000), thus causing sedimentation in those reservoirs due to unscientific agricultural practices. Land use patterns and exploitation in different conditions such as farming, industrial factory and domestic household wastes have poured excessive loads of nutrients (sulphates and phosphates), sewage, toxin, insecticides and pesticides (Strayer 2010; Smith 2003; Smol 2008) which cause eutrophication of the freshwater, unable to support the natural growth and survival of biotic communities. For example, these pollutants as causal factors have led to removal and elimination of all fishes from 5% of the rivers (Dudgeon 1999).

Wetlands are either permanently waterlogged areas or lands that are flooded or saturated for longer periods with a regular frequency (Tiner 2014). For instance, besides nutrient supply, settling of sediment-laden floodwater in wetlands of Jammu and Kashmir is one of the major reasons responsible for sedimentation. Furthermore, rapidly growing population and fast urbanisation processes led to the encroachment on those wetlands for agriculture, road and house construction (Pandit and Qadri 1990; Kumar and Acharya 2016; Kumar 2016).

This study focuses on advances in remote-sensing technology to prepare an inventory and to perform the change detection of surface water of wetlands in the middle Ghaghara River basin. The modified normalised difference water index (MNDWI) (Xu 2006) and an appropriate threshold value of shortwave infrared (SWIR) bands of Landsat 5 Thematic Mapper (TM) (1989) and Landsat 8 Operational Land Imager-Thermal Infrared Sensor (OLI-TIRS) (2015) images have been applied to map the open water bodies. This study also analyses impacts of land use change and climatic parameters such as precipitation and temperature on the open water bodies of wetlands.

10.2 Study Area

The study area is located in the middle part of Ghaghara River basin. Its longitudinal extent is between 81°23′19.61″ E and 83°06′ 17.45″ E, while the latitudinal extent is from 26°33′33.715″ N to 28°10′32.286″ N (Fig. 10.1). The middle Ghaghara River basin covers an area of 21,134 km^2 out of the total area (1,207,950 km^2) of the entire Ghaghara basin. The middle Ghaghara River basin is drained by Ghaghara and its tributaries (Sarju, Kuwano and Rapti) and sub-tributary (Burhi Rapti). The Sarju, Kuwano, Rapti and Burhi Rapti rivers have sinuous channels, while the Ghaghara River has braided channels. Permanently flooded wetlands are the Ghaghara, Rapti, Burhi Rapti, Sarju and Kuwano rivers, while the saturated or temporarily flooded wetlands are situated in the floodplains of these rivers and its surrounding flood-free regions. The general slope of the study area is from northwest (NW) to southeast (SE), following the general direction of lineaments (Singh 1996).

The elevation ranges from 41 to 260 m, above mean sea level (amsl). An average elevation of the study area is about 109 m, amsl. Annual precipitation varies from 101 (in Basti) to 109 cm (in Bahraich). Normally, high amount of rainfall occurs during monsoon season (June to September). The annual mean temperature ranges between 25.3° (Bahraich) and 25.8 °C (Basti). According to Varghese et al. (2008), major flora of the wetlands of the study area is aquatic weeds like water hyacinth (*Eichornia crassipes*), marsh glory (*Ipomoea aquatica*), water fern (*Salvinia molesta*) and the aquaculture farms of water chestnut (*Trapa natans*), while the major fauna includes catfishes like bighead carp (*Ariyocthys nobilis*) and northern African catfish (*Clarias gariepinus*). The administrative units, covering the study area, are Bahraich, Shrawasti, Balrampur, Siddharthnagar, Basti, Gonda, Sant Kabir Nagar, Ambedkar Nagar, Faizabad, Barabanki and Sitapur districts of Uttar Pradesh (Fig. 10.1).

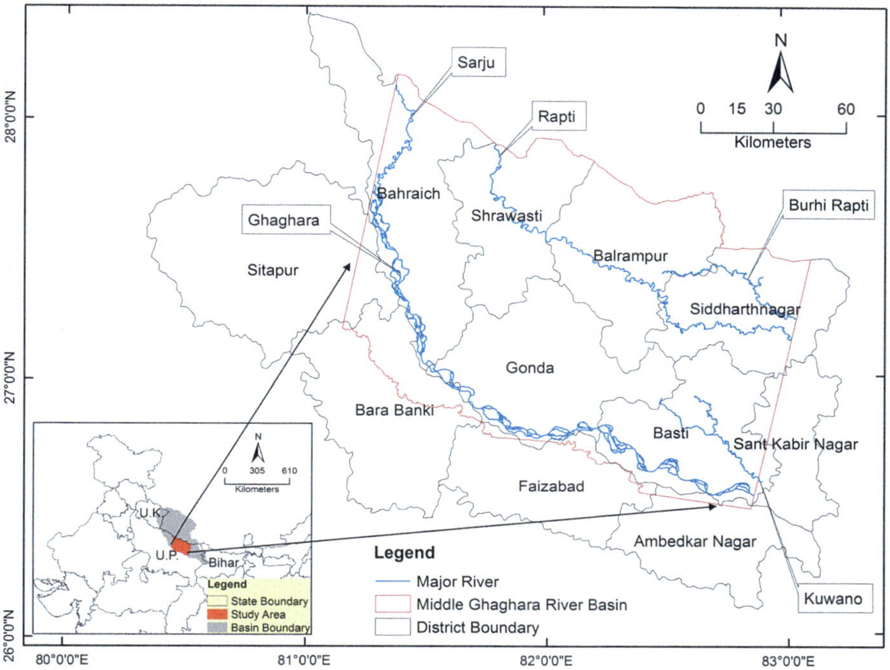

Fig. 10.1 Location map of the study area. Name of the major rivers of the region is given in callouts

10.3 Materials and Methods

10.3.1 Data Used

This study utilises different types of data sets, collected and analysed from different sources. Remotely sensed multispectral satellite images (30 m spatial resolution) from different sensors like Landsat 5 Thematic Mapper (TM) and Landsat 8 Operational Land Imager-Thermal Infrared Sensor (OLI-TIRS) were obtained from the US Geological Survey EarthExplorer (USGSEE 2016) (Table 10.1). Total annual rainfall (in mm) and annual mean temperature (in °C) for Bahraich, Shrawasti, Balrampur, Siddharthnagar, Basti and Gonda districts were downloaded from India Water Portal (2017). Google Earth image of the period November 19, 2015, was also used as a reference map to check the accuracy of open water bodies, extracted from Landsat 8 OLI-TIRS image. Data on cropland and built-up areas were obtained from Bhuvan Geo-Platform of ISRO (NRSC 2006 and 2014). District-wise total population data were collected from District Sankhikiya Patrika (2015).

Table 10.1 Details of satellite images

Satellite/sensor	Date of acquisition	Scan time (GMT)	Sun azimuth (°)	Sun elevation (°)	Cloud cover (%)	Earth-sun distance (in astronomical units)
Landsat 5 TM	November 14, 1989	04:24:55.6110810Z	146.3829196	37.03540569	12.45	0.9913869
Landsat 8 OLI-TIRS	November 6, 2015	05:00:49.4526352Z	155.2753964	43.12107867	1	0.9893152

10.3.2 Atmospheric Correction of Landsat (5 TM and 8 OLI-TIRS) Images

The digital number (DN) of images acquired from the Landsat 5 and 8 sensors was converted into top of atmosphere (TOA) radiance using Eq. (10.1) (USGS 2015):

$$R = M_{\mathrm{L}}{}^{*}(\mathrm{DN}) + A_{\mathrm{L}} \tag{10.1}$$

where R denotes TOA radiance (Watts/(m^2 * srad * μm)), M_{L} represents band-specific multiplicative rescaling factor and A_{L} stands for band-specific additive rescaling factor, obtained from the metadata files of the Landsat images.

TOA radiance values of these images were converted into surface reflectance that is crucial for the calculation of modified normalised difference water index (MNDWI) (Xu 2006) and for determining an appropriate threshold value of short-wave infrared (SWIR) band to extract water.

The radiance images were converted into band-interleaved-by-line (BIL) format and divided by 10 to obtain the radiance values in μW/(cm^2 * srad * nm) that are further used as an input image in the Fast Line-of-sight Atmospheric Analysis of Spectral Hypercubes (FLAASH) module of ENVI 5.1 software. FLAASH is a MODTRAN®-based radiation transfer code that is widely used to correct scattering from atmospheric water vapour, oxygen, carbon dioxide, methane, ozone, molecule and aerosols on the basis of scene centre elevation, viewing and solar angles of the scenes (Jensen 2005). The central location of each scene, sensor altitude, flight date and time (GMT) were also put into the FLAASH model as input variables. In this study, the selected aerosol model was tropospheric, while the atmospheric model was midlatitude summer.

The Kaufman–Tanre aerosol retrieval method was based on two bands that use dark pixels of the input scene (Kaufman et al. 1997). If there is no dark pixel in the scene, the value of initial visibility (50 km) is used. The input for the FLAASH model is basically the selection of channels to run the K–T aerosol retrieval. Landsat 5 and 8 sensors have inappropriate wavelengths for water retrieval; therefore, in this study, the overland retrieval alternate method was selected, considering the band 7 of Landsat 5 and the shortwave infrared 2 (SWIR2) of Landsat 8 image as K–T upper channels. Further, band 1 of Landsat 5 and the coastal aerosol band of Landsat 8 were by default selected as K–T lower channels. The maximum upper channel reflectance was 0.1, while the reflectance ratio between upper and lower channel was 0.25. The retrieved mass of water molecules in the atmosphere was 2.77 g/cm^2.

Aerosol scale height and CO_2 mixing ratio were 1.5 km and 390 ppm, respectively. The scene centre elevation was 0.109 km. However, the satellite images used in this study were sensed by the multispectral sensors; therefore, the MODTRAN resolution of 15 cm^{-1} was selected. The MODTRAN multiscatter model was DISORT with 16 directions.

10.3.3 Open Water Body Extraction

The open water bodies were extracted using MNDWI and density slicing of surface reflectance values of SWIR and SWIR1 band of Landsat 5 TM and 8 OLI-TIRS images, respectively. The formula of MNDWI computation for Landsat 5 TM image is given in Eq. (10.2), while for Landsat 8 OLI-TIRS is given in Eq. (10.3).

$$[(\text{Band2} - \text{Band5})/(\text{Band2} + \text{Band5})] \tag{10.2}$$

$$[(\text{Green} - \text{SWIR1})/(\text{Green} - \text{SWIR1})] \tag{10.3}$$

The MNDWI values range between -1 and $+1$. The negative MNDWI values show non-water objects on the image, while the positive values denote clear and turbid water pixels (Xu 2006). In the study area, the presence of weeds and algae in the wetlands causes problems in water pixel detection using MNDWI. Therefore, in this study, appropriate thresholds were applied on SWIR bands of Landsat 5 and SWIR1 band of Landsat 8 images to map the open water bodies of the wetlands. The defined thresholds for atmospherically corrected SWIR bands of Landsat (5 and 8) images were 0.095 and 0.08, respectively. These threshold values surpass the effects of weeds and algae and enhance the open water bodies of wetlands. Finally, water body layers, derived using MNDWI and density slicing of SWIR bands, were combined to compute the areal extent of surface water bodies of wetlands under different districts using ARCGIS 10. The selected coordinate system was World Geodetic System (WGS) Universal Transverse Mercator (UTM) zone 44 north for the calculation of areal extent of open water bodies.

10.3.4 Trend Analysis of Total Annual Precipitation and Annual Mean Temperature

A non-parametric Mann–Kendall (MK) test was applied to statistically analyse if there is a monotonic upward or downward trends in temperature and precipitation over six different stations of UP for 100 years starting from 1903 to 2002. The trend was considered at a minimum of 99% confidence interval. The sign of *Z*-value provides the increasing or decreasing trends of the time series (positive increasing and negative decreasing trend). The Sen's slope provides a rate of change (Q) in precipitation and temperature (Sen 1968).

10.3.5 Percentage Change Computation

Percentage change in the open water bodies, cropland and built-up areas and a total population of two different periods were computed using Eq. (10.4):

$$\text{Percentage change} = [(T_2 - T_1)/T_1] \times 100 \tag{10.4}$$

where T_2 denotes a variable of recent time and T_1 shows a variable of the previous time.

10.4 Results and Discussion

The total area under open water bodies has come down from 472 km^2 in 1989 to 317 km^2 in 2015. Thus, a decrease of 33% is observed in the last 26 years (Fig. 10.2). Regarding district-wise areal coverage, 91% area of the study region is distributed among Bahraich, Shrawasti, Balrampur, Siddharthnagar, Basti and Gonda districts. In these districts, a decrease in open water bodies is observed in the last 26 years (Table 10.2). Highest decrease (51%) is observed in Shrawasti while the lowest (12%) in Basti District. The rest of the districts, namely Barabanki, Sitapur, Faizabad, Ambedkar Nagar and Sant Kabir Nagar, only cover 9% area of the study region. Among these districts, open water bodies in Faizabad District show an increase of 6% due to the formation of new lakes and channel widening and a decrease in all other districts. The highest decrease is observed in Sant Kabir Nagar while the lowest in Sitapur District (Table 10.2).

Google Earth image dated November 19, 2015, has been used as a base map to perform accuracy assessment of the open water bodies, extracted from Landsat 8 OLI-TIRS image. Overall accuracy of the extracted open water bodies is 89.4% with a kappa coefficient of 0.79 (Table 10.3). The accuracy assessment of open water bodies of the period 1989, extracted from Landsat 5 TM image, could not be performed due to unavailability of base map of water bodies of the same period.

The open water bodies (oxbow lakes, cutoffs and abandoned channels) along the Ghaghara, Rapti, Burhi Rapti, Sarju and Kuwano rivers are temporary in nature, depending upon the sediments supply from the mountainous and fragile upper reaches (Nepal) as well as from the plain (UP). This also depends upon the frequency of inundation. Depleting forest cover in mountainous reaches as well as in UP further increased the sediment load in these rivers (Rana et al. 2009; Kumar et al. 2013a, b; Kumar et al. 2016). Hence, sedimentation in wetlands is occurring in this region during monsoon season.

The mean spectral profiles of open water bodies of 1989 and 2015 show a decrease in reflectance in NIR and an increase in green band (Fig. 10.3). A reduction in reflectance in NIR band denotes a decrease in weeds and algal content (Fig. 10.4a and b). An increase in reflectance in green band signifies high suspended sediment

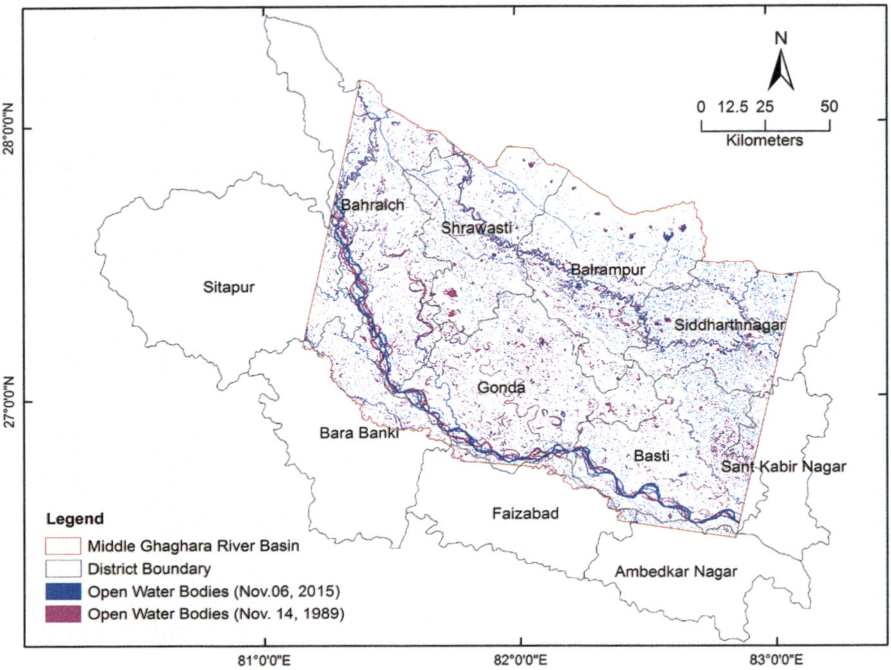

Fig. 10.2 Distribution of open water bodies of 1989 and 2015

Table 10.2 District-wise areal extent of open water bodies (in km^2) and percentage change during 1989–2015

District	1989	2015	Percentage change
Gonda	106.1	63.7	−40
Bahraich	99.5	82.1	−18
Basti	77.6	68.3	−12
Balrampur	64.6	32.8	−49
Siddharthnagar	52.7	30.3	−43
Shrawasti	43.4	21.2	−51
Barabanki	16.1	9.7	−40
Sitapur	4.9	4.6	−8
Faizabad	2.2	2.3	6
Ambedkar Nagar	2.0	1.4	−32
Sant Kabir Nagar	2.9	0.6	−80
Total	472	317	−33

concentrations (e.g. Han 1997) (Fig. 10.4c). Deposition of sediments in open water bodies is detrimental to many aquatic flora and fauna (e.g. Varghese et al. 2008; Pandit and Qadri 1990). Furthermore, such changes in mean spectral profiles in the last 26 years are indicating towards land use change and encroachments on the wetlands in the middle Ghaghara River basin (Fig. 10.4d).

Table 10.3 Accuracy assessment of non-water areas and open water bodies (November 19, 2015)

Class	Commission (%)	Omission (%)	Commission (pixels)	Omission (pixels)
Non-water areas	20.9	0	9/43	0/34
Open water bodies	0	17.6	0/42	9/51
	Producer accuracy	User accuracy	Producer accuracy	User accuracy
	(%)	(%)	(Pixels)	(Pixels)
Non-water areas	100	79.1	34/34	34/43
Open water bodies	82.4	100	42/51	42/42

Overall accuracy = 89.4%; kappa coefficient = 0.79

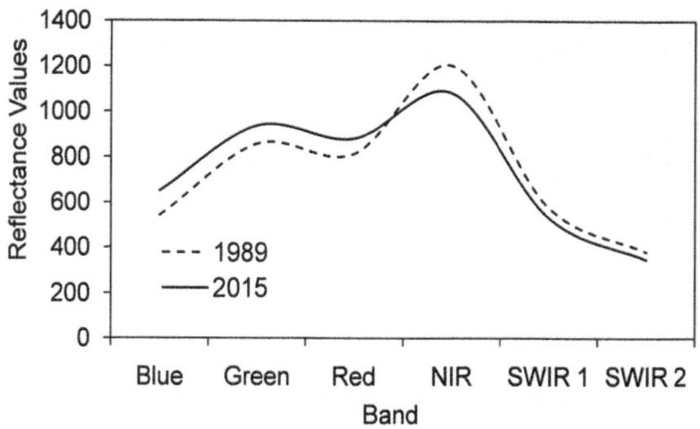

Fig. 10.3 Mean spectral curves for open water bodies of 1989 and 2015

The encroachment is mainly due to rapidly growing population, which needs land for living, agriculture, commercial and industrial activities. More than 50% increase in population is observed in Siddharthnagar, Balrampur and Gonda districts from 1991 to 2011(Table 10.4).

In Bahraich and Basti districts, an increase in total population is 42 and 47% from 1991 to 2011, respectively. This is evident from the statistics that there is an increase in the areal extent of build-up areas and cropland. Overall the built-up areas increase by 42.6% between 2005–2006 and 2011–2012 in the selected districts. An increase in built-up areas is highest in Siddharthnagar, while the lowest is observed in Gonda district (Table 10.5).

On the other hand, the overall increase in cropland is 12.1% from 2005–2006 to 2011–2012. The highest expansion in cropland is observed in Balrampur while the lowest in Basti (Table 10.6). Such expansion in built-up areas and cropland has caused a loss in the areal extent of open water bodies. For instance, Yousefi et al. (2017) reported that the channel area of Talar River of Iran is decreasing due to increase in built-up areas along the river.

Precipitation and temperature are two major factors that nurture the water bodies of any region (Sinha 2011). Besides anthropogenic causes, a decreasing total annual

Fig. 10.4 (**a** and **b**) Weeds and algal contents in open water bodies (82.09798 E and 27.1039 N and 82.07782 E and 27.11352 N), (**c**) turbid water due to sediments mobilised from the nearby agricultural land (82.26267 E and 27.029554 N) and (**d**) construction of houses in the desiccated open water body (82.35455 E and 27.004051 N) (Photographs taken by the authors)

precipitation and an increasing annual mean temperature are also the main causal factors for desiccation of open water bodies in the study region. The MK test for the total annual precipitation for all stations is significant at the 0.01 level. The negative

Table 10.4 Percentage change in total population

District	1991	2001	2011	1991–2001	2001–2011	1991–2011
Bahraich	2,090,852	2,701,478	2,961,861	29	10	42
Shrawasti	N.A.	855,864	1,116,840	N.A.	30	N.A.
Balrampur	1,368,630	1,682,350	2,148,795	23	28	57
Siddharthnagar	1,618,932	2,040,085	2,557,742	26	25	58
Basti	1,675,359	2,084,814	2,463,939	24	18	47
Gonda	2,204,445	2,765,586	3,405,376	25	23	54

Source: District Sankhikiya Patrika (2015)

Table 10.5 District-wise areal extent of built-up areas (in km^2) and percentage change between 2005–2006 and 2011–2012

District	2005–2006[a]	2011–2012[b]	Percentage change
Gonda	42.1	45.7	8.7
Balrampur	41.5	46.8	12.6
Shrawasti	49.4	64.3	30.2
Bahraich	34.3	67.2	95.8
Siddharthnagar	47.5	126.6	166.5
Basti	152.3	172.9	13.5
Total	367.2	523.5	42.6

Source: [a]NRSC (2006) and [b]NRSC (2014)

Table 10.6 District-wise areal extent of crop land (in km^2) and percentage change between 2005–2006 and 2011–2012

District	2005–2006[a]	2011–2012[b]	Percentage change
Shrawasti	1739.2	1928.8	10.9
Basti	2093.8	2099.5	0.3
Siddharthnagar	2242.8	2484.1	10.8
Balrampur	1946.7	2512.2	29.0
Bahraich	2907.6	2942.5	1.2
Gonda	3051.6	3709.8	21.6
Total	13981.7	15676.8	12.1

Source: [a]NRSC (2006) and [b]NRSC (2014)

Z-values for all station show that the total annual precipitation is decreasing significantly over all the stations (Table 10.7).

The rate of decrease (*Q*) in total annual precipitation is highest over Siddharthnagar and least over Gonda district. The rate of decrease is 0.045 mm/year over Bahraich and 0.046 mm/year over Balrampur. Over Basti, Gonda, Shrawasti and Siddharthnagar, the rate of decrease is 0.026, 0.012, 0.014 and 0.052 mm/year, respectively. The MK test for annual mean temperature is significant at the 0.01 level. All stations exhibit an increasing trend in annual mean temperature, showing positive *Z*-values (Table 10.8).

Table 10.7 MK test output for total annual precipitation (1903–2002)

Stations	*Z* statistics	Level of significance	*Q* value (mm/year)
Bahraich	−2.61	0.01	0.045
Balrampur	−3.02	0.01	0.046
Basti	−2.96	0.01	0.026
Gonda	−2.73	0.01	0.012
Shrawasti	−2.76	0.01	0.014
Siddharthnagar	−3.22	0.01	0.052

Table 10.8 MK test output for mean annual temperature (1903–2002)

Stations	*Z* statistics	Level of significance	*Q* value (°C/year)
Bahraich	2.59	0.01	0.0036
Balrampur	3.26	0.01	0.0046
Basti	2.8	0.01	0.0037
Gonda	2.66	0.01	0.0037
Shrawasti	2.52	0.01	0.0035
Siddharthnagar	2.58	0.01	0.0033

Table 10.9 Trend in rainfall and temperature in Uttar Pradesh (1951–2010)

Month/season	Parameters	Trend	Rate
November	Mean temperature	Increasing[a]	+0.02 °C/year
	Monthly rainfall	Increasing[a]	+0.01 mm/year
Post-monsoon	Mean temperature	Increasing[a]	+0.01 °C/year
	Rainfall	Decreasing	−0.33 mm/year
Annual	Mean temperature	No trend	0 °C/year
	Rainfall	Decreasing[a]	−4.42 mm/year
Monsoon	Mean temperature	No trend	0 °C/year
	Rainfall	Decreasing[a]	−3.52 mm/year

[a]Trend is significant at 95%
Source: Rathore et al. (2013)

The rate of increase (*Q*) in annual mean temperature is highest in Balrampur and least in Siddharthnagar district. The rate of increase in annual mean temperature in Bahraich, Balrampur, Basti, Gonda, Shrawasti and Siddharthnagar districts is 0.0036, 0.0046, 0.0037, 0.0037, 0.0035 and 0.0033 °C/year, respectively. According to Rathore et al. (2013), an overall decreasing trend in annual precipitation over entire Uttar Pradesh is significant at 95% level during 1951–2010. The rate of change in annual precipitation is −4.42 mm/year. There is no observable trend in the annual mean temperature. In other words, the annual mean temperature is static between 1951 and 2010 (Table 10.9). During the post-monsoon season, a significant increasing trend in the mean temperature is observed with a rate of +0.01 °C/year, while an insignificant decreasing trend in rainfall is observed with a rate of −0.33 mm/year. During monsoon season, the trend in mean temperature mimics the trend in annual mean temperature, while a significant decreasing trend in monsoonal rainfall is observed with a rate of −3.52 mm/year. In the month of November, significant increasing trends in mean temperature and rainfall are noted. The rate of change in mean temperature and rainfall is +0.02 °C/year and + 0.01 mm/year, respectively (Table 10.9). Besides physical and anthropogenic factors, lack of management and good governance are also responsible for a decrease in the areal extent of open water bodies of the middle Ghaghara basin (e.g. Kumar et al. 2013a, b).

10.5 Conclusions

Physical (sedimentation, precipitation and temperature) and anthropogenic factors (land use change) are responsible for the decrease in the areal extent of open water bodies of the middle Ghaghara River basin. If the same rate of depletion in open water bodies continues, the region will face severe ground as well as surface water crisis. The encroachment of built-up areas and cropland on the desiccated water bodies should be restricted by law enforcement to minimise the damage, caused by backwaters from these water bodies during the monsoon period, and also to maintain the hydrological processes, operating therein. The data on open water bodies discussed in this study are of immense value to the planners and environmentalists, working on the restoration of wetlands in the middle Ghaghara River basin.

Conflict of Interest The authors declare no conflict of interest.

References

Allan JD, Palmer M, Poff NL (2005) Climate change and freshwater ecosystems. In: Lovejoy TE, Hannah L (eds) Climate change and biodiversity. Yale University Press, New Haven, pp 274–290

Ashizawa D, Cole JJ (1994) Long-term temperature trends of the Hudson River – a study of the historical data. Estuaries 17:166–171

Banarescu PM (1990) Zoogeography of fresh waters. General distribution and dispersal of freshwater animals, vol 2. Aula Verlag, Wiesbaden

Bassi N, Kumar MD, Sharma A, Pardha-Saradhi P (2014) Status of wetlands in India: a review of extent, ecosystem benefits, threats and management strategies. J Hydrol 2:1–19. https://doi.org/10.1016/j.ejrh.2014.07.001

Brinson MM, Malvárez AI (2002) Temperate freshwater wetlands: types, status, and threats. Environ Conserv 29(2):115–133

Burridge CP, Craw D, Waters JM (2006) River capture, range expansion, and cladogenesis: the genetic signature of freshwater variance. Evolution 60(3):1038–1049

Carlton JT (2000) Global change and biological invasions in the oceans. In: Mooney HA, Hobbs RJ (eds) Invasive species in a changing world. Island Press, Covelo, pp 31–53

Deutsch CA, Tewksbury JJ, Huey RB, Sheldon KS, Ghalambor CK, Haak DC, Martin PR (2008) Impacts of climate warming on terrestrial ectotherms across latitude. Proc Natl Acad Sci U S A 105:6668–6672

District Sankhyikiya Patrika (2015), Decadal no. of inhabited villages, population and decadal percent difference according to census in the district 1901–2011 (Table 9). http://updes.up.nic.in/spatrika/engspatrika/select_distt_yr.asp. Accessed 23 Jan 2017

Dudgeon D (1999) Tropical Asian streams: zoobenthos, ecology and conservation. Hong Kong University Press, Hong Kong

Dudgeon D, Arthington AH, Gessner MO, Kawabata Z, Knowler DJ, Lévêque C, Naiman RJ, Prieur-Richard AH, Soto D, Stiassny ML, Sullivan CA (2006) Freshwater biodiversity: importance, threats, status and conservation challenges. Biol Rev 81:163–182

Firth P, Fisher SG (eds) (1992) Global climate change and freshwater ecosystems. Springer, New York

Han L (1997) Spectral reflectance with varying suspended sediment concentrations in clear and algae-laden waters. Photogramm Eng Remote Sens 63(6):701–705

Heino J, Virkkala R, Toivonen H (2009) Climate change and freshwater biodiversity: detected patterns, future trends and adaptations in northern regions. Biol Rev 84:39–54

Hirsch RM, Archfield SA (2015) Flood trends not higher but more often. Nat Clim Chang 5:198–199

Indiawaterportal (2017) Met data. http://www.indiawaterportal.org/met_data/. Accessed 23 Jan 2017

Intergovernmental Panel on Climate Change (IPCC) (2007) Climate change 2007: a synthesis report. Intergovernmental panel on climate change. Valencia, Spain

Jackson RB, Carpenter SR, Dahm CN et al (2001) Water in a changing world. Ecol Appl 11:1027–1045

Jensen JR (2005) Introductory digital image processing: a remote sensing perspective, 3rd edn. Pearson Prentice Hall, Upper Saddle River

Kaufman YJ, Wald AE, Remer LA, Gao B-C, Li R-R, Flynn L (1997) The MODIS 2.1-mm channel-correlation with visible reflectance for use in remote sensing of aerosol. IEEE Trans Geosci Remote Sens 35(5):1286–1298

Kumar R (2016) Flood hazard assessment of 2014 floods in Sonawari sub-district of Bandipore district (Jammu and Kashmir): an application of geoinformatics. Remote Sens Appl Soc Environ 4:188–203

Kumar R, Acharya P (2016) Flood hazard and risk assessment of 2014 floods in Kashmir Valley: a space-based multisensor approach. Nat Hazards 84(1):437–464

Kumar MD, Panda R, Niranjan V, Bassi N (2013a) Technology choices and institutions for improving economic and livelihood benefits from multiple uses tanks in western Orissa. In: Kumar MD, Sivamohan MVK, Bassi N (eds) Water management, food security and sustainable agriculture in developing economies. Routledge, Oxford

Kumar R, Kamal V, Singh RK (2013b) Geomorphic effects of 2011 floods on channel belt parameters of Rapti River: a remote sensing and GIS approach, Corona. J Sci Technol 2(2):4–12

Kumar R, Kumar S, Pandey PC (2016) Delineation and zonation of flood prone area using geo-hydrological parameters: a case study of lower Ghaghara River valley. In: Srivastava PK, Pandey PC, Kumar P, Raghubanshi AS, Han D (eds) Geospatial technology for water resource applications. CRC Press/Taylor & Francis Group, Boca Raton, pp 78–100

Magnuson JJ, Robertson DM, Benson BJ, Wynne RH, Livingstone DM, Arai T, Assel RA, Barry RG, Card V, Kuusisto E, Granin NG, Prowse TD, Stewart KM, Vuglinski VS (2000) Historical trends in lake and river ice cover in the northern hemisphere. Science 289:1743–1746

Mitsch W, Gosselink JG (2000) The value of wetlands: importance of scale and landscape. Ecol Econ 35:25–33

Nakano S, Kitano F, Maekawa K (1996) Potential fragmentation and loss of thermal habitats for charrs in the Japanese archipelago due to global warming. Freshw Biol 36:711–722

NRSC (2006) Land Use/Land Cover database on 1:50,000 scale. Natural Resources Census Project, LUCMD, LRUMG, RS & GIS AA, National Remote Sensing Centre, ISRO, Hyderabad
NRSC (2014) Land Use/Land Cover database on 1:50,000 scale. Natural Resources Census Project, LUCMD, LRUMG, RSAA, National Remote Sensing Centre, ISRO, Hyderabad
Olden JD, Kennard MJ, Leprieur F, Tedesco PA, Winemiller KO, García-Berthou E (2010) Conservation biogeography of freshwater fishes: recent progress and future challenges. Divers Distrib 16(3):496–513
Pandit AK, Qadri SS (1990) Floods threatening Kashmir valley. J Environ Manag 31:299–311
Parmesan C (2006) Ecological and evolutionary responses to recent climate change. Annu Rev Ecol Evol Syst 37:637–669
Pekel J-F, Cottam A, Gorelick N, Belward AS (2016) High-resolution mapping of global surface water and its long-term changes. Nature 540:418–436. https://doi.org/10.1038/nature20584
Poff NL, Olden JD, Strayer DL (2009) Climate change and freshwater extinction risk. In: Hannah L (ed) Extinction risk from climate change. Island Press, Washington DC
Rahel FJ, Olden JD (2008) Assessing the effects of climate change on aquatic invasive species. Conserv Biol 22(3):521–533
Rana NK, Kumar R, Kumar D (2009) Nature of channel shift of a foothill-fed river in the alluvial setting: a case study of river Rapti India. Indian J Geomor 13(1&2):83–98
Rathore L S, Attri S D, Jaswal A K (2013) State level climate change trends in India. Meteorological Monograph No. ESSO/IMD/EMRC/02/2013. Ministry of earth Sciences, Earth System Science organization, Meteorological Department, Govt. of India
Sen PK (1968) Estimates of the regression coefficient based on Kendall's tau. J Am Stat Assoc 63:1379–1389
Singh IB (1996) Geological evolution of ganga plain: an overview. J Palaeontol Soc India 41:99–137
Sinha CP (2011) Climate change and its impacts on the wetlands of North Bihar, India. Lakes Reserv Res Manag 16:109–111
Smith VH (2003) Eutrophication of freshwater and coastal marine ecosystems: a global problem. Environ Sci Pollut Res 10:126–139
Smol JP (2008) Pollution of lakes and rivers: a paleoecological perspective, 2nd edn. Blackwell, Malden
Strayer DL (2010) Alien species in fresh waters: ecological effects, interactions with other stressors, and prospects for the future. Freshw Biol 55(s1):152–174
Strayer DL, Dudgeon D (2010) Freshwater biodiversity conservation: recent progress and future challenges. J N Am Benthol Soc 29(1):344–358
Tiner RW (2014) Wetlands: An Overview. In: Tiner RW, Lang MW, Klemas VV (eds) Remote sensing of wetlands: applications and advances. CRC Press Taylor & Francis Group, Boca Raton, pp 3–18
U.S. EPA (Environmental Protection Agency) (2008) Effects of climate change on aquatic invasive species and implications for management and research. EPA, Office of Research and Development, Washington, DC
USGS. 2015. Landsat 8 (L8) data users handbook (version 1.0). http://www.greenpolicy360.net/mw/images/Landsat8DataUsersHandbook.pdf. Accessed 20 Dec 2015
USGSEE (2016) EarthExplorer web portal. https://earthexplorer.usgs.gov/. Accessed 23 Jan 2017
Varghese S, Narayanan SP, Prashob Raj VM, Hari Prasad V, Narendra Prasad S (2008) Analyses of wetland habitat changes and its impacts on avifauna in selected districts of the indo-Gangetic plains of Uttar Pradesh, India, between 1972 and 2004. In: Sengupta M, Dalwani R (eds) Proceedings of Taal 2007: the 12th world lake conference, pp 2045–2055
Verma M, (2001) Economic Valuation of Bhoj Wetlands for Sustainable Use. [EERC Working Paper Series: WB-9]. Indian Institute of Forest Management, Bhopal
Vörösmarty C, Sahagian D (2000) Anthropogenic disturbance of the terrestrial water cycle. Bioscience 50(9):753–765

Whittaker RJ, M B A'j, Jepson P Ladle RJ, Watson JEM, Willis KJ (2005) Conservation biogeography: assessment and prospect. Divers Distrib 11(1):3–23
Woodward G (2009) Biodiversity, ecosystem functioning and food webs in fresh waters: assembling the jigsaw puzzle. Freshw Biol 54(10):2171–2187
Woodward G, Perkins DM, Brown LE (2010) Climate change and freshwater ecosystems: impacts across multiple levels of organization. Philos Trans R Soc, B 365(1549):2093–2106
Xu H (2006) Modification of normalised difference water index (NDWI) to enhance open water features in remotely sensed imagery. Int J Remote Sens 27:3025–3033. https://doi.org/10.1080/01431160600589179
Yousefi S, Moradi HR, Keesstra S, Pourghasemi HR, Navratil O, Hooke J (2017) Effects of urbanization on river morphology of the Talar River, Mazandaran Province, Iran. Geocarto Int:1–7. https://doi.org/10.1080/10106049.2017.1386722

Chapter 11
Climate Change, Water and Wastewater Treatment: Interrelationship and Consequences

Suruchi Singh and Supriya Tiwari

Abstract Climate change has become an inevitable phenomenon in the recent years and is expected to gain momentum with the coming time. One of the important processes of the society which is affected by present-day climate change scenario is the wastewater management. Wastewater includes municipal sewage effluents, industrial effluents and urban and agricultural runoff, which without appropriate management could be harmful to human health. Changing climate impacts wastewater systems through changes in temperature, precipitation patterns, sea level rise and storm-related changes. It is estimated that climate change has a dual effect on water resources and wastewater treatment plants. Climate change events like extreme weather range will lead to more untreated wastewater. As such, the necessity for wastewater management is continuously increasing. However, emission of certain greenhouse gases like CO_2, CH_4 and N_2O during the process of wastewater treatment further contributes to the already aggravated problem of climate change. As such, there exists a nonending scuffle between the phenomenon of climate change and wastewater management, each intensifying the other problem.

Keywords Climate change · Greenhouse gases · Weather · Water · Wastewater treatment plants

11.1 Introduction

The world population is projected to increase by more than one billion within the next 15 years (UN 2015). With the exponential population growth in different regions of the world, urban areas are also expanding. Urbanization will continue both in the developed and developing nations, and by 2050, 86 and 64% of total population of developed and developing nations, respectively, will be urban population. Overall, the world population is expected to be 67% urban in 2050. This

S. Singh · S. Tiwari (✉)
Department of Botany, Institute of Science, Banaras Hindu University, Varanasi, India

R. P. Singh et al. (eds.), *Water Conservation, Recycling and Reuse: Issues and Challenges*, https://doi.org/10.1007/978-981-13-3179-4_11

urbanization accompanied with industrialization comes with unavoidable consequences like increased industrial activity and release of greenhouse gases (GHGs) (Dedenko et al. 2017). There is general agreement that human industrial activity has released vast quantities of GHGs, about 900 billion tonnes of CO_2, of which 450 have stayed in the atmosphere. About 80% of CO_2 emission is caused by industrialization and the remaining by land use changes (Stephenson et al. 2010). Population, which is the most neglected dimension of climate change, is the main contributing factor. Increased human activities are not only contributing in GHGs emissions but also generating wastes which need to be treated before being discharged into the environment for reuse (Scovronick et al. 2017). Poor waste management causes air, water and soil contamination in return. Open and unsanitary landfills contribute in contamination of drinking water and in spread of infection and transmit diseases.

Many studies have been conducted on climate change and water resources, but there is dearth of information available on potential effect of climate change on water and wastewater treatment. So, this chapter focuses on highlighting the influences of climate change variables on water resources and wastewater treatment.

11.2 Climate Change and Water Resources

Anthropogenically released GHGs are the most significant forces that drive climate change (IPCC 2013). The greenhouse gases trap energy which then warms the planet. The warming potential of these gases depends on their ability to change the energy balance of the planet. The main greenhouse gases are carbon dioxide (CO_2), methane (CH_4) and nitrous oxide (N_2O). The sources of different greenhouse gases are given in Table 11.1 and the anthropogenic contribution in release of different gases is given in Fig. 11.1.

Owing to global warming, the chances of extreme precipitation events have increased across the world (Donat et al. 2016). Alterations in spatial precipitation characteristics are among the most obvious aspects of climate change. Scientists are of the notion that dry region would be more drier and wet would be more wetter (Donat et al. 2016), taking into consideration the difference between precipitation and evaporation, as atmosphere moisture convergence and divergence are expected to increase in magnitude with increasing atmosphere moisture content in warmer regions (Held and Soden 2006). Warmer temperature increases the rate of evaporation of water into the atmosphere, which in turn increases the atmosphere's capacity to hold water (Georgakakos et al. 2014). Increased evaporation may dry out some areas and fall as excess precipitation on other areas. As temperatures rise, people and animal need more water to maintain their health and thrive. Many important economic activities, like producing energy to power plants, raising livestock and growing food crops, also require water. The amount of water available for these activities may be reduced as Earth warms and if competition for water resources increases.

Table 11.1 Different greenhouse gases, their sources, potency with respect to CO_2 and residence time

Greenhouse gases	Sources	Potency/residence time
CO_2	Burning of fossil fuels	–
	Solid waste	
	Trees and wood products	
CH_4	Production and transport of coal, natural gas and oil	21 times more potent than CO_2
	Livestock and other agriculture practices	12 years atmospheric life span
	Decay of organic waste in municipal solid waste landfills	
N_2O	Agriculture and industrial waste	310 times more potent than CO_2
	Combination of fossil fuels and solid waste	210 years life span
	Wastewater treatment plants during the nitrification and denitrification processes	
Fluorinated gases	Industrial process	HFCs: 140–11,700 times more potent than CO_2; 1 to 260 years life span
HFCs		PFCs: 6500–9200 times more potent than CO_2; several thousand years life span
PFCs		SF_6: 23,900 times more potent; extremely long lived with few sinks
SF_6		

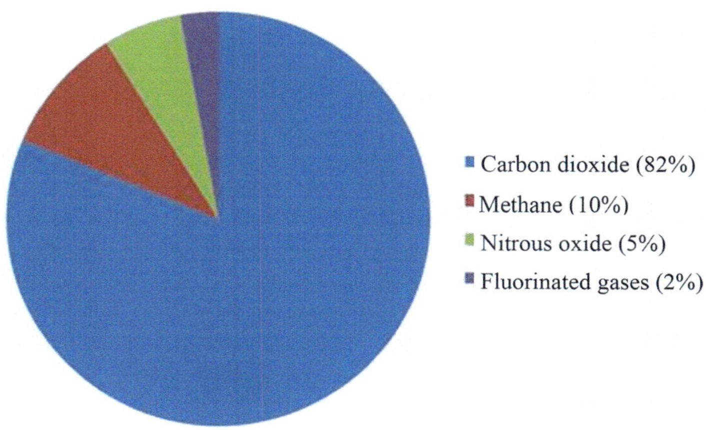

Fig. 11.1 Anthropogenic contribution (%) in the release of different greenhouse gases (GHGs)

Water quality could suffer in areas experiencing increases in rainfall. Heavy precipitation events could cause problems for the water infrastructure, as sewer systems and water treatment plants are overwhelmed by the increased volume of water. Higher temperature also facilitates biodegradation of chemicals in aqueous state, and thus many reduce the toxic levels, but other way round, it may increase the toxicity by other means. Precipitation may affect the movement and distribution of chemicals, e.g. increased leaching of applied pesticides. Through surface erosion, the water flow inputs untreated sewage effluent and diffuse pollutants in water bodies. Flooding due to climate change has implications for the inundation of land that is contaminated. Thus, there is a higher risk on remobilization of chemicals through floodwater to river or marine ecosystem. Climate change also led to changes in farming practices which required more input of specific chemicals. Changing climate affects land use and may also influence mobilization and fate of chemicals applied to land, thus increasing discharge to surface and groundwater. Higher temperature will stimulate the volatilization and degradation of pesticides in soil and surface waters.

Wetter soils (as a result of heavy precipitation) are known to have high hydraulic conductivities, and so, pesticide-rich water may move rapidly through soils, but they will also show enhanced degradation rates. But heavy precipitation may also increase the bypass flow, thus increasing the chances of pesticide movement to drains, surface waters and soil layers. Changes in the land cover and land management practices have been regarded as the key influencing factors behind the alteration of the hydrological system, which led to change in the runoff as well as the water quality (Tong and Chen 2002). Groundwater contaminants mainly include arsenic, fluoride, nitrate, iron, phosphates, bacteria and heavy metals.

11.3 Climate Change and Water Cycle

Increasing temperature will intensify Earth's water cycle by intensifying evaporation. Increased evaporation will lead to more storms but will also contribute in drying over some land mass. Consequently, storm-stricken regions are expected to receive more rainfall and, thus, will face flooding, while regions not aligning along the storm tracks are expected to face drought-like situation. During the previous decade, temperature was observed to have increased by about 0.75 °C. Temperature increase is leading to a rise in sea levels, glacier melt and alterations in precipitation patterns (WHO 2008). Also, the climate change and the hydrological cycle are solar radiation driven. Changes in greenhouse gases (GHGs), both natural and anthropogenic, affect the absorption and reflection of solar radiation in the atmosphere, causing changes in the Earth's temperature, thus impacting climate (Fig. 11.2).

In natural hydrological cycle, solar energy enters the Earth's system leading to the process of evaporation, formation of clouds and then precipitation in the form of rain, hail or snow (Bates et al. 2008). Once the precipitation has initiated, water

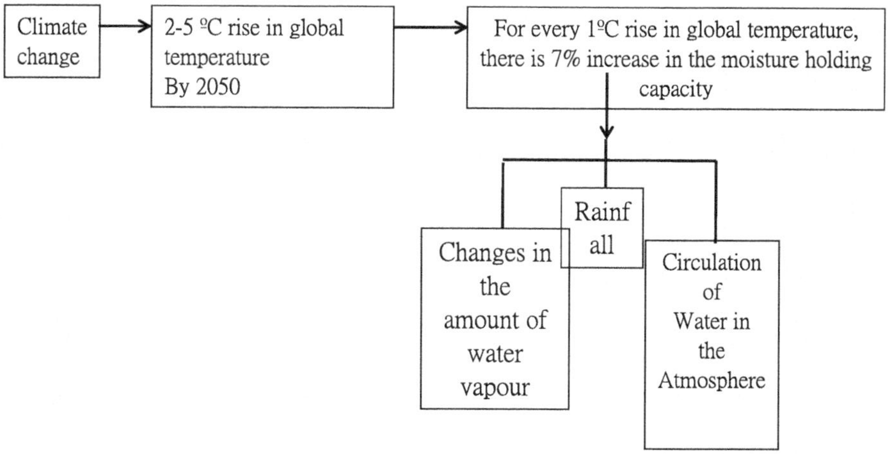

Fig. 11.2 Flowchart depicting how changes in greenhouse gases (GHGs) affects the absorption and reflection of solar radiation in the atmosphere which causes changes in the Earth's temperature thus impacting climate

evaporates or transpires back into the atmosphere, or it can be a runoff. It can also percolate down the surface to become a part of groundwater. This means water resides within the hydrological cycle and is thus considered as a renewable resource. There are many other pressures that exert on the water cycle, firstly, because of over-exploitation of water resources for drinking, agriculture, industries and other purposes, thus restricting the chances for replenishing. Secondly, building of structures like dam on water bodies or diverting rivers may unevenly distribute ample water at places where dam is built while lesser water where the water is diverted, thus altering the local water cycle and the microclimate of the region. Thirdly, climate can also affect hydrological cycle (World Bank 2009).

Water is a common component of all climate system, i.e. atmosphere, hydrosphere, cryosphere and biosphere. This obviously means that any change in the climate would have impacts on water through different means. As the temperature increases, saturation of vapour pressure in air increases, and it is expected that with a warming climate (increasing temperature), the amount of water vapour suspended in the air will increase (IPCC 2013). Due to GHGs, the temperature of lower atmospheric layer will rise, then the evaporation of water will increase, hence more water vapour will get into the atmosphere, and this additional water vapour will also absorb more heat. Most of the solar energy received by the Earth is used by the hydrological cycle; thus more and more of solar energy will be trapped which will lead to intensification of hydrological cycle, resulting in changes of precipitation pattern. Such changes will bring about more floods and droughts and thus will have drastic impact on the availability of freshwater. These impacts on freshwater will be compounded by rising sea levels and melting of glaciers (Mimura 2013).

Speculations for the changes in hydrological cycle across the globe are:

- Water vapour and precipitation will increase but not in uniform manner (Ye et al. 2014).
- With temperature, average atmospheric water vapour tends to increase (Ye et al. 2014).
- Heavy precipitation for mid-latitudes is expected (Fischer and Knutti 2016).
- Especially in tropics and subtropics, soil moisture will decrease, and droughts will be more intense and for longer duration (Fischer and Knutti 2016).
- Occurrence of cyclone will be more common in tropics (Pfahl and Wernli 2012).

Frequency as well as amount of average precipitation in a much warmer world will not be uniform, with some parts experiencing increases and others decreases or no change. Owing to higher carrying capacity of warmer troposphere, the higher latitudes could experience more precipitation, while many mid-latitude, arid and semiarid regions are likely to experience less precipitation.

There are many factors that can deviate the hydrological cycle from regular course, and also, this change can harm human health in various ways:

- Increase in temperature of water bodies will alter many biochemical properties like oxygen solubility.
- Assimilative property of river to break organic waste will be disturbed.
- On the one hand, increased precipitation will increase dilution, while on the other hand, erosion will be enhanced and thus will increase the sediments.
- Decreased precipitation may decrease the dissolved oxygen and increase water bloom and thus would make the water unfit for human consumption.
- Rise in sea level can increase salt water intrusion in estuaries and coastal aquifers.

11.4 Wastewater Treatment and Climate Change

Effects of climate change are observed on municipalities of both urban and rural nature. The degree of damage on wastewater treatment plant depends on geography, economics, administrative capacity, demography and climate change (Major et al. 2011). The immediate problem due to climate change induced malfunctioning of treatment plant because of heavy precipitation which is combined with sewage overflows leading to environmental pollution, contamination of drinking water, percolation of sewer water into groundwater, intrusion of pollutants in wastewater treatment plants, flooding in pumping station, surcharging of manholes causing surface flooding, increased sediments loading in water collection systems and decreased infiltration capacity in the infiltration basin (Berggren et al. 2007).

Urban water supplies majorly come from surface water and groundwater. The associated structures and facilities are vulnerable to adverse effects of climate change (Case 2008). Numerous urban water systems that are already under pressure may face additional challenges by growing demands of population that come by population growth and climate change. Climate change imposes problems on various aspects of wastewater treatment plants (WWTPs) like treatment, distribution, disposal as well as reuse by incurring high energy cost and by increasing the volume of wastewater by adding additional storm/floodwater and also through increasing the needs for reuse where droughts become more prevalent (Major et al. 2011). The rise in temperature may also impose direct effects on WWTPs by raising the likelihood of sewer corrosion and odour problems. Combined sewer overflows need to be checked for protecting the surface water quality (PREPARED 2010). Temperature otherwise also plays a crucial role as it regulates treatment processes especially natural and/or non-mechanized one. Warm temperature has manyfold advantages like decrease in land requirements, enhanced conversion processes and increased removal efficiencies. In general, outcomes of climate change can affect wastewater treatment in the following ways:

1. Rainfall: Wastewater infrastructure is affected by increased rainfall intensity aggravating flooding, and combined sewer overflows affected WWTP efficiency.
2. Temperature: Temperature increase will rise the likelihood of sewer corrosion and odour problems.
3. Sea level rise can decrease the hydraulic capacity of downstream sewer and increase salt water intrusion. Rising downstream water levels may make pumping effluent requirements increasing energy need.
4. Storm increase creates flooding, which can be harmful to infrastructure when WWTPs are built in coastal areas and can cause pollutants to directly enter the waterways and contaminate water supplies (EPA 2012).

Anaerobic reactors utilize diluted wastewater in warm temperature, although, in low-temperature regions, stabilization ponds are utilized, which occupy larger areas. Other processes like activated sludge and aerobic biofilm reactors depend relatively less on temperature (Von Sperling and de Lemos Chemicharo 2005). More exacerbated weather owing to extreme warmer temperature includes urban heat islands which can in addition cause convective thunderstorms, hail, storms, cyclonic events and higher winds that may exceed the design capacity (Major et al. 2011). Flooded WWTPs can release the untreated waste into the ecosystem, thus can incur financial loss and is a threat to public health (Langeveld et al. 2013). Thus, wastewater and sewage network need to accommodate more intense precipitation (Kleidorfer et al. 2009).

11.5 Processes Affected Under Climate Change in a WWTP and Solutions Utilized

Wastewater treatment gets affected by more water, less water and poor water quality. These are the four basic steps utilized for treating wastewater:

- Preliminary treatment: Involves removal of floating material and settled inorganic solids.

- Primary treatment: Involves removal of fine suspended organic solids. It utilizes processes of sedimentation or settling.
- Secondary or biological treatment: Involves removal of dissolved or fine colloidal organic matter. This process involves the use of microorganisms that decompose the unstable organic matter to stable inorganic forms. The biological treatment processes of sewage are broadly classified as aerobic, anaerobic and pond processes.
- Tertiary treatment: Needed for the removal of suspended or dissolved substances, after the conventional primary and secondary treatments.

11.6 Various Processes of WWTP Affected by Climate Change (Zouboulis and Tolkore 2015)

- *Sedimentation*
  - Warm wastewater increases the bacterial reaction and thus would decrease the density of settled sludge.
  - Rate of biological reactions is temperature dependent according to $k = k_{20}\Theta^{T\text{-}20}$ where k = reaction rate constant at temperature, T; k_{20} = reaction rate constant at 20 °C; Θ = temperature coefficient; and T = temperature of biological reaction.
- *Biological aeration of warm wastewater*
  - BOD increased
  - Activated sludge aeration system working at high temperature to support nitrification
- *Stabilization ponds*
  - Land demand is high.
  - High emission of foul odours.
- *Chlorination*
  - High chlorine will be utilized to disinfect highly infected sewage which is lethal to fish and invertebrate species.

There are propositions from urban planners and water managers to government for including the adaptive strategies with respect to infrastructure design, operation and maintenance for dealing with climate change. Historic design-based WWTPs will become obsolete; rather reconstruction and rehabilitation will become a necessity. In many parts of the world, WWTPs are made more effective even under the consequences of climate change by following these directives:

- Adaptive capacity: The degree to check a municipality is able to deal with the impacts of climate change is often referred to as adaptive capacity.
- Assessment of tool: An impact assessment tool was created for assessing the effects of a facility flooding into the surrounding community. Ratio between the average flow rate and design flow rate of the plant is the crucial factor for wastewater facilities. This rate measures how close to maximum capacity a wastewater treatment facility operates. A facility that operates close to maximum capacity will be less able to handle an increase in flow. Facilities that have an average flow rate up to 50% of their design capacities were rated as low impact, between 50 and 70% as medium impact and above 70% as high impact (Blumenau et al. 2011).
- Monitoring of wastewater treatment plants: Continuous and long-term monitoring of urban wastewater infrastructure is increasingly applied on wastewater treatment plants. This monitoring is done with the help of sensors, data communication and data handling capacity which is useful in depicting the performance of wastewater infrastructure over a long period of time. Combining this data along with the meteorological data, it is possible to anticipate the impact of climate changes on treatment plants. The development of innovative monitoring of combined sewer overflows and an early warning system for facial contamination in recreational waters will allow wastewater utilities to be better prepared and to respond faster to any contamination due to combined sewer overflows and uncontrolled runoff caused by more frequent and heavier rainfall (PREPARED 2010).
- Vulnerability analysis: Helps in identifying the prior vulnerable issues related to climate change and lays foundation for follow-up actions.
- Membrane treatment processes: Many water suppliers in over-constrained settlings have also turned to energy-intensive membrane treatment processes to enable desalination of water sources and reuse of highly treated wastewater effluent (Cromwell et al. 2007).

11.7 Wastewater Treatment Contribution to Climate Change

Greenhouse gases like CO_2, CH_4 and N_2O are emitted by WWTPs (Law et al. 2012; Kampschreur et al. 2008), through three main mechanisms, i.e. direct, indirect internal and external (Campos et al. 2016), and these sources are also referred to as scope I, II and III emissions, respectively (LGOP 2008). CO_2 production is attributed to two main factors, treatment process and electricity consumption. During anaerobic reactions, the BOD of wastewater meets two fates, (1) incorporate into biomass and (2) convert into CO_2 and CH_4. Via endogenous respiration, fraction of biomass is further converted into CO_2 and CH_4. In aerobic process, CO_2 is produced

Table 11.2 Ways to check the greenhouse gases emissions from wastewater treatment plants

Greenhouse gases	Procedure
N_2O	Biological wastewater treatment plants should be operated at high solid retention times to maintain low NH_3 and nitrite concentration
	Large bioreactor volume is recommended to buffer loadings and reduce the transient O_2 depletion
CH_4	Collection of CH_4 and its use as a biogas
CO_2	Decrease of sludge production
	Applying shortest solid retention time

through the breakdown of organic matter in the activated sludge process and some through the primary clarifiers.

The magnitude of CH_4 production depends on the amount of degradable organic matter, temperature and the type of treatment system. With the rise in temperature, the rate of CH_4 production increases. This reaction is especially important in warm climates. As per Daelman et al. (2012), about 1% of the incoming chemical oxygen demand to the WWTPs was emitted as CH_4. Release of N_2O is associated with the degradation of nitrogen components, e.g. urea, nitrate and protein, in the wastewater. Nitrification is an aerobic process converting ammonia and other nitrogen compounds into nitrate, while denitrification occurs under anoxic condition and involves the biological conversion of nitrate into nitrogen gas (N_2). N_2O though is an intermediate product of both the reactions but it is more associated with denitrification. In anoxic conditions, both ammonia- and nitrite-oxidizing bacteria are able to produce N_2O, while in aerobic condition only ammonia-oxidizing bacteria can produce it. However, according to some, N_2O production occurs predominantly in aerobic tank (Ahn et al. 2010). Ammonia-oxidizing bacteria have been identified as the main N_2O producers, while heterotrophic denitrifying bacteria contribution is only relevant when nitrite and/or oxygen are present in the anoxic stage (Wunderlin et al. 2012). Hydroxylamine oxidation pathway can be the main process responsible for the emissions of N_2O at high ammonia and low nitrite concentration, when a high metabolic activity of ammonia-oxidizing bacteria occurs.

GHGs emission can also be checked by modifying operational conditions of WWTPs which is not very feasible way owing to its own operational limitations. The common procedures for checking the emission of these three gases are given in Table 11.2.

11.8 Conclusions

Human activities induced emissions of GHGs which through climate forcing alter the energy budget of the globe, thus affecting the weather patterns. Extreme weather events like temperature rise and heavy precipitation are not only altering hydrological cycle but also deteriorating the water quality. Consequently, wastewater treatment plants (WWTPs) are facing many direct and indirect challenges. GHGs are not only affecting proper functioning of WWTPs but also making it a contributor of GHGs.

References

Ahn JH, Kim S, Park H, Rahm B, Pagilla K, Chandran K (2010) N_2O emissions from activated sludge processes, 2008–2009: results of a national monitoring survey in the United States. Environ Sci Technol 44(12):4505–4511

Bates BC, Kundzewicz ZW, Wu S, Palutik JP (2008) Climate change and water, intergovernmental panel on climate change. Technical Paper VI, Geneva, pp 33–48

Berggren, K., Olofsson, M., Viklander, M., Svensson, G. (2007) Tools for measuring climate change impacts on urban drainage systems. In: Proceedings of the 6th NOVATECH international conference: sustainable techniques and strategies in urban waste water management. Lyon, pp 239–246

Blumenau M, Barnoush A, Thomas I, Hofmann H, Vehoff H (2011) Impact of selective oxidation during inline annealing prior to hot-dip galvanizing on Zn wetting and hydrogen-induced delayed cracking of austenitic FeMnC steel. Surf Coat Technol 206(2–3):542–552

Campos JL, Valenzuela-Heredia D, Pedrouso A, Val del Rio A, Belmonte M, Mosquera-Corral A (2016) Greenhouse gases emissions from wastewater treatment plants: minimization, treatment and prevention. J Chem 2016:1–12

Case T (2008) Climate change and infrastructure issues. AWWA Research Foundation, drinking water research. Clim Change Spec Issue 18:15–17

Cromwell JE, Smith JB, Raucher RS (2007) No doubt about climate change and its implications for water suppliers. J Am Water Works Assoc 99(9):112–117

Daelman MRJ, van Voorthuizen EM, van Dongen UGJM, Volcke EIP, van Loosdrecht MCM (2012) Methane emission during municipal wastewater treatment. Water Res 46(11):3657–3670

Dedenko NI, Skripnuk DF, Mirolyubova OV (2017) Urbanization and greenhouse gas emissions from industry. IOP Conf Ser Earth Environ Sci 72:1–8

Donat MG, Lowry AL, Alexander LV, Gorman PAO, Maher N (2016) More extreme precipitation in the worlds dry and wet regions. Nature Clim Change 6:508–513

EPA (2012) Watershed academy web. The effect of climate change on water resources and programs. http://www.epa.gov/watertrain

Fischer EM, Knutti R (2016) Observed heavy precipitation increase confirms theory and early models. Nat Clim Change 6:986–991

Georgakakos KP, Graham NE, Modrick TM, Murphy MJ, Shamir E, Spencer CR, Sperfslage JA (2014) Evaluation of real-time hydrometeorological ensemble prediction on hydrologic scales in Northern California. J Hydrol 519:2978–3000

Held IM, Soden BJ (2006) Robust responses of the hydrological cycle to global warming. J Clim 19 (21):5686–5699

International Panel for Climate Change (2013) Fifth assessment report for climate change

Kampschreur MJ, van der Star WRL, Wielders HA, Mulder JW, Jetten MSM, van Loosdrecht MCM (2008) Dynamics of nitric oxide and nitrous oxide emission during full-scale reject water treatment. Water Res 42(3):812–826

Kleidorfer M, Mőderl M, Sitzenfrei R, Urich C, Rauch W (2009) A case independent approach on the impact of climate change effects on combined sewer system performance. Water Sci Tech 60 (6):1555–1564

Langevelt JG, Schilperoort RPS, Weijers SR (2013) Climate change and urban wastewater infrastructure: there is more to explore. J Hydrol 476:112–119

Law LY, Pan Ye Y, Yuan Z (2012) Nitrous oxide emissions from wastewater treatment processes. Philos Trans R Soc B Biol Sci 367(1593):1265–1277

LGOP (2008) Local Government Operations Protocol. For the quantification and reporting of greenhouse gas emissions inventories (Version 1.0). Available at: www.theclimateregistry.org/resources/protocols/local-givernment-operations-protocol

Major DC, Omojola A, Dettinger M, Hanson RT, Sanchez-Rodriguez R (2011) Climate change, water, and wastewater in cities. In: Rosenzweig C, Solecki WD, Hammer SA, Mehrotra S (eds) Climate change and cities: first assessment report of the urban climate change research network. Cambridge University Press, Cambridge, pp 113–143

Mimura N (2013) Sea-level rise caused by climate change and its implications for society. Proc Japan Acad Ser B Phys Biol Sci 89(7):281–301

Pfahl S, Wernli H (2012) Quantifying the relevance of cyclones for precipitation extremes. J Clim 25:6770–6781

PREPARED Enabling Change (2010) www.prepared-fp7.eu. Accessed 4 Mar 2014

Scovronick N, Budolfson MB, Dennig F, Fleurbaey M, Siebert A, Socolow RH, Spears D, Wagner F (2017) Impact of population growth and population ethics on climate change mitigation policy. PNAS 114(46):12338–12343

Stephenson J, Newman K, Mayhew S (2010) Population dynamics and climate change: what are the links? J Public Health 32(2):150–156

Tong Susanna T, Chen W (2002) Modelling the relationship between land use and surface water quality. J Environ Manag 66:377–393

UN (2015) World urbanization prospects: the 2015 revision. United Nations, New York

Von Sperling M, de Lemos Chernicharo CA (2005) Biological wastewater treatment in warm climate regions. IWA Publishing, Padstow

WHO (2008) Safer water, better health: costs, benefits and sustainability of interventions to protect and promote health. World Health Organization, Geneva

World Bank (2009) Water and climate change: impacts on groundwater resources and adaptation options. Water Unit Energy, Transport, and Water Department, Washington, DC, p 98

Wunderlin P, Mohn J, Joss A, Emmenegger L, Siegrist H (2012) Mechanisms of N_2O production in biological wastewater treatment under nitrifying and denitrifying conditions. Water Res 46 (4):1027–1037

Ye H, Fetzer EJ, Behrangi A, Olsen ET, Cohen J, Lambrigtsen BH, Chen L (2014) Impact of increased water vapour on precipitation efficiency over northern Eurasia. Geophys Res Lett 41 (8):2941–2947

Zouboulis A, Tolkore A (2015) Effect of climate change in wastewater treatment plants: reviewing the problems and solutions. In: Shrestha S, Anal AK, Salam PA, Valk MV (eds) Managing water resources under climate uncertainty. Springer, Cham, pp 197–220

Chapter 12
Treatment of Wastewater Using Vermifiltration Technology

M. M. Manyuchi, N. Mupoperi, C. Mbohwa, and E. Muzenda

Abstract This chapter focuses on an alternative biological method for wastewater treatment using the vermifiltration system. The vermifiltration system must meet the desired process and design parameters for optimum wastewater treatment. A case study for swine wastewater treatment using a 3-stage vermifiltration process was used. Three vermifilters with media which is comprised of *Eisenia fetida* earthworms, garden soil, sand and quartz stones were used as the filtration media. The swine wastewater chemical oxygen demand (COD), biological oxygen demand (BOD_5), total suspended solids (TSS), total dissolved solids (TDS), electrical conductivity (EC) and dissolved oxygen (DO) values were measured before and after treatment with the vermifiltration at each stage. The parameters were measured using standard methods. Treatment using a 3-stage vermifilter connected in series resulted in 99.2% reduction in COD, 99.4% in BOD_5, 99.2% in TSS, 80.2% in TDS and 86.9% in EC. The DO concentration increased by >345.5%. Application of the vermifiltration technology in swine wastewater treatment allows for effective

M. M. Manyuchi (✉)
BioEnergy and Environmental Technology Centre, Department of Quality and Operations Management, Faculty of Engineering, University of Johannesburg, Johannesburg, South Africa

Department of Chemical and Processing Engineering, Faculty of Engineering, Manicaland State University of Applied Sciences, Mutare, Zimbabwe

N. Mupoperi
Department of Chemical and Process Systems Engineering, Harare Institute of Technology, Harare, Zimbabwe

C. Mbohwa
BioEnergy and Environmental Technology Centre, Department of Quality and Operations Management, Faculty of Engineering, University of Johannesburg, Johannesburg, South Africa

E. Muzenda
BioEnergy and Environmental Technology Centre, Department of Quality and Operations Management, Faculty of Engineering, University of Johannesburg, Johannesburg, South Africa

Department of Chemical, Materials and Metallurgical Engineering, Faculty of Engineering and Technology, Botswana International University of Science and Technology, Palapye, Botswana

R. P. Singh et al. (eds.), *Water Conservation, Recycling and Reuse: Issues and Challenges*, https://doi.org/10.1007/978-981-13-3179-4_12

biological contaminants, and the technology is easily adoptable in developing countries due to its simplicity and treats water to acceptable standards.

Keywords Earthworms · Vermifiltration · Vermicompost · Wastewater treatment

12.1 Introduction

Wastewater treatment using vermifiltration is increasingly becoming popular. Vermifiltration is the process of using earthworms and aerobic bacteria to treat wastewater (Sinha et al. 2009; Adugna 2016). During this process, wastewater comes in contact with a biofilter which contains the earthworms and bacteria. These earthworms then consume the organic matter and bacteria, leaving behind effluent with lower total suspended solids (TSS), total dissolved solids (TDS), biochemical oxygen demand (BOD_5) and chemical oxygen demand (COD) (Sinha et al. 2009). The vermifiltration system is comprised of earthworms, different sized layers of gravel with soil bedding on top. During vermifiltration, the wastewater is uniformly distributed onto the biofilter surface. This setup is very similar to a conventional trickle filter system with earthworms added on top. The results of the experiment showed that the system with earthworms present reduced BOD_5, COD, TDS and TSS more than the conventional systems. A schematic of the vermifiltration process is shown in Fig. 12.1.

The vermifiltration process is a low-energy, low-cost and efficient way to treat wastewater as a system when it is functioning properly. The wastewater which is produced by these systems can be up to a secondary quality (Suthar 2012). The earthworms' body works as a biofilter, and they have been found to remove BOD_5 by over 90%, COD by 80–90%, TDS by 90–92% and the TSS by 90–95% from wastewater by general (Sinha et al. 2008). Earthworms such as *Perionyx excavatus*, *Eudrilus eugeniae* and *Eisenia fetida* are commonly used in vermifiltration

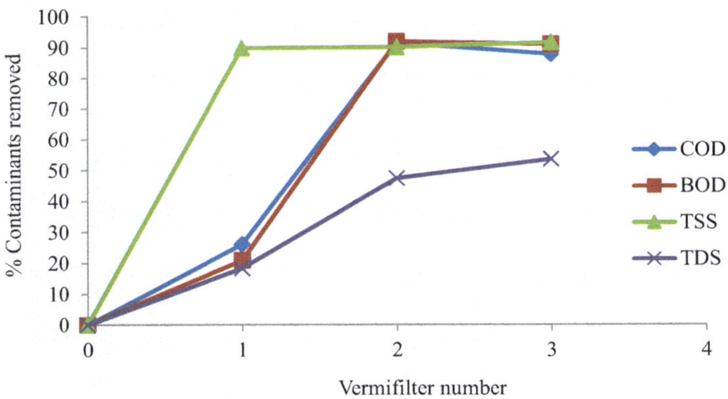

Fig. 12.1 Comparison of single-stage and 3-stage contaminants removal in swine wastewater

(Kharwade and Khedikar 2011). Vermifiltration wastewater treatment systems promise to provide a low-cost and efficient means of treating domestic wastewater, especially the one with high biological contaminants to a prescribed standard, and can easily be adoptable in developing countries.

12.2 Vermifilter Process Design Considerations

The vermifiltration process is a very sensitive process since we deal with living organisms. Key process design for a vermifiltration system includes temperature, moisture content, pH, dissolved oxygen availability, earthworm type and performance, the earthworm loading rate as well as retention time in the vermifiltration bed.

12.2.1 Temperature

The temperature is a significant factor in the worm survival (Neuhauser et al. 1988). Studies have shown that earthworms can survive temperatures ranging from 10 °C to 30 °C but are most active between 15 °C and 25 °C (Sinha et al. 2009). When temperatures are <10 °C, earthworms are significantly inactive but are less likely to die during these times; at the same time earthworms will not live in parts of the filter which are too warm (Baumgartner 2013). This also applies to vermifiltration beds in direct sunlight. The removal of the top layer revealed earthworms were living further down the vermifilter where temperatures were cooler. This layer of the vermifilter where there were no earthworms living effectively decreases the depth of the vermifilter which may lead to decreased wastewater treatment.

12.2.2 Moisture Content

Moisture content is an important factor for earthworm survival. While earthworms have been known to survive in both high and low moisture levels, the moisture level at which earthworms are most active is between 60% and 75% (Baumgartner 2013). High moisture levels can indicate that there are problems within the vermifiltration bed such as blockages or overloading.

12.2.3 pH

The pH within the vermifilter is expected to change with depth during treatment. Earthworms are most sensitive to pH outside of the range of 4.5–9.0 (Baumgartner 2013). Although earthworms are most productive at a neutral pH of 7, the wastewater that the vermifiltration bed will be treating is unlikely to have a neutral pH and hence must be continually monitored. The vermifiltration system's pH can be maintained by not using chemicals which will significantly alter the pH of the wastewater.

12.2.4 Dissolved Oxygen and Aerobic Conditions

Earthworms and the other microorganisms that break down the waste in wastewater are aerobic organisms. Earthworms acquire this oxygen through contact with oxygen-rich water, and oxygen levels need to be sustained to maintain the earthworm and bacteria population. Wastewater has a high biochemical oxygen demand, and it is important that areas of the vermifilter system do not become anaerobic, as this causes offensive odours and earthworm deaths (Tchobanoglous et al. 1981). It has been noted that the dissolved oxygen level (DO) in vermifiltration systems should be greater than 5 mg/L, with a lower boundary of 3 mg/L. It is expected that DO levels will be lower at the bottom of the filter however must still be above 5 mg/L. Earthworms are aerobic organisms which require oxygenated environments to survive and thrive (Sinha et al. 2009). It is important that the vermifilter does not have areas which become anaerobic. Earthworms will die in these areas and aerobic treatment will cease. Aerobic digestion is significantly faster than anaerobic digestion which also creates foul odours. The termination of aerobic digestion will also lead to a backlog of untreated biowaste in the wastewater, which may cause the system to overflow and fail. Once this happens, the vermifiltration system will need to be pumped out and rebuilt internally. The operating conditions of the vermifilter are shown in Table 12.1.

Table 12.1 Operational conditions of the vermifiltration bed

Parameter	Conditions
Temperature	10–25 °C
Moisture	60–75%
pH	4.5–9.0
Dissolved oxygen	>5 mg/L
Odour	No offensive odours must come from the vermifiltration system

Sinha et al. (2008)

12.2.5 Earthworm Growth and Performance

For efficient wastewater treatment, vermifiltration systems should be started with a large population of earthworms: at least 15,000–20,000/m^3 (Sinha et al. 2008). The population of the earthworms should increase after commissioning the vermifiltration system and fluctuate with the loading of the system. A significant decline in earthworm population can indicate that process conditions within the vermifiltration system are not ideal. The population of earthworms should consist of adults, juveniles and cocoons so that the system remains steady.

The estimation of the earthworm population is an important feature in the operation of the vermifiltration bed. As earthworms are the key means of treatment in these vermifiltration beds, it is important that the population remains steady or increases (Meiyan et al. 2010). The earthworm population is an extremely good indicator of how the vermifiltration bed performed during the treatment period. It is expected that if the vermifiltration bed provided a suitable environment, earthworm numbers would dramatically increase from the initial population. There are a number of techniques used in the field to estimate earthworm population: from hand sorting soil to using chemical irritants to lure earthworms to the surface (Chan and Munro 2001). However, with all of these techniques come advantages and disadvantages (Bartlett et al. 2010). For example, hand sorting is very time-consuming, but it is good for estimating population dynamics, and chemical irritants are more time efficient but are not as good for estimating cocoon numbers (Jiménez et al. 2006; Čoja et al. 2008).

12.2.6 Hydraulic Loading Rate

The hydraulic loading rate (HRT) of a vermifiltration system is characterized as the volumetric flow rate of wastewater through the vermifilter media as defined by Eq. 12.1. The HRT is important to maintain correct temperature, aerobic conditions and moisture levels.

$$\text{Hydraulic loading rate}\left(\frac{\text{m}}{\text{h}}\right) = \frac{\text{Volumetric low rate }(\text{m}^3)}{\text{Area }(\text{m}^2) \times \text{Time to flow through profile (h)}} \tag{12.1}$$

12.2.7 Hydraulic Retention Time

Hydraulic retention time (HRT) is characterized as the time which the wastewater is in contact with the filter media and is defined by Eq. 12.2. HRT is significant because it is the amount of time that the vermifilter is treating the wastewater. For BOD_5 loads between 200 mg/L and 400 mg/L, an acceptable HRT for significant reduction is 30–40 min (Sinha et al. 2009). This amount of time will allow earthworms and bacteria to reduce BOD_5, COD and TSS and facilitate biological treatment. However, for domestic wastewater treatment, it has been recommended that the HRT is increased to between 1 h and 2 h. This is due to the need to reduce pathogens, toxic chemicals and heavy metal concentrations found in sewage (Sinha et al. 2009).

$$\text{Hydraulic retention time(h)} = \frac{\text{Porosity} \times \text{soil profile volume} \left(\text{m}^3\right)}{\text{Flowrate through filter} \left({}^{\text{m}^3}/_{\text{h}}\right)} \tag{12.2}$$

12.3 Vermifilter Systems Design Considerations

By taking aspects of biological wastewater treatment system design and vermifiltration into account, a number of factors must be satisfied and become apparent for vermifiltration-based wastewater treatment systems. Many of these are general wastewater treatment factors which have to be tailored, to take into account the presence and continuing survival of earthworms and bacteria. Other systems design for the vermifilter is shown in Table 12.2.

12.3.1 Depth of the System

The depth of the vermifilter refers to the depth in which earthworms are expected to be found. In conventional vermicomposting systems, most species inhabit the top 100–200 mm of the compost material (Edwards and Fletcher 1988). It has also been suggested that when the depth of the vermifilter reaches over 450 mm in depth that compaction and anaerobic conditions are present, this will slow down the performance of the vermifilter. In vermifiltration systems, the biofilter has a substrate material which should give the filter a structure which allows for more voided areas. However, if the loading rate of the vermifiltration system is such that to facilitate a buildup of a layer of vermicompost on the surface, then there may be problems with areas becoming anoxic and ultimately anaerobic. It can then be seen that the waste application method and proper sizing of the vermifiltration system are important to the viability of the system. The overall depth of a vermifiltration system is dependent on the substrate material, aeration rate and internal temperature of the system.

Table 12.2 Vermifiltration systems design parameters consideration

Design parameter	Comment
Flow control	Flow control is required to maintain the duration of time which the wastewater is in contact with the filter, which will allow for sufficient treatment. The minimum hydraulic retention time is 1 h; however any increase in this time will result in enhanced treatment outcomes
Substrate material	The substrate is the material which earthworms live and wastewater trickles through. This material should be relatively inert and not degrade significantly or leach after extensive use. This substrate may need to be replaced periodically and if so should be easily removed
Servicing	Servicing of the system should be undertaken at regular intervals at approximately 6-month interval by appropriate technicians
Integrity of the system	The vermifiltration system has to be constructed to the appropriate standard
Backflow prevention	The vermifiltration system should prevent backflow through the inlet and outlet
External inflow prevention	The vermifiltration system should only receive input through specific inlets. Storm water should not be able to flow into the system as it puts more pressure on the system
Disinfection	The use of disinfection is dependent on the end use of the treated wastewater, and if used it is probably only a viable option for post-treatment
Odour	Vermifiltration systems should not produce an unpleasant odour. A vermifiltration system which is producing an odour is most likely not operating properly and has anaerobic pockets present
Removal of vermicompost	The by-product of worm treatment is vermicompost. In a vermifiltration system, it needs to be clear where and how the castings are going to be removed over a known period
Vermifilter processing capacity	The processing capacity must maintain a sufficient hydraulic retention time

12.3.2 Wastewater Application Technique

The technique in which wastewater is applied to the vermifilter needs to ensure a consistent application over the entire surface. This will reduce flooding in the vermifiltration system, which can also result in anaerobic conditions. These systems may include spraying and dripping.

12.3.3 Prevention of Clogging

Clogging within a wastewater system can occur in a number of areas such as membranes, pipes and filters. The vermifiltration system needs to demonstrate measures that will prevent clogging from occurring in filters and pipes. This can be done in a few ways such as pre-agitating the influent wastewater. It is also necessary to be able to detect where a system is clogged to be able to remediate the problem. A change in pressure drop in the vermifiltration system can indicate clogging.

12.3.4 Aeration

Aeration is a significant process aspect of a vermifiltration system necessary for the continual survival of the earthworms and bacteria. The vermifiltration system needs to demonstrate that it can provide sufficient aeration and maintain oxygen levels.

12.3.5 Alarms

A number of alarms could be fitted to a vermifiltration system including temperature, inlet pressure, water level, backflow, power failure and instrument failure. An alarm system should detect a malfunction quickly and alert the user. It needs to be clear what conditions will cause the alarm to sound, how long the system can be used until it fails after an alarm sounds and what is the contingency plan if there are failures.

12.3.6 Vermifilter Tank Location

Wastewater treatment systems can be above or below ground. However, a below-ground tank is not affected by fluctuations in air temperature as much as above-ground tanks. Vermifiltration-based systems require a relatively constant temperature so belowground tanks are probably best suited to this application.

12.4 Case Study: Treatment of Swine Wastewater Using the Vermifiltration Technology

Swine wastewater was treated using a 3-stage vermifiltration system as a biological wastewater technique.

12.4.1 Materials

The swine wastewater was obtained from a local swine farm and 5 L were used as the vermifiltration sample. *Eisenia fetida* earthworms used for the vermifiltration process were obtained from the local earthworm farmers. The sand and gravel were obtained from a local building supplier. The 3-stage vermifilter is comprised of three 20 L/day vermifiltration beds arranged in series comprising of 40% garden soil and earthworms, 30% sand and 15% 10 mm and 30 mm quartz stones. The garden soil facilitated in trapping of the solids in the swine wastewater for conversion into vermicompost (Kharwade and Khedikar 2011). A 15% voidage was maintained for free movement of the earthworms as well as proper sprinkling of the wastewater in the vermifilters. An optimal earthworm loading density of 40 g/L was used in each of the three vermifilters (Malek et al. 2013).

12.4.2 *Methods*

The raw swine wastewater was tested for pH, total nitrogen (TKN), ammonia, nitrite, oil and grease, chemical oxygen demand (COD), biochemical oxygen demand (BOD_5), dissolved oxygen (DO), sulphate ions concentration (SO_4^{2-}), chloride ions concentration (Cl^-), iron ions concentration (Fe^{2+}), manganese ions concentration (Mn^{2+}), total suspended solids (TSS), total dissolved solids (TDS) and electrical conductivity (EC). All parameters were measured in milligrams per litre (mg/L) except for pH and the EC which was measured in microsiemens per centimetre (μS/cm). The TKN, ammonia, nitrite, oil and grease, Fe^{2+}, Mn^{2+}, BOD_5, DO, COD, Cl^-, SO_4^{2-} values were measured using titration methods in accordance to the APHA Standards Methods of determination (APHA 2005). TSS and TDS were measured through a 20 μm filter. The EC and pH were measured using a Hanna HI electrode probe. Three readings of the swine wastewater parameters were measured at each vermifilter and an average value was noted. The studies were conducted over a period of 3 months and vermicompost was removed monthly for use as bio-fertilizer.

12.4.3 *Results and Discussion*

12.4.3.1 Raw Swine Wastewater Characteristics

The COD, BOD_5, TSS, TDS, EC and the DO were out of specification in accordance to the Environmental Management Agency (EMA) Guidelines for effluent disposal in Zimbabwe. The parameters that were off-specification are italicized. The 3-stage vermifiltration process was then employed in a bid to reduce these off-limit parameters to acceptable limits. In addition, the raw swine wastewater had a COD/BOD_5 ratio of 1.17 and BOD_5/TKN ratio of 264 indicating the high biodegradability potential of the swine wastewater, hence the use of the vermifiltration technology necessary (Deng et al. 2008).

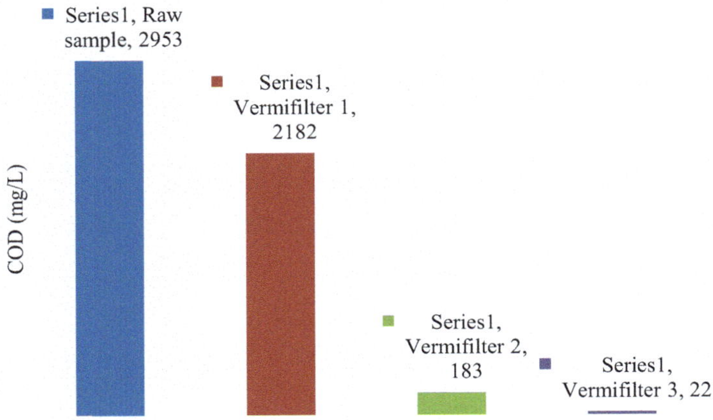

Fig. 12.2 COD concentration reduction in swine wastewater using the 3-stage vermifilter

12.4.3.2 Application of Vermifiltration Technology on the Swine Wastewater

12.4.3.2.1 The 3-Stage Vermifiltration Process Description

The excess earthworms and the vermicompost obtained from the vermifiltration process were sent to the vermicomposting section for further processing.

12.4.3.2.2 Single-Stage vs. 3-Stage Swine Biological Contaminants Removal

Removal of swine wastewater bio-contaminants using the 3-stage vermifiltration process increased significantly from the raw sample to vermifilter number 3 for the COD, BOD_5, TSS and TDS (Fig. 12.1). However the increase in removal efficiency became almost constant between vermifilters number 2 and 3 since most of the contaminants in the wastewater would have been removed. The COD removal rate increased from 26.1% to 91.6% and then a slight dip to 87.9% in the third vermifilter, whereas the BOD_5 removal rate increased from 20.9% to 91.9% and then a dip to 91.1% in the third vermifilter. TSS removal rate increased from 90.1% to 90.1% and then a slight dip at 91.8% in the third vermifilter (Fig. 12.2). However, the TDS removal rate increased throughout from 18.4% to 53.6% in the three vermifilters (Fig. 12.1). This is a clear indication that the 3-stage vermifiltration process is a necessity in the swine wastewater treatment industry.

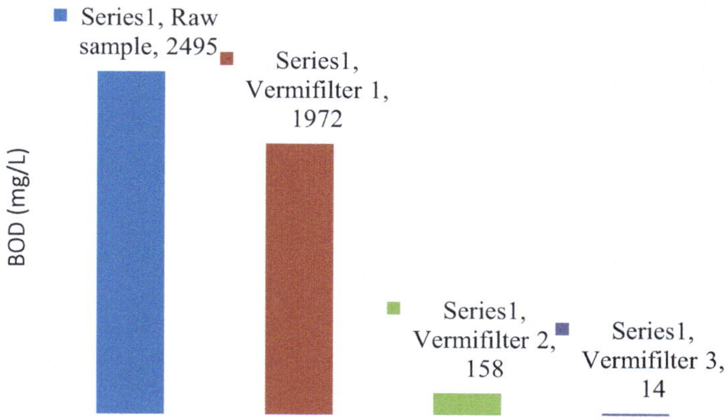

Fig. 12.3 BOD_5 concentration reduction in swine wastewater using the 3-stage vermifilter system

12.4.3.3 Effect of 3-Stage Vermifiltration on Swine Wastewater Physicochemical Properties

12.4.3.3.1 Effect on Chemical Oxygen Demand

The swine wastewater COD concentration decreased significantly as the vermifiltration process progressed through the various vermifiltration system (Fig. 12.2). COD values decreased by 99.2% resulting in an effluent with COD concentration of 22 mg/L which are within the acceptable disposal limits of $\leq$60 mg/L according to EMA standards. COD reduction values by more than 90.8% were reported for the gelatine industry wastewater treatment using a 3-tier vermifiltration process (Ghatnekar et al. 2010). This is a clear indication of the effectiveness of the 3-stage vermifiltration technology that can be adopted for swine wastewater treatment.

12.4.3.3.2 Effect on Biological Oxygen Demand

The swine wastewater BOD_5 concentration decreased significantly as the vermifiltration process progressed through the various vermifiltration system (Fig. 12.3). BOD_5 values decreased by 99.4% resulting in an effluent with BOD_5 concentration of 22 mg/L which is within the acceptable disposal limits of $\leq$30 mg/L according to EMA standards. Ghatnekar et al. (2010) also reported an 89.24% reduction in BOD_5 when a 3-stage vermifilter was used for gelatine industry wastewater.

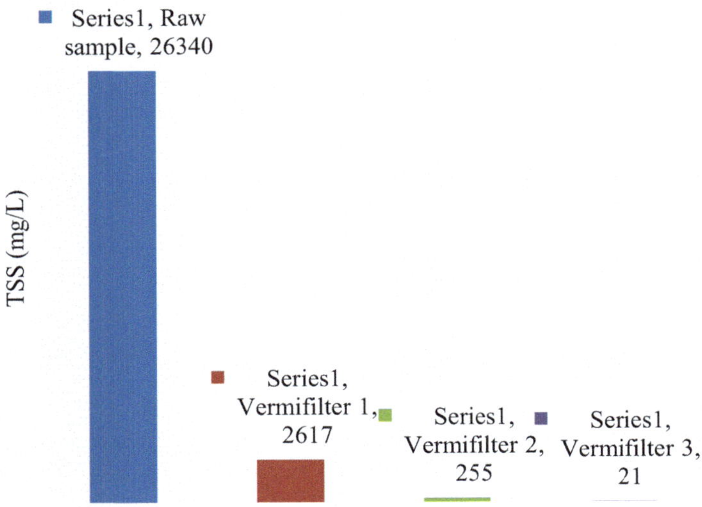

Fig. 12.4 TSS concentration reduction in swine wastewater using the 3-stage vermifilter system

Table 12.3 Raw swine wastewater characteristics in comparison to EMA disposal guidelines

Parameter	Raw swine wastewater	EMA standards
pH at 25 °C	7.38 ± 0.1	6.0–9.0
EC at 25 °C (μS/cm)	1250 ± 37.5	1000
Fe^{2+} (mg/L)	0.05 ± 0.001	≤1
Mn^{2+} (mg/L)	2.0 ± 0.04	
SO_4^{2-} (mg/L)	12 ± 0.36	
Ammonia (mg/L)	<0.01	≤0.5
DO (mg/L)	18.4 ± 0.46	≥60
Cl^- (mg/L)	284 ± 7.1	
Oil and grease (mg/L)	<0.01	
COD (mg/L)	2960 ± 88.8	60
TKN (mg/L)	9.5 ± 0.01	≤10
BOD_5 at 20 °C (mg/L)	2510 ± 75.3	30
Nitrite (mg/L)	<0.01	≤3
TSS (mg/L)	26,300 ± 657	25
TDS (mg/L)	800 ± 24	500

12.4.3.3.3 Effect on Total Suspended Solids

The swine wastewater TSS decreased significantly as the vermifiltration process progressed through the 3-stage vermifilter system (Fig. 12.4). TSS values decreased by 99.9% resulting in an effluent with TSS concentration of 21 mg/L which is within the acceptable disposal limits of ≤25 mg/L according to EMA standards (Table 12.3). Xing et al. (2010) reported a 57.18–77.90% decrease in a single-

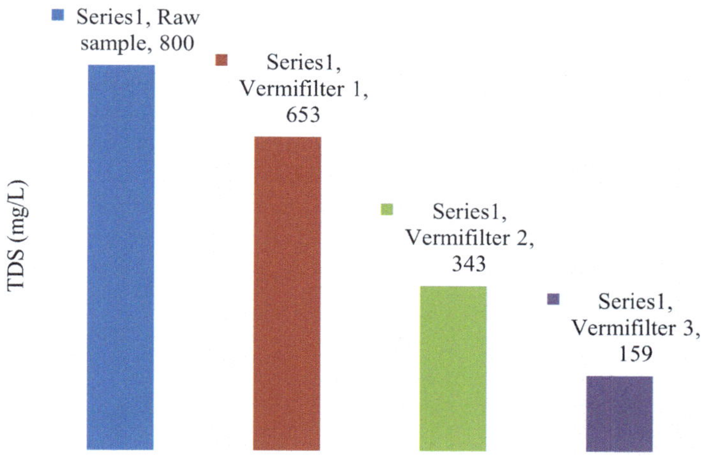

Fig. 12.5 TDS concentration reduction in swine wastewater using the 3-stage vermifilter system

stage vermifiltration system for domestic, in which the removal efficiency could be improved by a 3-stage system as in this case. The significant decrease is attributed to earthworm activity in the bioconversion of the solid material to vermicompost.

12.4.3.3.4 Effect on Total Dissolved Solids

The swine wastewater TDS concentration decreased significantly as the vermifiltration process progressed through the 3-stage vermifilter system (Fig. 12.5). TDS values decreased by 80.2% resulting in an effluent with TDS concentration of 159 mg/L which is within the acceptable disposal limits of $\leq$600 mg/L according to EMA standards. Earthworms have a tendency to feed on the dissolved solids as food resulting in the significant TDS decrease. Furthermore, as the wastewater progressed through the vermifilters, some of the dissolved particles were captured onto the filtering media.

12.4.3.3.5 Effect on Electrical Conductivity

The swine wastewater EC decreased significantly as the vermifiltration process progressed through the 3-stage vermifilter system (Fig. 12.6). EC values decreased by 86.9% resulting in an effluent with EC concentration of 165 µS/cm which are within the acceptable disposal limits of $\leq$25 µS/cm according to EMA standards. The EC decrease was also due to the decrease in TSS and TDS.

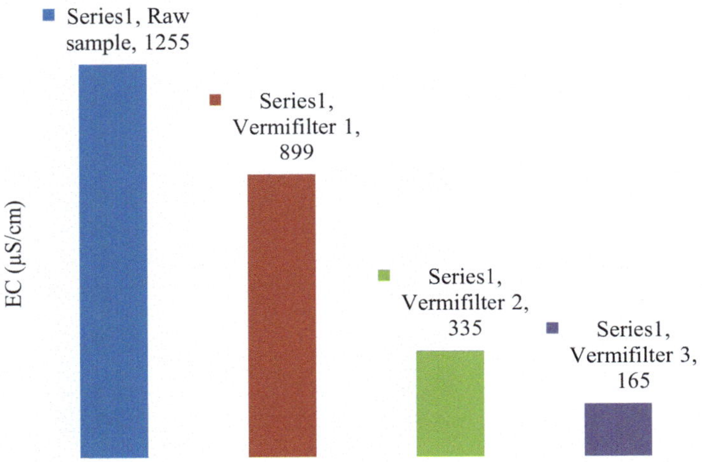

Fig. 12.6 EC reduction in swine wastewater using the 3-stage vermifilter

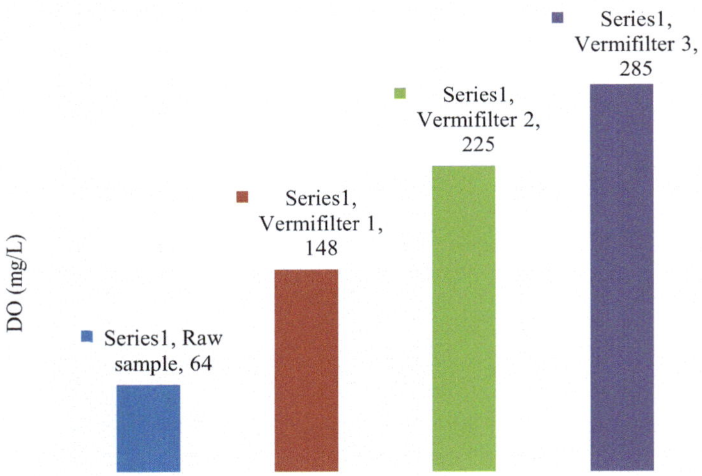

Fig. 12.7 DO concentrations increase in swine wastewater using the 3-stage vermifilter

12.4.3.3.6 Effect on Dissolved Oxygen

The swine wastewater DO concentration increased significantly as the vermifiltration process progressed through the 3-stage vermifiltration system (Fig. 12.7). DO values increased by >345.5% resulting in an effluent with DO concentration of 285 mg/L which are within the acceptable disposal limits of $\geq$60 mg/L according to EMA standards. The DO concentration increase was due the high removal of COD and BOD_5 by the earthworms during the vermifiltration process. High DO values at disposal will result in booming aquatic life as plants and

Table 12.4 Swine waste vermicompost composition

Nutrient	Composition (%)
Nitrogen	2.4 ± 0.07
Phosphorous	3.9 ± 0.01
Potassium	0.5 ± 0.01

animals will have adequate oxygen to thrive if the water is disposed of in existing water bodies.

12.4.3.4 Vermifiltration By-Products

Earthworms and vermicompost were produced as by-products during the vermifiltration process. The earthworms can be reused back in the process or can be used as feedstock for fish or poultry. The vermicompost produced had nitrogen, phosphorous and potassium (NPK) nutrient composition indicated in Table 12.4 and can be utilized as a bio-fertilizer. The vermicompost also contained traces of magnesium (0.5 ± 0.01%) and iron (0.03 ± 0.01%).

12.5 Conclusions

Vermifiltration of wastewater is an attractive wastewater treatment option in developing countries. Vermifiltration must meet the adequate process and systems design for proper operation. Vermifiltration can be applied either as multistage or single-stage systems, and multistage systems show a high degree of bio-contaminants removal with more than 90% removal. A case study with a 3-stage vermifilter indicated that the use of a 3-stage vermifiltration system enhances swine wastewater treatment to acceptable disposal values. Vermifiltration wastewater treatment is essential as a measure of promoting aquatic life in water bodies and promoting environmentally friendly technologies.

References

Adugna AT (2016) Concentrated greywater treatment by vermifiltration for Sub-Saharan urban poor. International Institute for Water and Environmental Engineering Thesis for the grade of: doctor in sciences and Technologies of Water, Energy and Environment Specialty: Water

APHA (2005) Standard methods for the examination of water and wastewater, 21st edn. America Public Health Association, American Water Works, Association, Water Environment Federation, Washington, DC

Bartlett MD, Briones MJI, Neilson R, Schmidt O, Spurgeon D, Creamer RE (2010) A critical review of current methods in earthworm ecology: from individuals to populations. Eur J Soil Biol 46(2):67–73

Baumgartner R (2013) Developing a testing protocol for Vermifiltration-based Onsite Wastewater Treatment Systems (VOWTS), pp 1–60

Chan K-Y, Munro K (2001) Evaluating mustard extracts for earthworm sampling. Pedobiologia 45 (3):272–278

Čoja T, Zehetner K, Bruckner A, Watzinger A, Meyer E (2008) Efficacy and side effects of five sampling methods for soil earthworms (Annelida, Lumbricidae). Ecotoxicol Environ Saf 71 (2):552–565

Deng L, Zheng P, Chen Z, Mamood Q (2008) Improvement in post-treatment of digested piggery wastewater. Bioresour Technol 99(8):3136–3145

Edwards CA, Fletcher KE (1988) Interactions between earthworms and micro-organisms in organic-matter breakdown. Agric Ecosyst Environ 24:235–247

Ghatnekar SD, Kavian MF, Sharma SM, Ghatnekar GS, Ghatnekar AV (2010) Application of vermifilter based effluent treatment plant (pilot scale) for bio-management of liquid effluents for the gelatin industry. Dyn Soil Dyn Plant 4(1):83–88

Jiménez JJ, Lavelle P, Decaens T (2006) The efficiency of soil hand-sorting in assessing the abundance and biomass of earthworm communities. Its usefulness in population dynamics and cohort analysis studies. Eur J Soil Biol 42(1):S225–S230

Kharwade AM, Khedikar IP (2011) Laboratory scale studies on domestic grey water through vermifilter and non-vermifilter. J Eng Res Stud 2(4):35–39

Malek TEBUA, Ismali SA, Ibrahim MH (2013) A study on hydraulic loading rate and worm density in vermifiltration of palm oil mill effluent. J Ind Res Technol 3(1):1–5

Meiyan X, Xiaowei L, Yang L (2010) Treatment performance of small-scales vermifilter for domestic wastewater and its relationship to earthworm growth, reproduction and enzymatic activity. Afr J Biotechnol 9(44):7513–7520

Neuhauser EF, Loehr RC, Malecki MR (1988) The potential of earthworms for managing sewage sludge. In: Edward CA, Neuhauser EF (eds) Earthworms in waste and environmental management. SPB Academic Publishing, The Hague ISBN:90-5103-017-7

Sinha RK, Bharambe G, Chaudhari U (2008) Sewage treatment by vermifiltration with synchronous treatment of sludge by earthworms: a low-cost sustainable technology over conventional systems with potential for decentralization, vol 28. Springer Science, pp 409–420

Sinha RK, Herat S, Bharambe G, Patil S, Bapat PS, Chauhan K, Valani D (2009) Vermiculture biotechnology: the emerging cost-effective and sustainable technology of the 21st century for multiple uses from waste and land management to safe and sustained food production. Environ Res J. NOVA Science Publishers, Hauppauge 3(1):41–110

Suthar S (2012) Vermistabilization of wastewater sludge from milk processing industry. Ecol Eng 47:115–119

Tchobanoglous G, Burton FL, Stensel HD (1981) Wastewater engineering: treatment and reuse. Metcalf and Eddy, New York

Xing M, Xiaowei L, Jian Y (2010) Treatment performance of small scale vermifilter for domestic wastewater and its relationship to earthworms' growth, reproduction and enzymatic activity. Afr J Biotechnol 9(44):7513–7520

Chapter 13
Reuse of Wastewater in Agriculture

M. P. Tripathi, Yatnesh Bisen, and Priti Tiwari

Abstract Reuse of water is defined as "water which is used twice or more time before it returns back to the natural water cycle". Treated wastewater is reused for beneficial purposes which include domestic use as toilet flushing, agricultural and landscape irrigation, industrial processes and replenishing/recharging a groundwater basin. water reuse defined as the dependency on the use of groundwater and surface water sources and can reduce the diversion of water from susceptible ecological systems. Moreover, water reuse may decline the nutrient concentration from wastewater flows into waterways, thereby decreasing and controlling pollution. This chapter built in the different sources and possibilities of reuse wastewater, their advantages, disadvantages and possible risks. The Environmental Protection Agency (EPA), the World Health Organization (WHO) health guidelines for the reuse of wastewater and the Food and Agricultural Organization (FAO) water quality guidelines for irrigation are integrated in this chapter. On the basis of these guidelines, recommendations and policy implementations for safe reuse of wastewater in irrigation and various purposes are suggested in this chapter. Further, the issue of wastewater reclamation is given and discussed properly in this chapter, which can be taken into consideration before implementing the reuse of wastewater for agriculture in India and abroad. The prospective reuse of wastewater depends on the hydraulic and biochemical individuality of wastewater, which determines the systems and extent of treatment required. Irrigation usually requires a lower quality of treatment of wastewater. However, properly designed and adequately implemented wastewater reuse system is an environmental protection measure which is superior to discharging treated wastewater into surface waters. It is the authors' hope that the content of this chapter will facilitate the consideration of reuse as an integral part of water management strategies in development projects. In this context the aim of the authors is to convey the message that the wastewater irrigation is to maximize the benefits to the poor (who depend on the resource) while minimizing the risks. The integrated guidelines will help task managers and development agency staff to

M. P. Tripathi (✉) · Y. Bisen · P. Tiwari
Department of Soil and Water Engineering, SVCAET & RS, FAE, IGKV,
Raipur, Chhattisgarh, India

R. P. Singh et al. (eds.), *Water Conservation, Recycling and Reuse: Issues and Challenges*, https://doi.org/10.1007/978-981-13-3179-4_13

prepare wastewater reuse projects. The information and material about wastewater reuse in agriculture shows that an integrated planning approach, considering economic as well as environmental and health issues, related to water reuse, is essentially a guaranty for the success.

13.1 Introduction

The amount of water is fixed and its form is changing as per atmospheric changes. Water never destroyed, its circulation is a continuous process and only its distribution is changing. Its quantity and quality both should be preserved carefully. It will be the most scarce commodity in the future and has to be dealt carefully. Seventy percent of world water use, including all the water diverted from rivers and pumped from underground, is utilized for irrigation, 20% is used by industry, and 10% goes to residences. The freshwater withdraw in agriculture and the industry worldwide is shown in Fig. 13.1. India utilized its 84% water for agriculture, 4% for drinking purposes and 12% for industry and others. Hence, efficient management in agriculture is very important.

The major objective of wastewater treatment is usually to allocate domestic and industrial effluents to be removed/disposed without any risk to human health or unacceptable harm to the natural environment. In fact, irrigation with wastewater is an effective form of wastewater disposal (as in slow-rate land treatment). It can be completely untreated municipal or industrial wastewater, mechanically purified wastewater or particularly or fully purified wastewater treated biologically (Donta 1997). The amount of treated effluent used in agriculture/horticulture has a great control on the performance and function of the wastewater-soil-plant or aquacultural systems. In the case of irrigation, the required amount of effluent will depend on the different crops, soil conditions and distribution system of effluent to be adopted. Through crop constraint and selection of irrigation methods, which minimize human health, the extent of pre-application wastewater treatment can be decreased.

India represents about 16% of the world's population. It accounts only about 2.5% of land area and 4% of water resources of the world. The total utilizable water resources of the country are estimated to be about 1123 BCM (690 BCM from surface and 433 BCM from ground), which was found to be 28% of precipitation. About 85% (688 BCM) of water for irrigation is being diverted (Fig. 13.2), and it was estimated to be increased up to 1072 BCM by the year 2050. Major water resource for irrigation is groundwater in India. Annual groundwater recharge in India is about 433 BCM, out of which about 213 BCM is used for irrigation and about 18 BCM for domestic and industrial purposes (CGWB 2011). Water demand for domestic and industrial purposes will be increased up to about 29 BCM till the year 2025. Thus availability of water for irrigation is estimated to reduce up to about 162 BCM.

The population in India is expected to be more than 1.5 billion by the end of the year 2050 as per the present rate of population growth (1.9% per annum). In India,

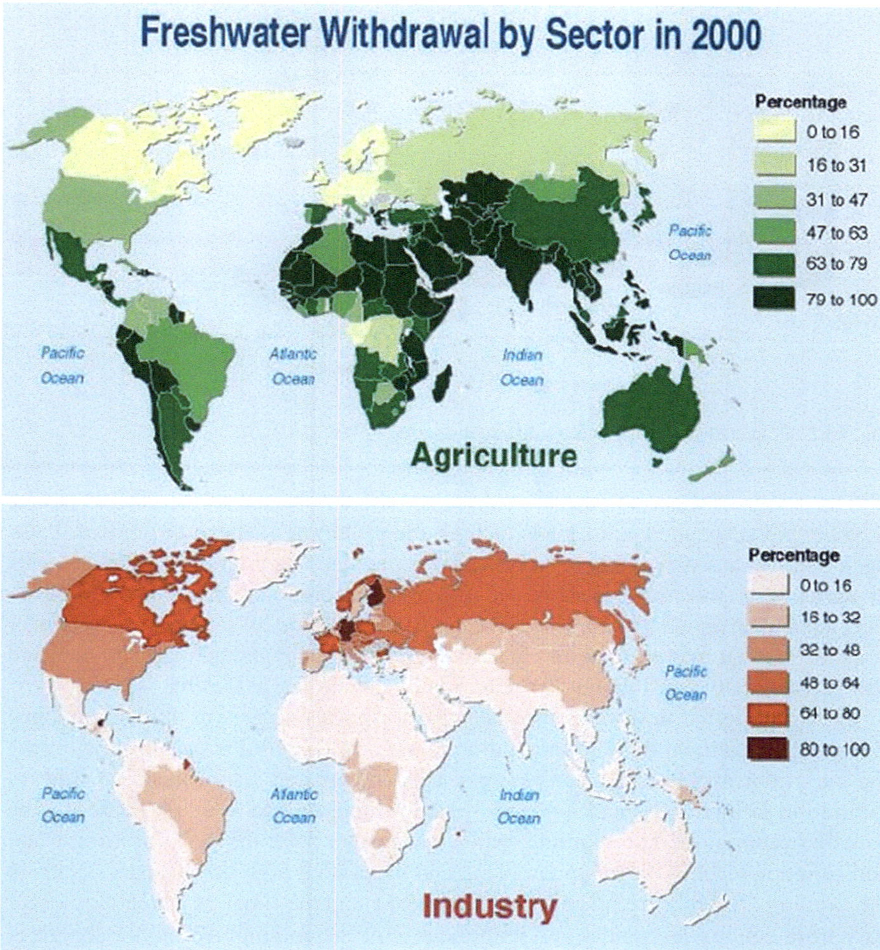

Fig. 13.1 Freshwater withdraw worldwide by sector in 2000: Source World Resources 2000–2001. People and Ecosystems, Washington, DC 2000

the average annual freshwater availability per capita has gone down due to increase in population, overall development and change in standard of living. Per capita availability of freshwater per annum in India was reported to be as 5177 m^3, 1869 m^3 and 1588 m^3 for the years 1951, 2001 and 2010, respectively. The effective and efficient water resource management through adopting wastewater treatment process and recycling system is urgently needed to meet the water requirement of the country. The project-wise water required/needed by the various sectors is shown in Fig. 13.2 (CWC 2010). There is an alarming situation of the disposal of wastewater and effluents because of deficit capacity of treatment plants and their increasing rate of sewage production.

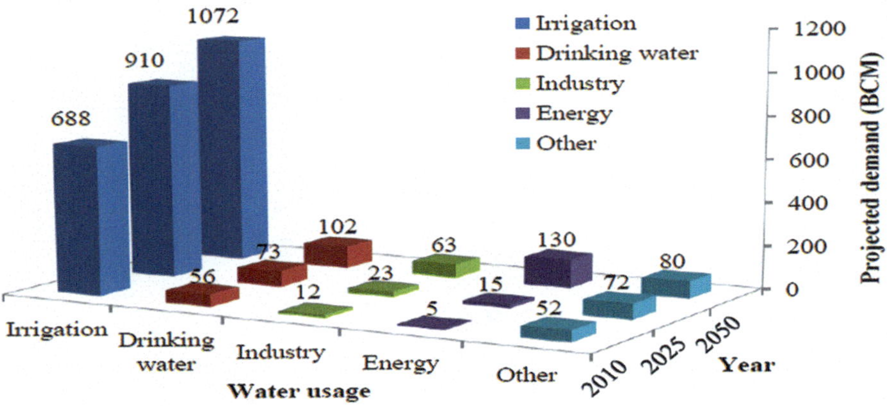

Fig. 13.2 Projected water demand by different sectors. (CWC 2010)

The design of wastewater treatment plants is usually based on the need to reduce organic and suspended solid loads to limit the pollution of the environment. Pathogen removal has very rarely been considered an objective, but, for reuse of effluents in agriculture, this must now be of primary concern, and processes should be selected and designed accordingly (Hillman 1988). Treatment to remove wastewater constituents that may be toxic or harmful to crops, aquatic plants (macrophytes) and fish is technically possible but is not normally economically feasible.

The short-period deviations in wastewater disposals noticed at public/municipal wastewater treatment units follow a diurnal pattern. Disposal is usually low during the early-morning hours, when water use is lowest and when the base flow of infiltration-inflow and small amount of sanitary wastewater. A first peak of flow usually occurs in the late morning, when wastewater from the peak morning water consumption reaches the treatment plant, and the second high flow usually occurs in the evening. The relative magnitude of the peaks and the times at which they occur vary from country to country and with the size of the community and the length of the sewers. Small communities with small sewer systems have a much higher ratio of peak flow to average flow than do large communities. Even though the extent of peaks is satisfied as wastewater passes through a treatment plant, the daily differences in flow from a public/municipal treatment unit make it impracticable, in most cases, to irrigate with effluent directly from the treatment unit. The value of reclaimed water in crop irrigation has long been recognized, particularly where fresh water resources are limited (Webster 1954; Mertz 1956; Sepp 1971). Some form of flow equalization or short-term storage of treated wastewater is important to provide a relatively continuous/uniform supply of treated wastewater for efficient irrigation, although extra benefits result from storage.

Wastewater is composed of 99.9% water and 0.1% of other materials like suspended, colloidal and dissolved solids (International reference centres for wastes disposal, 1985). Water resources are so scarce that there is often a major conflict between urban (domestic and industrial) and agricultural demands for water. This conflict can usually only be resolved by the agricultural use of wastewater: the cities must use the fresh water first; urban wastewater after proper treatment must be used for crop irrigation. If such a sequence of water resource utilization is not followed, both urban and agricultural developments may be seriously constrained with consequent adverse effects on national economic development. Moreover, the knowledge of fertilizer value of the effluent is necessary. All of the nitrogen and much of the phosphorus and potassium normally required for agricultural crop production would be supplied by the effluent (Al-Salem 1987). Further, other valuable micronutrients and the organic matter contained in the effluent will provide additional benefits (Abdel-Ghaffar et al. 1988).

13.2 Challenges

Insufficient capacity of wastewater treatment and increasing sewage generation pose a big question of the disposal of wastewater (Bhamoriya 2004). Thus, if the world is facing a water shortage, it is also facing a food shortage. Water deficits, which are already spurring heavy grain imports in numerous smaller countries, may soon do the same in larger countries, such as China and India (Earth Policy Institute 2002). Water supply for agriculture and sanitation will be one of the main future challenges in growing population and industrialization. The growing awareness of water resource scarcity, the competition for water resources and the negative impact of contaminated water on human health and the environment demand the development of adequate strategies in water management. The development of new management strategies to supply fresh water and the issue of treating and recycling wastewater will play an important role in tackling the existing and occurring problems.

It has been reported that application of sewage or sewage mixed with industrial effluents can save 25–50% of N and P fertilizer and produce 15–27% higher crop yield as compared to the normal water (Anonymous 2004). As per an estimate, 73,000 ha of peri-urban agriculture in India is focused to provide irrigation with wastewater (Strauss and Blumenthal 1990). Minhas and Samra (2004) reported that farmers of peri-urban areas generally take up, round the year, rigorous vegetable production system with 300–400% cropping intensity. They also reported that perishable commodity like fodders irrigated with wastewater farmers earn four times more per unit area as compared to normal water. Arar (1988) stated that the most suitable treatment of wastewater to be used prior to effluent application in agriculture is that which will turn out as an effluent meeting the suggested microbiological and chemical quality guidelines both at low cost and at minimum operational and maintenance requirements.

A biological treatment processes alone are not sufficient to meet tightening environmental regulations (Pant and Adholeya 2007). Presently there are no separate regulations/guidelines for safe handling, transport and disposal of wastewater in the country. The existing policies for regulating wastewater management are based on certain environmental laws and certain policies and legal provisions, viz.

Constitutional Provisions on sanitation and water pollution; National Environment Policy, 2006; National Sanitation Policy, 2008; Hazardous Waste (Management and Handling) Rules, 1989; Municipalities Act; District Municipalities Act; etc. Water Act 1974 also emphasizes utilization of treated sewage in irrigation, but this issue has been ignored by the State Governments. With the improper design, poor maintenance, frequent electricity breakdowns and lack of technical manpower, the facilities constructed to treat wastewater do not function properly and remain closed most of the time (CPCB 2007). In addition to setting up treatment plants, the Central Government, the State Government and the Board have given financial incentive to industries/investors to encourage them to invest in pollution control (CPCB 2005).

13.3 Irrigation with Wastewater

This section includes (1) conditions for successful irrigation, (2) strategies for managing treated wastewater on the farm, (3) selection of irrigation methods, (4) field management practices in wastewater irrigation and (5) planning for wastewater irrigation.

13.3.1 Conditions for Successful Irrigation

Irrigation is the application of water to plants as per the requirement. It helps to grow agricultural/horticultural crops, maintain landscapes and revegetate disturbed soil in dry areas during the period of insufficient rainfall. In arid and semiarid region of the country, irrigation is important for enhancing the production, where it is usually required on supplementary basis in humid and semi-humid regions.

At the farm level, the following basic conditions should be met to make irrigated farming a success:

- The required *amount* of water should be applied.
- The water should be of acceptable *quality*.
- Water application should be properly *scheduled*.
- Appropriate irrigation *methods* should be used.
- Salt accumulation in the root zone should be prevented by means of *leaching*.
- The rise of water table should be controlled by means of appropriate *drainage*.
- Plant *nutrients* should be managed in an optimal way.

The above requirements are equally applicable when the source of irrigation water is treated wastewater. Nutrients in municipal wastewater and treated effluents are a particular advantage of these sources over conventional irrigation water sources, and supplemental fertilizers are sometimes not necessary. However,

Table 13.1 Water requirements, sensitivity to water supply and water utilization efficiency of some selected crops

Crop	Water requirements (mm/growing period)	Sensitivity to water supply (ky)	Water utilization efficiency for harvested yield, Ey, kg/m^3 (% moisture)
Alfalfa	800–1600	Low to medium-high (0.7–1.1)	1.5–2.0 hay (10–15%)
Banana	1200–2200	High (1.2–1.35)	Plant crop: 2.5–4
			Ratoon: 3.5–6
			Fruit (70%)
Bean	300–500	Medium-high (1.15)	Lush: 1.5–2.0 (80–90%)
			Dry: 0.3–0.6 (10%)
Cabbage	380–500	Medium-low (0.95)	12–20 head (90–95%)
Citrus	900–1200	Low to medium-high (0.8–1.1)	2–5 fruit (85%, lime: 70%)
Cotton	700–1300	Medium-low (0.85)	0.4–0.6 seed cotton (10%)
Groundnut	500–700	Low (0.7)	0.6–0.8 unshelled dry nut (15%)
Maize	500–800	High (1.25)	0.8–1.6 grain (10–13%)
Potato	500–700	Medium-high (1.1)	4–7 fresh tuber (70–75%)
Rice	350–700	High	0.7–1.1 paddy (15–20%)
Safflower	600–1200	Low (0.8)	0.2–0.5 seed (8–10%)
Sorghum	450–650	Medium-low (0.9)	0.6–1.0 grain (12–15%)
Wheat	450–650	Medium high (spring: 1.15; winter: 1.0)	0.8–1.0 grain (12–15%)

Source: FAO (1979)

additional environmental and health requirements must be considered when treated wastewater is the source of irrigation water.

13.3.1.1 Amount of Water to be Applied

It is well known that more than 99% of the water absorbed by plants is lost by transpiration and evaporation from the plant surface. Thus, for all practical purposes, the water requirement of crops is equal to the evapotranspiration requirement, ET_c. Crop evapotranspiration is mainly determined by climatic factors and hence can be estimated with reasonable accuracy using meteorological data. An extensive review of this subject and guidelines for estimating ET_c, prepared by Doorenbos and Pruitt, are given in Irrigation and Drainage Paper 24 (FAO 1977). A computer program, called CROPWAT, is available in FAO to determine the water requirements of crops from climatic data. Table 13.1 presents the water requirements of some selected crops, reported by Doorenbos and Kassam (FAO 1979). It should be kept in mind that the actual amount of irrigation water to be applied will have to be adjusted for effective rainfall, leaching requirement, application losses and other factors.

13.3.1.2 Quality of Water to be Applied

Irrigation water quality requirements from the point of view of crop production will have to be adjusted depending on the local climate, soil conditions and other factors. In addition, farm practices, such as the type of crop to be grown, irrigation method and agronomic practices, will determine to a great extent the quality suitability of irrigation water. Some of the important farm practices aimed at optimizing crop production when treated sewage effluent is used as irrigation water. The suitability of water for irrigation will greatly depend on the climatic conditions, the physical and chemical properties of the soil, the salt tolerance of the crop grown and the management practices. Thus, classification of water for irrigation will always be general in nature and applicable under average use conditions. Ayers and Westcot (FAO 1985) classified irrigation water into three groups based on salinity, sodicity, toxicity and miscellaneous hazards, as shown in Table 13.2. These general water quality

Table 13.2 Guidelines for interpretation of water quality for irrigation

		Degree of restriction on use		
Potential irrigation problem	Units	None	Slight to moderate	Severe
Salinity				
Ec_w[a]	dS/m	< 0.7	0.7–3.0	> 3.0
or				
TDS	mg/l	< 450	450–2000	> 2000
Infiltration				
SAR[b] = 0–3 and EC_w		> 0.7	0.7–0.2	< 0.2
3–6		> 1.2	1.2–0.3	< 0.3
6–12		> 1.9	1.9–0.5	< 0.5
12–20		> 2.9	2.9–1.3	< 1.3
20–40		> 5.0	5.0–2.9	< 2.9
Specific ion toxicity				
Sodium (Na)				
Surface irrigation	SAR	< 3	3–9	> 9
Sprinkler irrigation	me/l	< 3	> 3	
Chloride (Cl)				
Surface irrigation	me/l	< 4	4–10	> 10
Sprinkler irrigation	m^3/l	< 3	> 3	
Boron (B)	mg/l	< 0.7	0.7–3.0	> 3.0
Trace elements				
Miscellaneous effects				
Nitrogen (NO_3-N)[c]	mg/l	< 5	5–30	> 30
Bicarbonate (HCO_3)	me/l	< 1.5	1.5–8.5	> 8.5
pH	Normal range 6.5–8			

Source: FAO (1985)

[a]EC_w means electrical conductivity in deciSiemens per metre at 25 °C

[b]SAR means sodium adsorption ratio

[c]NO_3-N means nitrate nitrogen reported in terms of elemental nitrogen

classification guidelines help to identify potential crop production problems associated with the use of conventional water sources. The guidelines are equally applicable to evaluate wastewaters for irrigation purposes in terms of their chemical constituents, such as dissolved salts, relative sodium content and toxic ions.

Municipal wastewater effluents may contain a number of toxic elements, including heavy metals, because under practical conditions wastes from many small and informal industrial sites are directly discharged into the common sewer system. These toxic elements are normally present in small amounts, and hence, they are called trace elements. Some of them may be removed during the treatment process, but others will persist and could present phytotoxic problems. Thus, municipal wastewater effluents should be checked for trace element toxicity hazards, particularly when trace element contamination is suspected. Table 13.3 presents phytotoxic threshold levels of some selected trace elements.

13.3.1.3 Scheduling of Irrigation

To obtain the maximum yields, water should be applied to crops before the soil moisture potential reaches a level at which the evapotranspiration rate is likely to be reduced below its potential. The relationship of actual and maximum yields to actual and potential evapotranspiration is illustrated in the following equation:

$$\left(1 - \frac{Y_a}{Y_m}\right) = ky\left(1 - \frac{ET_a}{ET_m}\right) \tag{13.1}$$

where

Y_a = actual harvested yield
Y_m = maximum harvested yield
ky = yield response factor
ET_a = actual evapotranspiration
ET_m = maximum evapotranspiration

Several methods are available to determine optimum irrigation scheduling. The factors that determine irrigation scheduling are available water-holding capacity of the soils, depth of root zone, evapotranspiration rate and amount of water to be applied per irrigation, irrigation method and drainage conditions.

13.3.1.4 Irrigation Methods

Many different methods are used by farmers to irrigate crops. They range from watering individual plants from a can of water to highly automated irrigation by a centre pivot system. However, from the point of wetting the soil, these methods can be grouped under five headings, namely:

Table 13.3 Threshold levels of trace elements for crop production

	Element	Recommended maximum concentration (mg/l)	Remarks
Al	(Aluminium)	5.0	Can cause nonproductivity in acid soils (pH < 5.5), but more alkaline soils at pH > 7.0 will precipitate the ion and eliminate any toxicity
As	(Arsenic)	0.10	Toxicity to plants varies widely, ranging from 12 mg/l for Sudan grass to less than 0.05 mg/l for rice
Be	(Beryllium)	0.10	Toxicity to plants varies widely, ranging from 5 mg/l for kale to 0.5 mg/l for bush beans
Cd	(Cadmium)	0.01	Toxic to beans, beets and turnips at concentrations as low as 0.1 mg/l in nutrient solutions. Conservative limits recommended due to its potential for accumulation in plants and soils to concentrations that may be harmful to humans
Co	(Cobalt)	0.05	Toxic to tomato plants at 0.1 mg/l in nutrient solution. Tends to be inactivated by neutral and alkaline soils
Cr	(Chromium)	0.10	Not generally recognized as an essential growth element. Conservative limits recommended due to the lack of knowledge on its toxicity to plants
Cu	(Copper)	0.20	Toxic to a number of plants at 0.1–1.0 mg/l in nutrient solutions
F	(Fluoride)	1.0	Inactivated by neutral and alkaline soils
Fe	(Iron)	5.0	Not toxic to plants in aerated soils but can contribute to soil acidification and loss of availability of essential phosphorus and molybdenum. Overhead sprinkling may result in unsightly deposits on plants, equipment and buildings
Li	(Lithium)	2.5	Tolerated by most crops up to 5 mg/l; mobile in soil. Toxic to citrus at low concentrations (<0.075 mg/l). Acts similarly to boron
Mn	(Manganese)	0.20	Toxic to a number of crops at a few tenths to a few mg/l but usually only in acid soils
Mo	(Molybdenum)	0.01	Not toxic to plants at normal concentrations in soil and water. Can be toxic to livestock if forage is grown in soils with high concentrations of available molybdenum
Ni	(Nickel)	0.20	Toxic to a number of plants at 0.5 mg/l to 1.0 mg/l; reduced toxicity at neutral or alkaline pH
Pd	(Lead)	5.0	Can inhibit plant cell growth at very high concentrations
Se	(Selenium)	0.02	Toxic to plants at concentrations as low as 0.025 mg/l and toxic to livestock if forage is grown in soils with relatively high levels of

(continued)

Table 13.3 (continued)

	Element	Recommended maximum concentration (mg/l)	Remarks
			added selenium. As essential element to animals but in very low concentrations
Sn	(Tin)		
Ti	(Titanium)	–	Effectively excluded by plants; specific tolerance unknown
W	(Tungsten)		
C	(Vanadium)	0.10	Toxic to many plants at relatively low concentrations
Zn	(Zinc)	2.0	Toxic to many plants at widely varying concentrations; reduced toxicity at pH > 6.0 and in fine-textured or organic soils

Source: Adapted from National Academy of Sciences (1972) and Pratt (1972)
Note: The maximum concentration is based on a water application rate which is consistent with good irrigation practices (10,000 m^3 per hectare per year). If the water application rate greatly exceeds this, the maximum concentrations should be adjusted downwards accordingly. No adjustment should be made for application rates less than 10,000 m^3 per hectare per year. The values given are for water used on a continuous basis at one site

(i) *Flood irrigation* – water is applied over the entire field to infiltrate into the soil (e.g. wild flooding, contour flooding, borders, basins, etc.).
(ii) *Furrow irrigation* – water is applied between ridges (e.g. level and graded furrows, contour furrows, corrugations, etc.). Water reaches the ridge, where the plant roots are concentrated, by capillary action.
(iii) *Sprinkler irrigation* – water is applied in the form of a spray and reaches the soil very much like rain (e.g. portable, semiportable and solid set sprinklers, travelling sprinklers, spray guns, centre pivot systems, etc.). The rate of application is adjusted so that it does not create ponding of water on the surface.
(iv) *Subirrigation* – water is applied beneath the root zone in such a manner that it wets the root zone by capillary rise (e.g. sub-surface irrigation canals, buried pipes, etc.). Deep surface canals or buried pipes are used for this purpose.
(v) *Localized irrigation* – water is applied around each plant or a group of plants so as to wet locally and the root zone only (e.g. drip irrigation, bubblers, micro-sprinklers, etc.). The application rate is adjusted to meet evapotranspiration needs so that percolation losses are minimized. Table 13.4 presents some basic features of selected irrigation systems as reported by Doneen and Westcot (FAO 1988).

13.3.1.5 Leaching

Under irrigated agriculture, a certain amount of excess irrigation water is required to percolate through the root zone so as to remove the salts which have accumulated as

Table 13.4 Basic features of some selected irrigation systems

Irrigation method	Topography	Crops	Remarks
Widely spaced borders	Land slopes capable of being graded to less than 1% slope and preferably 0.2%	Alfalfa and other deep rooted close-growing crops and orchards	The most desirable surface method for irrigating close-growing crops where topographical conditions are favourable. Even grade in the direction of irrigation is required on flat land and is desirable but not essential on slopes of more than 0.5%. Grade changes should be slight and reverse grades must be avoided. Cross slops is permissible when confined to differences in elevation between border strips of 6–9 cm. Water application efficiency 45–60%
Graded contour furrows	Variable land slopes of 2–25% but preferable less	Row crops and fruit	Especially adapted to row crops on steep land, though hazardous due to possible erosion from heavy rainfall. Unsuitable for rodent-infested fields or soils that crack excessively. Actual grade in the direction of irrigation 0.5–1.5%. No grading required beyond filling gullies and removal of abrupt ridges. Water application efficiency 50–65%
Rectangular checks (levees)	Land slopes capable of being graded so single or multiple tree basins will be levelled within 6 cm	Orchard	Especially adapted to soils that have either a relatively high or low water intake rate. May require considerable grading. Water application efficiency 40–60%
Subirrigation	Smooth flat	Shallow rooted crops such as potatoes or grass	Requires a water table, very permeable subsoil conditions and precise levelling. Very few areas adapted to this method. Water application efficiency 50–70%
Sprinkler	Undulating 1–> 35% slope	All crops	High operation and maintenance costs. Good for rough or very sandy lands in areas of high production and

(continued)

Table 13.4 (continued)

Irrigation method	Topography	Crops	Remarks
			good markets. Good method where power costs are low. May be the only practical method in areas of steep or rough topography. Good for high rainfall areas where only a small supplementary water supply is needed. Water application efficiency 60–70%
Localized (drip, trickle, etc.)	Any topographic condition suitable for row crop farming	Row crops or fruit	Perforated pipe on the soil surface drips water at base of individual vegetable plants or around fruit trees. Has been successfully used in Israel with saline irrigation water. Still in the development stage. Water application efficiency 75–85%

Source: FAO (1988)

a result of evapotranspiration from the original irrigation water. This process of displacing the salts from the root zone is called leaching, and that portion of the irrigation water which mobilizes the excess of salts is called the leaching fraction, LF.

$$\text{Leaching Fraction (LF)} = \frac{\text{depth of water leached below the root zone}}{\text{depth of water applied at the surface}} \quad (13.2)$$

Salinity control by effective leaching of the root zone becomes more important as irrigation water becomes more saline.

13.3.1.6 Drainage

Drainage is defined as the removal of excess water from the soil surface and below so as to permit optimum growth of plants. Removal of excess surface water is termed surface drainage, while the removal of excess water from beneath the soil surface is termed sub-surface drainage. The importance of drainage for successful irrigated agriculture has been well demonstrated. It is particularly important in semiarid and arid areas to prevent secondary salinization. In these areas, the water table will rise with irrigation when the natural internal drainage of the soil is not adequate. When the water table is within a few metres of the soil surface, capillary rise of saline groundwater will transport salts to the soil surface. At the surface, water evaporates, leaving the salts behind. If this process is not arrested, salt accumulation will continue, resulting in salinization of the soil. In such cases, sub-surface drainage can control the rise of the water table and hence prevent salinization.

13.3.2 Strategies for Managing Treated Wastewater on the Farm

Success in using treated wastewater for crop production will largely depend on adopting appropriate strategies aimed at optimizing crop yields and quality, maintaining soil productivity and safeguarding the environment. Several alternatives are available, and a combination of these alternatives will offer an optimum solution for a given set of conditions. The user should have prior information on effluent supply and its quality, as indicated in Table 13.3, to ensure the formulation and adoption of an appropriate on-farm management strategy.

Basically, the components of an on-farm strategy in using treated wastewater will consist of a combination of:

- Crop selection
- Selection of irrigation method
- Adoption of appropriate management practices

Furthermore, when the farmer has additional sources of water supply, such as a limited amount of normal irrigation water, he will then have an option to use both the effluent and the conventional source of water in two ways, namely:

- By blending conventional water with treated effluent
- By using the two sources in rotation

These are discussed briefly in the following sections (Table 13.5).

13.3.3 Selection of Irrigation Methods

Under normal conditions, the type of irrigation method selected will depend on water supply conditions, climate, soil, crops to be grown, cost of irrigation method and the ability of the farmer to manage the system. However, when using wastewater as the source of irrigation, other factors, such as contamination of plants and harvested product, farm workers and the environment and salinity and toxicity hazards, will need to be considered. There is considerable scope for reducing the undesirable effects of wastewater use in irrigation through selection of appropriate irrigation methods.

Table 13.5 Information required on effluent supply and quality

Information	Decision on irrigation management
Effluent supply	
The total amount of effluent that would be made available during the crop growing season	Total area that could be irrigated
Effluent available throughout the year	Storage facility during non-crop growing period either at the farm or near wastewater treatment plant and possible use for aquaculture
The rate of delivery of effluent either as m^3 per day or litres per second	Area that could be irrigated at any given time, layout of fields and facilities and system of irrigation
Type of delivery: continuous or intermittent or on demand	Layout of fields and facilities, irrigation system and irrigation scheduling
Mode of supply: supply at farm gate or effluent available in a storage reservoir to be pumped by the farmer	The need to install pumps and pipes to transport effluent and irrigation system
Effluent quality	
Total salt concentration and/or electrical conductivity of the effluent	Selection of crops, irrigation method, leaching and other management practices
Concentrations of cations, such as Ca^{++}, Mg^{++} and Na^{+}	To assess sodium hazard and undertake appropriate measures
Concentration of toxic ions, such as heavy metals, Boron and Cl^{-}	To assess toxicities that are likely to be caused by these elements and take appropriate measures
Concentration of trace elements (particularly those which are suspected of being phytotoxic)	To assess trace toxicities and take appropriate measures
Concentration of nutrients, particularly nitrate-N	To adjust fertilizer levels, avoid over-fertilization and select crop
Level of suspended sediments	To select appropriate irrigation system and measures to prevent clogging problems
Levels of intestinal nematodes and faecal coliforms	To select appropriate crops and irrigation systems

The choice of irrigation method in using wastewater is governed by the following technical factors:

- The choice of crops
- The wetting of foliage, fruits and aerial parts
- The distribution of water, salts and contaminants in the soil
- The ease with which high soil water potential could be maintained
- The efficiency of application
- The potential to contaminate farm workers and the environment

Table 13.6 presents an analysis of these factors in relation to four widely practised irrigation methods, namely, border, furrow, sprinkler and drip irrigation.

A border (and basin or any flood irrigation) system involves complete coverage of the soil surface with treated effluent and is normally not an efficient method of

Table 13.6 Evaluation of common irrigation methods in relation to the use of treated wastewater

Parameters of evaluation	Furrow irrigation	Border irrigation	Sprinkler irrigation	Drip irrigation
1. Foliar wetting and consequent leaf damage resulting in poor yield	No foliar injury as the crop is planted on the ridge	Some bottom leaves may be affected, but the damage is not so serious as to reduce yield	Severe leaf damage can occur resulting in significant yield loss	No foliar injury occurs under this method of irrigation
2. Salt accumulation in the root zone with repeated applications	Salts tend to accumulate in the ridge which could harm the crop	Salts move vertically downwards and are not likely to accumulate in the root zone	Salt movement is downwards, and root zone is not likely to accumulate salts	Salt movement is radial along the direction of water movement. A salt wedge is formed between drip points
3. Ability to maintain high soil water potential	Plants may be subject to stress between irrigations	Plants may be subject to water stress between irrigations	Not possible to maintain high soil water potential throughout the growing season	Possible to maintain high soil water potential throughout the growing season and minimize the effect of salinity
4. Suitability to handle brackish wastewater without significant yield loss	Fair to medium. With good management and drainage acceptable yields are possible	Fair to medium. Good irrigation and drainage practices can produce acceptable levels of yield	Poor to fair. Most crops suffer from leaf damage and yield is low	Excellent to good. Almost all crops can be grown with very little reduction in yield

Source: Kandiah (1990b)

irrigation. This system will also contaminate vegetable crops growing near the ground and root crops and will expose farm workers to the effluent more than any other method. Thus, from both the health and water conservation points of view, border irrigation with wastewater is not satisfactory.

Furrow irrigation, on the other hand, does not wet the entire soil surface. This method can reduce crop contamination, since plants are grown on the ridges, but complete health protection cannot be guaranteed. Contamination of farm workers is potentially medium to high, depending on automation. If the effluent is transported through pipes and delivered into individual furrows by means of gated pipes, risk to irrigation workers will be minimum.

The efficiency of surface irrigation methods, in general, borders, basins and furrows, is not greatly affected by water quality, although the health risk inherent in these systems is most certainly of concern. Some problems might arise if the effluent contains large quantities of suspended solids, and these settle out and restrict flow in transporting channels, gates, pipes and appurtenances. The use of primary treated sewage will overcome many of such problems. To avoid surface ponding of stagnant effluent, land levelling should be carried out carefully, and appropriate land gradients should be provided.

Sprinkler, or spray, irrigation methods are generally more efficient in terms of water use, since greater uniformity of application can be achieved. However, these overhead irrigation methods may contaminate ground crops, fruit trees and farm workers. In addition, pathogens contained in aerosolized effluent may be transported downwind and create a health risk to nearby residents. Generally, mechanized or automated systems have relatively high capital costs and low labour costs compared with manually moved sprinkler systems. Rough land levelling is necessary for sprinkler systems, to prevent excessive head losses and achieve uniformity of wetting. Sprinkler systems are more affected by water quality than surface irrigation systems, primarily as a result of the clogging of orifices in sprinkler heads, potential leaf burns and phytotoxicity when water is saline and contains excessive toxic elements and sediment accumulation in pipes, valves and distribution systems. Secondary wastewater treatment has generally been found to produce an effluent suitable for distribution through sprinklers, provided that the effluent is not too saline. Further precautionary measures, such as treatment with granular filters or micro-strainers and enlargement of nozzle orifice diameters to not less than 5 mm, are often adopted.

Localized irrigation, particularly when the soil surface is covered with plastic sheeting or other mulch, uses effluent more efficiently, can often produce higher crop yields and certainly provides the greatest degree of health protection for farm workers and consumers. Trickle and drip irrigation systems are expensive, however, and require a high quality of effluent to prevent clogging of the emitters through which water is slowly released into the soil. Table 13.7 presents water quality requirements to prevent clogging in localized irrigation systems. Solids in the effluent or biological growth at the emitters will create problems, but gravel filtration of secondary treated effluent and regular flushing of lines have been found to be effective in preventing such problems in Cyprus (Papadopoulos and Stylianou

Table 13.7 Water quality and clogging potential in drip irrigation systems

		Degree of restriction on use		
Potential problem	Units	None	Slight to moderate	Severe
Physical				
Suspended solids	mg/l	< 50	50–100	> 100
Chemical				
pH		< 7.0	7.0–8.0	> 8.0
Dissolved solids	mg/l	< 500	500–2000	> 2000
Manganese	mg/l	< 0.1	0.1–1.5	> 1.5
Iron	mg/l	< 0.1	0.1–1.5	> 1.5
Hydrogen sulphide	mg/l	< 0.5	0.5–2.0	> 2.0
Biological	Maximum			
Bacterial populations	Number/ml	< 10,000	10,000–50,000	> 50,000

Source: Adapted from Nakayama (1982)

1988). Bubbler irrigation, a technique developed for the localized irrigation of tree crops, avoids the need for small emitter orifices, but careful setting is required for its successful application (Hillel 1987).

When compared with other systems, the main advantages of trickle irrigation seem to be:

(i) Increased crop growth and yield achieved by optimizing the water, nutrients and air regimes in the root zone.
(ii) High irrigation efficiency – no canopy interception, wind drift or conveyance losses and minimal drainage losses.
(iii) Minimal contact between farm workers and effluent.
(iv) Low-energy requirements – the trickle system requires a water pressure of only 100–300 k pa (1–3 bar).
(v) Low labour requirements – the trickle system can easily be automated, even to allow combined irrigation and fertilization (sometimes terms fertigation).

Apart from the high capital costs of trickle irrigation systems, another limiting factor in their use is that they are only suited to the irrigation of row crops. Relocation of sub-surface systems can be prohibitively expensive.

Clearly, the decision on irrigation system selection will be mainly a financial one, but it is to be hoped that the health risks associated with the different methods will be taken into account. Each measure will interact with others, and thus a decision on irrigation system selection will have an influence on wastewater treatment requirements, human exposure control and crop selection (e.g. row crops are dictated by trickle irrigation). At the same time, the irrigation techniques feasible will depend on crop selection, and the choice of irrigation system might be limited if wastewater treatment has already been decided before effluent use is considered.

13.3.4 Field Management Practices in Wastewater Irrigation

Management of water, soil, crop and operational procedures, including precautions to protect farm workers, plays an important role in the successful use of sewage effluent for irrigation.

13.3.4.1 Water Management

Most treated wastewaters are not very saline, salinity levels usually ranging between 500 and 200 mg/l ($EC_w = 0.7–3.0$ dS/m). However, there may be instances where the salinity concentration exceeds the 2000 mg/l level. In any case, appropriate water management practices will have to be followed to prevent salinization, irrespective of whether the salt content in the wastewater is high or low. It is interesting to note that even the application of a nonsaline wastewater, such as one containing 200–500 mg/l, when applied at a rate of 20,000 m^3 per hectare, a fairly typical

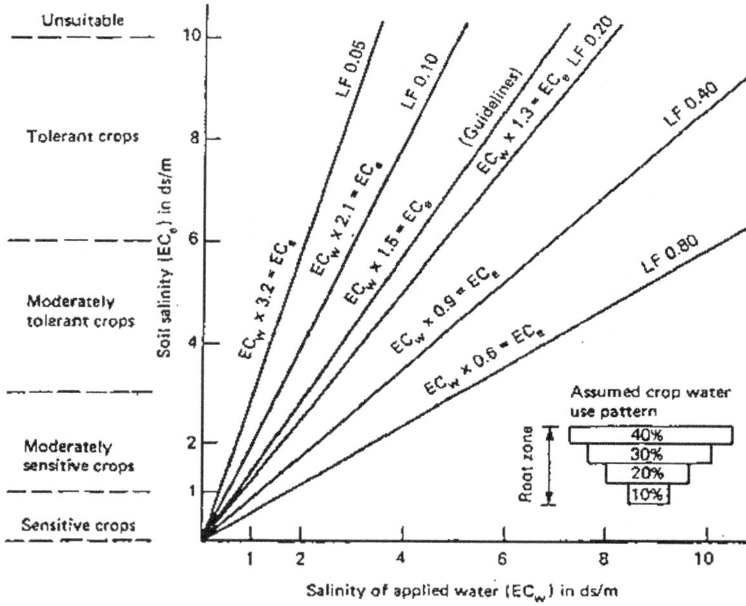

Fig. 13.3 Relationship between applied water salinity and soil water salinity at different leaching fractions (FAO 1985)

irrigation rate, will add between 2 and 5 tonnes of salt annually to the soil. If this is not flushed out of the root zone by leaching and removed from the soil by effective drainage, salinity problems can build up rapidly. Leaching and drainage are thus two important water management practices to avoid salinization of soils.

13.3.4.2 Leaching

The concept of leaching has already been discussed. The question that arises is how much water should be used for leaching, i.e. what is the leaching requirement? To estimate the leaching requirement, both the salinity of the irrigation water (EC_w) and the crop tolerance to soil salinity (EC_e) must be known. The necessary leaching requirement (LR) can be estimated from Fig. 13.3 for general crop rotations reported by Ayers and Westcot (FAO 1985). A more exact estimate of the leaching requirement for a particular crop can be obtained using the following equation:

$$\mathrm{LR} = \frac{\mathrm{EC_W}}{5(\mathrm{EC_e} - \mathrm{EC_W})} \tag{13.3}$$

where LR is the minimum leaching requirement needed to control salts within the tolerance (EC_e) of the crop with ordinary surface methods of irrigation, EC_w salinity of the applied irrigation water in dS/m and EC_e average soil salinity tolerated by the crop as measured on a soil saturation extract. It is recommended that the EC_e value that can be expected to result in at least a 90% or greater yield can be used in the calculation.

Figure 13.3 was developed using EC_e values for the 90% yield potential. For water in the moderate to high salinity range (>1.5 dS/m), it might be better to use the EC_e value for maximum yield potential (100%) since salinity control is critical in obtaining good yields. Further information on this is contained in Irrigation and Drainage Paper 29, Rev. 1 (FAO 1985).

Where water is scarce and expensive, leaching practices should be designed to maximize crop production per unit volume of water applied, to meet both the consumptive use and leaching requirements. Depending on the salinity status, leaching can be carried out at each irrigation, each alternative irrigation or less frequently, such as seasonally or at even longer intervals, as necessary to keep the salinity in the soil below the threshold above which yield might be affected to an unacceptable level. With good-quality irrigation water, the irrigation application level will almost always apply sufficient extra water to accomplish leaching. With high salinity irrigation water, meeting the leaching requirement is difficult and requires large amounts of water. Rainfall must be considered in estimating the leaching requirement and in choosing the leaching method.

The following practices are suggested for increasing the efficiency of leaching and reducing the amount of water needed:

(i) Leach during cool seasons instead of during warm periods, to increase the efficiency and ease of leaching, since the total annual crop water demand (ET, mm/year) losses are lower.
(ii) Use more salt-tolerant crops which require a lower leaching requirement (LR) and thus have a lower water demand.
(iii) Use tillage to slow overland water flow and reduce the number of surface cracks which bypass flow through large pores and decrease leaching efficiency.
(iv) Use sprinkler irrigation at an application rate below the soil infiltration rate as this favours unsaturated flow, which is significantly more efficient for leaching than saturated flow. More irrigation time but less water is required than for continuous ponding.
(v) Use alternate ponding and drying instead of continuous ponding as this is more efficient for leaching and uses less water, although the time required to leach is greater. This may have drawbacks in areas having a high water table, which allows secondary salinization between poundings.
(vi) Where possible, schedule leaching at periods of low crop water use or postpone leaching until after the cropping season.
(vii) Avoid fallow periods, particularly during hot summers, when rapid secondary soil salinization from high water tables can occur.
(viii) If infiltration rates are low, consider pre-planting irrigations or off-season leaching to avoid excessive water applications during the crop season.

(ix) Use one irrigation before the start of the rainy season if total rainfall is normally expected to be insufficient for a complete leaching. Rainfall is often the most efficient leaching method because it provides high-quality water at relatively low rates of application.

Drainage Salinity problems in many irrigation projects in arid and semiarid areas are associated with the presence of a shallow water table. The role of drainage in this context is to lower the water table to a desirable level, at which it does not contribute to the transport of salts to the root zone and the soil surface by capillarity. What is important is to maintain a downward movement of water through soils. Van Schilfgaarde (1984) reported that drainage criteria are frequently expressed in terms of critical water table depths; although this is a useful concept, prevention of salinization depends on the establishment, averaged over a period of time, of a downward flux of water. Another important element of the total drainage system is its ability to transport the desired amount of drained water out of the irrigation scheme and dispose it safely. Such disposal can pose a serious problem, particularly when the source of irrigation water is treated wastewater, depending on the composition of the drainage effluent.

Timing of Irrigation The timing of irrigation, including irrigation frequency, pre-planting irrigation and irrigation prior to a winter rainy season, can reduce the salinity hazard and avoid water stress between irrigations. Some of these practices are readily applicable to wastewater irrigation.

In terms of meeting the water needs of crops, increasing the frequency of irrigation will be desirable as it eliminates water stress between irrigations. However, from the point of view of overall water management, this may not always produce the desired results. For example, with border, basin and other flood irrigation methods, frequent irrigations may result in an unacceptable increase in the quantity of water applied, decrease in water use efficiency and larger amounts of water to be drained. However, with sprinklers and localized irrigation methods, frequent applications with smaller amounts may not result in decrease in water use efficiency and, indeed, could help to overcome the salinity problem associated with saline irrigation water.

Pre-planting irrigation is practised in many irrigation schemes for two reasons, namely, (i) to leach salts from the soil surface which may have accumulated during the previous cropping period and to provide a salt-free environment to germinating seeds (it should be noted that for most crops, the seed germination and seedling stages are most sensitive to salinity) and (ii) to provide adequate moisture to germinating seeds and young seedlings. A common practice among growers of lettuce, tomatoes and other vegetable crops is to pre-irrigate the field before planting, since irrigation soon after planting could create local water stagnation and wet spots that are not desirable. Treated wastewater is a good source for pre-irrigation as it is normally not saline and the health hazards are practically nil.

Blending of Wastewater with Other Water Supplies One of the options that may be available to farmers is the blending of treated sewage with conventional sources of water, canal water or ground water, if multiple sources are available. It is possible that a farmer may have saline ground water and, if he has nonsaline treated wastewater, could blend the two sources to obtain a blended water of acceptable salinity level. Further, by blending, the microbial quality of the resulting mixture could be superior to that of the unblended wastewater.

Alternating Treated Wastewater with Other Water Sources Another strategy is to use the treated wastewater alternately with the canal water or groundwater, instead of blending. From the point of view of salinity control, alternate applications of the two sources will be superior to blending. However, an alternating application strategy will require duel conveyance systems and availability of the effluent dictated by the alternate schedule of application.

13.3.5 Planning for Wastewater Irrigation

13.3.5.1 Central Planning

Government policy on effluent use in agriculture will have a deciding effect on what control measures can be achieved through careful selection of site and crops to be irrigated with treated effluent. A decision to make treated effluent available to farmers for unrestricted irrigation or to irrigate public parks and urban green areas with effluent will remove the possibility of taking advantage of careful selection of sites, irrigation techniques and crops in limiting the health risks and minimizing environmental impacts. However, if a government decides that effluent irrigation will only be applied in specific controlled areas, even if crop selection is not limited (i.e. unrestricted irrigation is allowed within these areas), public access to the irrigated areas will be prevented, and the control measures can be applied. Without doubt, the greatest security against health risk and adverse environmental impact will be achieved by limiting the effluent use to restricted irrigation on controlled areas to which the public has no access, but even imposing restrictions on effluent irrigation by farmers, if properly enforced, can achieve a degree of control.

A number of key issues or tasks were likely to have a significant effect on the ultimate success of effluent irrigation as follows:

(i) Organizational and managerial provisions made to administer the resource, to select the effluent use plan and to implement it
(ii) The importance attached to public health considerations and the levels of risk taken
(iii) The choice of single-use or multiple-use strategies
(iv) The criteria adopted in evaluating alternative reuse proposals
(v) The level of appreciation of the scope for establishing a forest resource

13.3.5.2 Desirable Site Characteristics

The features which are critical in deciding the viability of a land disposal project are the location of available land and public attitudes. Land which is far distant from the sewage treatment plant will incur high costs for transporting treated effluent to site and will generally not be suitable. Hence, the availability of land for effluent irrigation should be considered when sewerage is being planned, and sewage treatment plants should be strategically located in relation to suitable agricultural sites. Ideally, these sites should not be closed to residential areas, but even remote land might not be acceptable to the public if the social, cultural or religious attitudes are opposed to the practice of wastewater irrigation. The potential health hazards associated with effluent irrigation can make this a very sensitive issue, and public concern will only be mollified by the application of strict control measures. In arid areas, the importance of agricultural use of treated effluent makes it advisable to be as systematic as possible in planning, developing and managing effluent irrigation projects, and the public must be kept informed at all stages.

The ideal objective in site selection is to find a suitable area where long-term application of treated effluent will be feasible without adverse environmental or public health impacts. It might be possible in a particular instance to identify several potential sites within reasonable distance of the sewered community, and the problem will be to select the most suitable area or areas, taking all relevant factors into account. The following basic information on an area under consideration will be of value, if available:

- A topographic map
- Agricultural soil surveys
- Aerial photographs
- Geological maps and reports
- Groundwater reports and well logs
- Boring logs and soil test results
- Other soil and piezometric data

At this preliminary stage of investigation, it should be possible to assess the potential impact of treated effluent application on any usable aquifer in the area (s) concerned. The first ranking of sites should consider other factors, such as the cost and location of the land, its present use and availability and social factors, in addition to soil and groundwater conditions.

The characteristics of the soil profile underlying a particular site are very important in deciding on its suitability for effluent irrigation and the methods of application to be employed. Among the soil properties important from the point of view of wastewater application and agricultural production are physical parameters (such as texture, grading, liquid and plastic limits, etc.), permeability, water-holding capacity, pH, salinity and chemical composition. Preliminary observation of sites, which could include shallow hand auger borings and identification of vegetation, will often allow the elimination of clearly unsatisfactory sites. After elimination of marginal sites, each site under serious consideration must be investigated by on-site borings to ascertain the soil profile, soil characteristics and location of the water table. Piezometers should be located in each borehole, and these can be used for subsequent groundwater sampling. A procedure for such site assessment has been described by Hall and Thompson (1981) and, if applied, should not only allow the most suitable site among several possible to be selected but permit the impact of effluent irrigation at the chosen site to be modelled. When a site is developed, a long-term groundwater monitoring programme should be an essential feature of its management.

13.3.5.3 Economic, Institutional and Policy Issues

While the overall benefits of wastewater use in agriculture are obvious and the technology and expertise exist to allow it to be achieved without detriment to public health or the environment, governments must be prepared to control the process within a broader framework of a national effluent use policy forming part of the national plan for water resources. Lines of responsibility and cost allocation formulae have to be worked out between the various sectors involved: local authorities responsible for wastewater treatment and disposal, farmers who will benefit from any effluent use scheme and the state which is concerned with the provision of adequate water supplies, the protection of the environment and the promotion of public health. Sufficient attention must be given to the social, institutional and organizational aspects of effluent use in agriculture and aquaculture to ensure long-term sustainability.

13.3.5.4 Economic and Financial Implications

Although the responsibility for collecting, treating and disposing of urban wastewater will normally lie with a local water or sewerage authority or municipality, farmers wishing to take advantage of the effluent are often able and willing to pay for what they use but are not prepared to subsidize general disposal costs. They will base their decision on whether or not they will be better off paying for the effluent rather than doing without it, considering the quantity, timing, quality and cost of the treated effluent. The local sewerage authority should acknowledge their financial responsibility for the basic system to achieve environmental protection objectives and only charge farmers for any incremental costs associated with additional treatment or distribution required specifically for effluent use in agriculture or aquaculture. In practice, if the effluent use scheme is considered at the time the sewerage project is being planned, treatment costs might well be reduced over those normally required for environmental protection.

Payments by farmers might take the form of direct effluent use tariffs paid to the authority or contributions to the capital and/or operating costs of the wastewater treatment plant and effluent conveyance system. Cost sharing can be by cash payments or in-kind contributions, such as land for siting treatment or storage facilities and labour for operation and maintenance. Bartone (1986) has indicated that benefit-cost studies made in Peru showed that the irrigation components in effluent irrigation schemes were economically viable even if land costs and operation and maintenance for wastewater treatment were charged to farmers but not if the full cost of investment in treatment facilities was charged against the agricultural component. In the latter case, feasibility depended on the alternative minimum cost of treatment required for disposal without reuse.

Since wastewater treatment is a major cost in effluent use systems, accepting that local authorities are fully responsible for wastewater collection, it is essential that treatment process selection is made in conjunction with decisions on crop and irrigation system selection. Only in this way can a minimal investment in treatment be achieved without compromising the health risks of an effluent use scheme. Once a decision on effluent quality has been taken, the required standard must be achieved consistently, and the effluent treatment and conveyance system must be operated with complete reliability. Fluctuating production and demand for effluent created by seasonal and diurnal patters of water use, cropping and crop water needs must be accommodated at all times, even if the price of the effluent is varied, to be higher in the hot season.

13.3.5.5 Institutional Organization

The scope and success of any effluent use scheme will depend to a large extent on the administrative skills applied. Wastewater collection and treatment and effluent use in agriculture and aquaculture span a wide range of both urban-based and rural-based interests at both local and regional levels, and institutional responsibilities must be clearly defined. Decisions will have to be taken on:

- Allocation of effluent among competing uses
- Maintenance of quality standards and system reliability
- Investment in supporting resources, especially managerial and technical staff, required to administer each component of an effluent use scheme

Policy decisions should normally be taken by a national or regional body, with executive responsibilities in the hands of a regional organization. Such a regional organization would be responsible for project implementation and operation and would provide the criteria, framework and administrative mechanisms necessary for effective effluent utilization. However, they would also be responsible for effective monitoring and control of the crops irrigated, the quality of effluent and associated health and environmental impacts.

One of the most important features of a successful effluent use scheme is the supervision provided at all stages of the system. Strict control must be applied from the wastewater treatment plant, through the conveyance and irrigation systems, to the quality of the resulting products, whether they are of commercial or environmental value. The management, monitoring and public relations procedures are as important as the technological hardware involved in the system, and managers of regional organizations set up to administer effluent use projects must be firm if the schemes are to realize their full potential. Managerial and technical staff must be properly qualified and suitably trained to carry out their functions effectively. Treated effluent use in agriculture is a major resource development activity and requires an appropriate institutional structure, provided with adequate resources, to be successful.

13.3.5.6 Policy Issues

The legislative framework for effluent use in agriculture can have a significant influence on project feasibility. Bartone (1986) has indicated that the authorities in Mexico are able to impose effective crop restriction measures in irrigation districts because they are empowered to withhold effluent from farmers not observing the regulations, whereas in Chile the sanitary authorities have little leverage. Chilean water law vests water rights in the farmers (landowners), and the authorities have never been successful in imposing crop restrictions, even though lettuce and other vegetables being irrigated with raw sewage have been implicated in annual typhoid epidemics in Santiago.

A coherent national policy for wastewater uses in agriculture is essential. This must define the division of responsibilities among involved ministries and authorities and provide for their collaboration. Institutional mechanisms for implementation of the national policy must be established and legal backing provided for enforcement of regulations. Realistic standards must be adopted to safeguard public health and protect against adverse environmental impacts. Environmental issues associated with wastewater use are the main subject of a UNEP (1991) document. Provisions should be made to adequately staff and resource organizations charged with the responsibility for assessing, implementing, operating and monitoring effluent use schemes and enforcing compliance with regulations. A distinction between the upgrading of existing wastewater use schemes and the development of new schemes is drawn in Mara and Cairncross (1989). In addressing the former, it is stressed attention should be paid not only to the technical improvements required or feasible but also to the need for better management of existing schemes and to their improved operation and maintenance.

A national and/or regional consultative committee will often be of value in developing policy guidelines. Serving on this committee should be a representative of all the main interest groups, including water resources planning, public health, public works (municipalities), agriculture and forestry, environmental protection, trade and commercial interests (including farmers' representatives). Policies emanating from such a committee should be free of local or partisan influences but, nevertheless, should be pragmatic. In particular, enforcement legislation must be unequivocal, unambiguous and addressed to the main problem areas. The committee should also be charged with assessing the epidemiological and agricultural impacts of effluent use schemes.

References

Abdel-Ghaffar AS, El-Attar HA, Elsokkary IH (1988) Chapter 17: Egyptian experience in the treatment and use of sewage and sludge in agriculture. In: Pescod MB, Arar A (eds) Treatment and use of sewage effluent for irrigation. Butterworths, Sevenoaks

Al-Salem SS (1987) Evaluation of the Al Samra waste stabilization pond system and its suitability for unrestricted irrigation. Paper prepared for the Land and Water Development Division, FAO, Rome

Anonymous (2004) NATP – MM project report on 'Use of urban and industrial effluent in agriculture' CSSRI, Karnal 132001, India

Arar A (1988) Chapter 2: background to treatment and use of sewage effluent. In: Pescod MB, Arar A (eds) Treatment and use of sewage effluent for irrigation. Butterworths, Sevenoaks

Bartone CR (1986) Introduction to wastewater irrigation: extent and approach. Third annual World Bank irrigation and drainage seminar, Annapolis, Maryland. The World Bank, Washington, DC

Bhamoriya V (2004) Wastewater irrigation in Vadodara, Gujarat, India: economic catalyst for marginalized communities. In: Scott CA, Faruqui NI, Raschid-Sally L (eds) Wastewater use in irrigated agriculture: confronting livelihhod and environmental realities. CAB International in Association with IWMI/IDRC, Colmbo/Ottawa

CGWB (2011) Ground water year book – India 2010–11. Central Ground Water Board, Ministry of Water Resources, Government of India. http://www.cgwb.gov.in/documents/Ground%20Water%20Year%20Book-2010-11.pdf

CPCB (2005) Performance status of common effluent treatment plants in India. Central Pollution Control Board, India

CPCB (2007) Advance methods for treatment of textile industry effluents, Resource Recycling Series: RERES/&/2007. Central Pollution Control Board, India

CWC (2010) Water and related statistics. Water Planning and Project Wing, Central Water Commission, India

DONTA, ANTONIA ALKISTIS (1997) Der Boden als Bioreaktor bei der Aufbringung von Abwasser auf landwirtschaftlich genutzte Flächen. Veröffentlichungen des Institutes für Siedlungswasser-wirtschaft und Abfalltechnik der Universität Hannover, Heft 100

Earth Policy Institute (2002) http://www.earth-policy.org/Updates/Update15.htm

FAO (1977) Crop water requirements. In: Doorenbos J, Pruitt W (eds) Irrigation and drainage paper 24. FAO, Rome, 144 p

FAO (1979) Yield response to water. In: Doorenbos J, Kassam AH (eds) Irrigation and drainage paper 33. FAO, Rome, p 203

FAO (1985) Water quality for agriculture. In: Ayers RS, Westcot DW (eds) Irrigation and drainage paper 29 Rev. 1. FAO, Rome, p 174

FAO (1988) Irrigation practice and water management. In: Doneen LD, Westcot DW (eds) Irrigation and drainage paper 1, Rev. 1. FAO, Rome, p 71

Hall HC, Thompson SG (1981) Site selection techniques for land disposal of treated municipal wastewater. In: Proceedings water reuse symposium II: water reuse in the future. American Water Works Association (AWWA) Research Foundation, Denver, Colorado

Hillel D (1987) The efficient use of water in irrigation. World Bank technical paper no. 64. The World Bank, Washington, DC

Hillman PJ (1988) Chapter 5: health aspects of reuse of treated wastewater for irrigation. In: Pescod MB, Arar A (eds) Treatment and use of sewage effluent for irrigation. Butterworths, Sevenoaks

Kandiah A (1990) Criteria and classification of saline water. In: Kandiah A (ed) Water, soil and crop management practices relating to the use of saline water. AGL/MIC/16/90. FAO, Rome, pp 34–51

Mara DD, Cairncross S (1989) Guidelines for the safe use of wastewater and excreta in agriculture and aquaculture – measures for public health protection. World Health Organization, Geneva

Mertz RC (1956) Continued study of wastewater reclamation and utilization. California State Water Pollution Control Board Publication No. 15, Sacramento: California State Water Pollution Control Board

Minhas PS, Samra JS (2004) Wastewater use in Peri-urban agriculture: impacts and opportunities. Bulletin No. 2, CSSRI, Karnal 132001, India

Nakayama FS (1982) Water analysis and treatment techniques to control emitter plugging. In: Proceedings of irrigation association conference, 21–24 February 1982, Portland, Oregon

National Academy of Sciences and National Academy of Engineering (1972) Water quality criteria. US Environmental Protection Agency, Washington DC. Report No. EPA-R373–033, p 592

Pant D, Adholeya A (2007) Biological approaches for treatment of distillery wastewater: a review. Bioresour Technol 98:2321–2334

Papadopoulos I, Stylianou Y (1988) Ongoing research on the use of treated sewage effluent for irrigating industrial and fodder crops. In: Pescod MB, Arar A (eds) Treatment and use of sewage effluent for irrigation. Butterworths, Sevenoaks

Pratt PF (1972) Quality criteria for trace elements in irrigation waters. California Agricultural Experiment Station, p 46

Sepp E (1971) The use of sewage for irrigation – a literature review. State of California Department of Public Health, Sacramento

Strauss M, Blumenthal U (1990) Human waste use in agriculture and aquaculture: utilization practice and health perspectives. IRCWD report 09/90. International Reference Centre for Waste Disposal, Duebendorf, Germany

UNEP (1991) Environmental guidelines for municipal wastewater reuse in the Mediterranean region. In: Tedeschi S, Pescod MB (eds) Mediterranean action plan – priority actions programme. Regional Activity Centre, Split

Van Schilfgaarde J (1984) Drainage design for salinity control. In: Shainberg I, Shalheret J (eds) Proceedings of soil salinity under irrigation. Springer, New York, pp 190–197

Webster RA (1954) Sewage effluent disposal through crop irrigation. Sewage Ind Waste 26:133–137

Chapter 14
Application of the Ecological Network Analysis (ENA) Approach in Water Resource Management Research: Strengths, Weaknesses, and Future Research Directions

Ali Kharrazi and Tomohiro Akiyama

Abstract Population growth, climate change, and conflicting demand by industry and agriculture are increasingly straining our planet's water resources. In this light, there is a need to advance holistic approaches and objective tools which allow policymakers to better evaluate system-level properties and trade-offs of water resources. This chapter contributes to the expanding literature in this area by highlighting water resource management strategies based on the ecological network analysis (ENA) approach. This chapter overviews the theoretical underpinnings of the ENA approach and its application, limitations, and weaknesses for water resource management research. Furthermore, through the case study of the Heihe River Basin, this chapter demonstrates how to examine system-level properties and their trade-offs relevant to the resilience of water services. The ENA approach considers holistic trade-offs that may be used to evaluate alternative water recycling and saving scenarios. This approach can complement multiple criteria decision-making framework and scenario planning approaches and can be beneficial in developing new applicable water resource management strategies.

Keywords Resilience · River basin · Water network · Ecological network analysis · Water policy

---

A. Kharrazi (✉)
Advanced Systems Analysis Group, International Institute for Applied Systems Analysis, Laxenburg, Austria

The University of Tokyo, Tokyo, Japan
e-mail: kharrazi@iiasa.ac.at

T. Akiyama
Department of Socio-Cultural Environmental Studies, Graduate School of Frontier Sciences, The University of Tokyo, Tokyo, Japan
e-mail: akiyama@k.u-tokyo.ac.jp

R. P. Singh et al. (eds.), *Water Conservation, Recycling and Reuse: Issues and Challenges*, https://doi.org/10.1007/978-981-13-3179-4_14

14.1 Introduction

With the expansion of anthropocentric activities and climate change, management of scarce water resources has increasingly become one of the most critical challenges for sustainable development. Water scarcity not only exacerbates the vulnerability of food security but also increases the risk of geopolitical conflicts over water access and control. Given the transdisciplinary nature of the challenges associated with water supply and demand, the control and modification of individual sections may lead to unpredictable effects on other sections of the water system. Sustainable water resource management, therefore, requires an integrative evaluation of economic, social, and environmental dimensions. A holistic approach to a water system allows for a better understanding of the system-level dynamics of the water resource system and results in more effective and relevant sustainable decisions (Xue et al. 2015).

Traditionally, water resource management has focused on economic dimensions, while recently there has been a growing consensus on the necessity for more holistic concepts such as integrated water resource management (IWRM). Discussions surrounding IWRM began in the late 1980s and have evolved toward an overarching conceptual umbrella incorporating integrated sustainable social, environmental, and economic principles relevant to better water resources (Gallego-Ayala 2013). Despite its conceptual advantages, the IWRM is weak in practicality and application in the real world. Specifically, researchers have raised concern over the lack of successful evidence of this approach in the literature and the inability of the concept to offer analytical tools for highlighting cost-benefit trade-offs relevant to different water resource management approaches (Chikozho 2008; Garcia 2008). The inability to evaluate trade-offs is most critical with regard to groundwater resources where there is less physical visibility of its flow and recharge dynamics.

The ecosystem services concept has also been proposed by researchers seeking solutions relevant to economic and environmental dimensions (Garcia et al. 2016). In this avenue, a trade-off analysis is conducted with multiple ecosystem services and multiple stakeholders of a water system. This approach is beneficial toward integrating divergent managerial perspectives arising from different environmental and economic expenses and benefits. However, this approach may be overly complicated and would require large amounts of difficult-to-quantify data reflecting costs of each solution, detailed evaluations of the numerous stakeholders, and environmental resources of the water system (Hering et al. 2010; Hering and Ingold 2012).

The criticality of trade-off analysis is most evident with the increasing shift in recent years from consumption of surface water to consumption of less physically visible groundwater sources. While measuring groundwater resources continues to be a challenging endeavor, to a substantial extent conventional hydrological models are theoretically able to predict groundwater dynamics. However, these hydrological models are based on simplifying assumptions and are limited in reflecting network flows such as groundwater recharge, feedback, and water cycles (Goderniaux et al. 2009). Furthermore, these models are limited in their ability to directly examine system-level water network dynamics, most importantly, the resilience of the hydrological cycle. In this avenue, there is a critical need for advancing research on the system-level properties of water resource systems and more importantly develop objective tools to measure and communicate these properties for their application by sustainability practitioners.

In light of the complexities and limitations of previous approaches to water resource management, researchers have been exploring new directions arising from interdisciplinary network approaches. The main strength of holistic network approaches lies in their ability to illustrate system-level properties, trade-offs, and network properties such as the resilience of the system. For example, while the efficient extraction, transport, and consumption of water are commonly ascribed policy decisions, their system-level effects, especially on the resilience of water system, are not well understood (Scott et al. 2014). A resilient water system is defined as a system with the capacity to persist in its ability to deliver water services in the face of various disruptions and shocks, e.g., excessive water consumption, droughts, and climate change impacts. Network approaches are especially beneficial toward evaluating certain dynamics of the water system which are not easily and physically visible – this includes most importantly the dynamics of groundwater flow and recharge.

New insights toward evaluating system-level dynamics of water systems are arising from information-based network approaches such as the ecological network analysis (ENA). The ENA approach defines the dynamics of a resilient water system as arising from a balance between network redundancy and efficiency. In this view, a highly efficient water system maintains lower water flow redundancies and a weaker capacity for resilience. For example, highly efficient irrigation systems of a river basin may restrict the groundwater recharge flows and lower the system's ability to absorb water inflow shocks by relying on groundwater storage. Consequently, while the rate of pumping water may not be increasing, the net amount of what is pumped is increasing, leaving less residual to return to the aquifer. This has been referred elsewhere to the paradox of the "net water use" which is similar to Jevons paradox. Conversely, a highly redundant water system may result in excessive negligence, water loss, and detrimental to the resilience of the water system. The main value of the application of the ENA approach to water resource management research lies in its potential to overcome the limitations of previous approaches and hydrological models in reflecting system-level properties and trade-offs.

This article aims to introduce the ENA approach and discuss its strengths, limitations, and future research directions to a wide audience of researchers and practitioners in the area of water resource management. This chapter is organized as follows: Section 14.2 discusses the theoretical underpinning of the ENA approach and particularly focuses on network properties relevant to system resilience. Section 14.3 presents a review of the literature applying the ENA approach to water resource management research. Section 14.4 provides a case study example of the application of the ENA approach toward evaluating the trends of river basin network and applying relevant alternative scenario options. Finally, a discussion, conclusion, and future research directions follow in Sect. 14.5.

14.2 Ecological Network Analysis (ENA) Approach

The ecological network analysis (ENA) approach examines the emerging network structure from the flow of materials, e.g., information, money, electricity, nutrients, and water, among the node components of a system. This approach can be applied to systems that can be pictorially depicted as a web structure, i.e., a collection of compartment boxes connected by directed and weighted arrows that describe exchanges of materials that allow the functioning of a system. Rather than emphasizing the particular characteristics of nodes within a system, the ENA approach emphasizes the flow transfers between the nodes (Ulanowicz 1986). The ENA approach allows for a detailed examination of the system-level properties of the network and reveals network properties influencing the resilience of the system (Fath and Patten 1999). The underlying assumptions of this approach are, first, that growth and development are fundamentally two distinct system properties (Huang and Ulanowicz 2014). While growth reflects an extensive property of a system, e.g., the size of a system as quantified through the total system throughput (TST), development reflects an intensive property of organization within the system. More specifically, the development of the system is based on two dialectically related system-level network properties of efficiency and redundancy. In this avenue, the second underlying assumption of the ENA approach is the notion that the resilience of a networked system depends on the balance between network efficiency and redundancy arising from the topology and magnitudes of the pathways through which materials are circulated.

While in the long term, natural systems have been observed to increase their efficiency at the expense of their redundancy (Ulanowicz 1997), current levels of a trade-off between these two system variables depend on agency behavior, environmental constraints, and shocks or disruptions directed at the system. The efficiency of a system can be defined as the degree of articulation or constraints of flows in a network. In a water system, the efficiency of a system can be increased, for example, through the development of irrigation and drainage canals and aqueducts. In more objective and quantifiable terms, the average mutual information (Ulanowicz 2009; Ulanowicz and Norden 1990) is used to define the network efficiency of a system as:

$$\text{Efficiency} = \sum_{i,j} \frac{T_{ij}}{T_{..}} \log \frac{T_{ij} T_{..}}{T_{i.} T_{.j}} \tag{14.1}$$

Here, T_{ij} is a flow from agent i to agent j, $T_{i.} = \sum_j T_{ij}$ is the total flow leaving agent i, $T_{.j} = \sum_i T_{ij}$ is the total flow entering agent j, and the sum of all flows in the system, $T_{..} = \sum_{ij} T_{ij}$, is known as the total system throughput (TST).

The redundancy of a system, the countering variable to efficiency, can be defined as the degree of freedom or overhead of flows in a network. Network redundancy reflects the diversity of pathways in a system and is critical for a system's flexibility and capacity to adapt to changing environmental conditions arising from shocks and disruptions. In a water system, the redundancy of a system can be increased, for

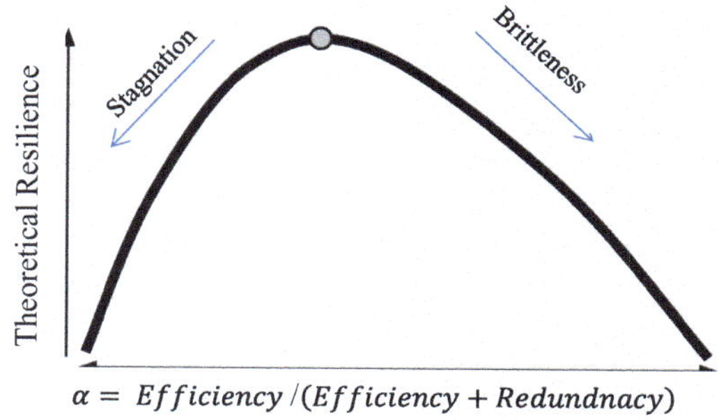

Fig. 14.1 The conceptual model of the ecological information-based approach

example, through the application of water reuse and recycling technologies or through the expansion of wetlands and aquifer recharge. In more objective and quantifiable terms, the conditional entropy is used to define the network redundancy (Ulanowicz 2009) of a system as follows:

$$\text{Redundancy} \quad = -\sum_{i,j}\frac{T_{ij}}{T_{..}}\log\frac{T_{ij}^2}{T_{i.}T_{.j}} \tag{14.2}$$

Both values of efficiency and redundancy are intensive, dimensionless, and based on units of information depending on the base logarithm used in their calculation, e.g., bits if the base 2 logarithm is used or nats if the natural logarithm is used. In the ENA literature, the natural logarithm is predominantly used in calculations.

Intuitively, following a disruption, networks with more diverse connections are more flexible in rerouting their flows and maintaining critical functions. Conversely, a more efficient network with a minimal number of well-organized connections can concentrate its capacity for growth and development. As illustrated in Fig. 14.1, overly redundant networks may be incoherent and lacking the efficiency to grow, while overly efficient networks may be brittle and prone to collapse when subjected to stress. To help determine a balance between constraint imposed by efficiency and the flexibility provided by redundancy, the relative order in the system is introduced as:

$$\alpha = \text{Efficiency}/(\text{Efficiency} + \text{Redundancy}), \quad \text{where } 0 \leq \alpha \leq 1 \tag{14.3}$$

The ratio α is a convenient way to express the degree of which property dominates the system at a given time. Departing from the relative order and invoking the Boltzmann measure (Boltzmann 1872) of its disorder, the level of a system's theoretical resilience can be expressed as (Ulanowicz et al. 2009):

$$\text{Theoretical Resilience} = -\alpha \, \log(\alpha) \tag{14.4}$$

From Eq. 14.4, a maximum value for theoretical resilience is derived as $1/e \approx 0.3704$ (independent of the logarithm's base). Empirical investigations have determined that natural networks, e.g., ecosystems and food webs, lie in close proximity to this maximum (Ulanowicz 2009), while economic systems indicate higher levels of redundancy (Kharrazi et al. 2013). The maximum resilience value, however, should be seen as a theoretical benchmark; optimal (minimal) vulnerability of real heterogeneous systems under various environmental conditions may be different from this value.

Theoretical resilience (Eq. 14.4) signifies the balance between efficiency and redundancy as a single metric and therefore is useful in exploring and comparing the configurations of heterogeneous networks. However, the analytical implications of the variable are limited and should be approached with caution. Firstly, the variable does not differentiate among shocks against which the network system might be judged to be resilient. Secondly, despite arguments of biomimicry, derived from evolutionary observations (Kharrazi et al. 2016a, b), it may be difficult to prescribe an optimal level of theoretical resilience to socioeconomic networks. Without a normative value, changes to the network configurations may be difficult to translate into actions, strategies, and practices.

14.3 Applying the ENA Approach in Water Resource Management Research

While the roots of the ENA approach lie within the ecological modeling and food web literature (Mukherjee et al. 2015), the approach has been gradually expanded to other economic (Goerner et al. 2009; Huang and Ulanowicz 2014) and environmental research areas (Chen et al. 2011, 2015). The ENA is a well-matched research approach for water system research as water systems encompass different inputs, outputs, and transformations between their compartments. Within the water resource management research domain, researchers have applied the ENA approach at different scales including, for example, at the level of urban, river basin, and virtual water systems.

At the level of urban water networks, the ENA approach has been applied in Albareto, Ravenna, and Sarmato in Italy (Bodini 2012; Bodini and Bondavalli 2002; Bodini et al. 2012). Within these studies, the ENA approach was used to illustrate water exchanges between the different sectors of the cities, and comparisons were made between present network configurations and network configurations arising from new water usage scenarios. In the same research vein, Pizzol et al. (2013) apply the ENA approach in examining the urban water management system of Hillerød, Denmark, using data from 2004 to 2008 and two future scenarios for 2015 and 2020. In this study, the authors compare the network properties found in the urban water system to natural ecosystems and discuss network-based strategies for increasing the resilience of the system to flooding and heavy rain events.

The ENA approach has been most frequently applied to the level of river basins, and most studies are based on case studies based in China. In one of the earliest studies in this avenue, Li et al. (2009) apply the ENA approach to six subsystems of the Yellow River Basin in China based on data from 1998 to 2006. This study develops new metrics based on the ENA approach and attempts to incorporate socio-environmental properties underlying the supply and demand of water. In a similar study, Li and Yang (2011) examine four subsystems of the Haihe River Basin in China based on data from 1999 to 2002 and 2005 to 2007. Within this study, the authors examine the optimal balance of network properties relevant to the resilience of the water system, i.e., efficiency against redundancy, in seven distinct scenarios. More recently, Hai et al. (2015) develop three new composite indicators based on the concept of optimality within ENA approach and in combination with conventional multidimensional social, economic, and environmental indicators. Using these composite indicators, the authors examine four subsystems of Huaihe River Basin in China from 2005 to 2011. The ENA approach has also been employed to examine the network configurations of the Baiyangdian River Basin in China from 2008 to 2013. Within this study, the authors attempt to advance the use of subsystem-level ENA indicators in examining different scenarios of the water system. Finally, Kharrazi et al. (2016a, b) examine the changes to system-level network configurations of the middle reaches of the Heihe River Basin in China from 2000 to 2009. In this study, the authors focus their discussion on the long-term resilience of the water system and more specifically the effects of changes in the groundwater body levels. The authors advance two hypothetical alternative scenarios, based on water recycling and saving strategies, to improve the long-term resilience of the water system.

The ENA approach has also been applied at the scale of virtual or embodied water flows. In this avenue, Fang and Chen (2015) construct virtual water network consisting of six economic sectors for Ganzhou District in the Heihe River Basin in China using data from 2002 to 2010. In addition to the system-level network properties reflecting the efficiency and redundancy of the water system, this study examines the dominant controlling sectors for water circulation and the utility relationships between pairwise sectors within the system.

14.4 Applying the ENA Approach in the Middle Reaches of the Heihe River Basin

Case studies are essential in advancing the application, practicality, and understanding of the strengths and limitations of the ecological network analysis (ENA) approach for water resource management research. Toward this end, we introduce a recent case study research by Kharrazi et al. (2016a, b) in examining the trends of the network properties of middle reaches of Heihe River Basin. The Heihe River Basin, China's second largest inland river basin, begins from the heights of the Qilian Mountains and ends in the perimeter of China's Gobi Desert (see Fig. 14.2). Given its bountiful water resources, this river basin has increased the region's capacity for agricultural development, especially in its middle reaches. The increasing anthropogenic activities

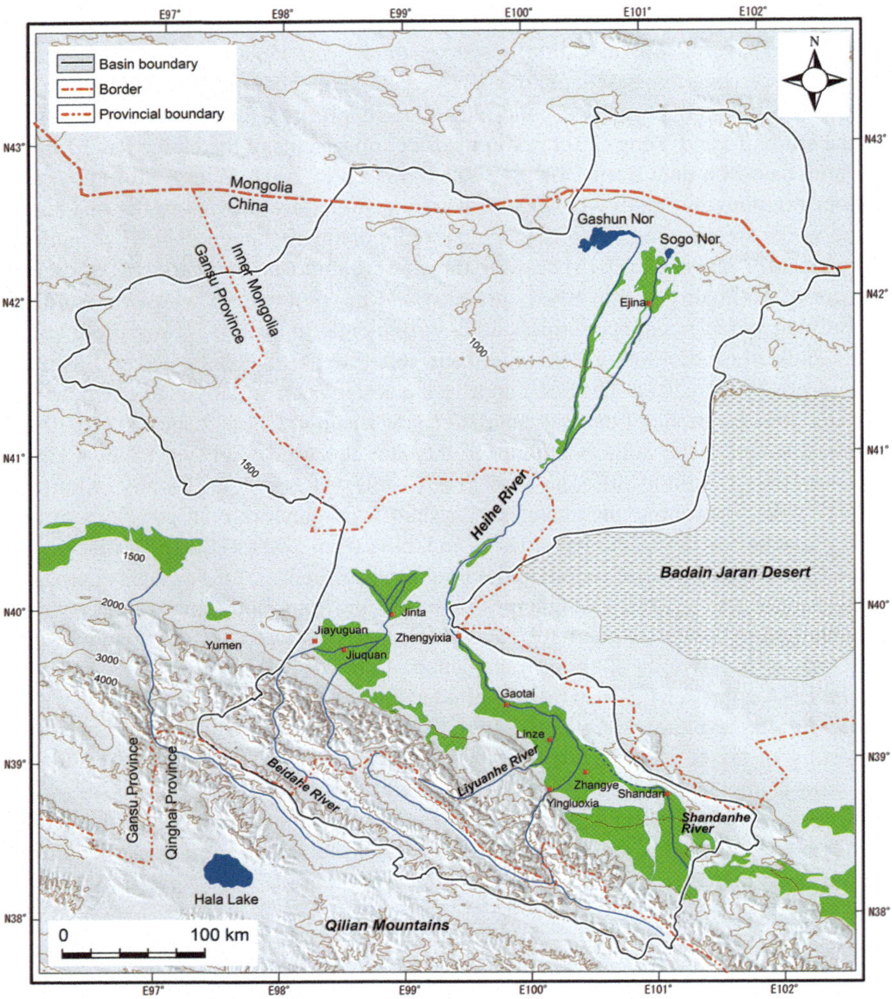

Fig. 14.2 A map of Heihe River Basin

require further exploitation of the surface and groundwater resources and in turn harming the ecology of the lower reaches. Realizing such critical trends, since the 2000s, the Chinese government has implemented regional water resource management plans, e.g., agricultural water quotas, more efficient water canals, and water conservation strategies (Cheng et al. 2006).

To examine the steady-state network (Jørgensen et al. 2007) of the Heihe River Basin, an eight-compartment model was constructed to represent the various hydrological and consumption flows between the compartments. This eight-compartment model represents hydrological inputs, outputs, and socioeconomic consumption

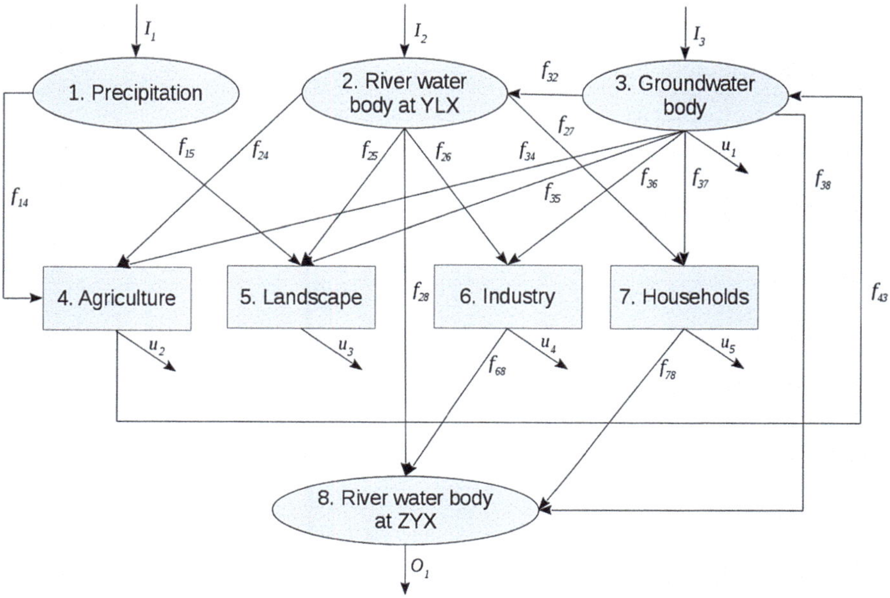

Fig. 14.3 The eight-compartment water network flows of Heihe River Basin

components of the river basin. The first compartment reflects rainwater, the second compartment represents river water from the upper reaches, and the third compartment reflects groundwater. The first three environmental compartments of the model circulate water to the social-economic compartments (4–7), i.e., agriculture, natural landscape, industry, and households. The output of the model to the lower reaches of the river basin is represented in the eighth compartment. Particular to the case study region, certain assumptions were made for this model. The groundwater recharge from the natural landscape f_{53} was not considered because annual rainfall below 200 mm usually results in negligible groundwater body recharge (Scanlon et al. 2006). All flows are static annual flows of water (10^8 m^3). Figure 14.3 illustrates the network model, and Table 14.1 illustrates the construction of the matrix of the network model, and a description of flows is given in Table 14.2. The data collected for this model is from the year 2000 to the year 2009.

14.4.1 Heihe River Basin Case Study Results

Through the ENA approach, the evolution of the network topology of the middle reaches of the Heihe River Basin from 2000 to 2009 can be revealed. As seen in Table 14.3, the extensive variable of TST, reflecting the total flows of the system, indicates a positive increase in the total size of the system during the 10-year period

Table 14.1 The matrix of the network model and corresponding row and column number, where (f) are internal flows between the compartments, (I) are boundary input flows, (u) are boundary output flows, and (o) is the water output to the lower reaches of the river basin

	1	2	3	...	...	...	...	n	$n+1$
1	–	–	–	f_{14}	f_{14}	–	–	–	–
2	–	–	–	f_{24}	f_{25}	f_{26}	f_{27}	f_{28}	–
3	–	f_{14}	–	f_{34}	f_{35}	f_{36}	f_{37}	f_{38}	u_1
...	–	–	f_{43}	–	–	–	–	–	u_2
...	–	–	–	–	–	–	–	–	u_3
...	–	–	–	–	–	–	–	f_{68}	u_4
...	–	–	–	–	–	–	–	f_{78}	u_5
n	–	–	–	–	–	–	–	–	o_1
$n+1$	I_1	I_2	I_3	–	–	–	–	–	–

Table 14.2 The flows and their description of the Heihe River Basin networks

Flows	Description
I_1	Precipitation
I_2	Annual river discharge at YLX
I_3	Groundwater discharge from mountainous areas
O_1	Annual river discharge at ZYX
u_1	Groundwater storage change
u_2	Evapotranspiration from the farmland
u_3	Evapotranspiration from the landscape
u_4	Consumptive water use of industrial sector
u_5	Consumptive water use of domestic sector
f_{14}	Precipitation to the farmland
f_{15}	Precipitation to the landscape
f_{32}	Groundwater discharge to surface water
f_{24}	Agricultural water use from surface water source
f_{25}	Landscape water use from surface water source
f_{26}	Industrial water use from surface water source
f_{27}	Domestic water use from surface water source
f_{28}	Surface water flowing from YLX directly to ZYX
f_{34}	Agricultural water use from groundwater source
f_{35}	Landscape water use from groundwater source
f_{36}	Industrial water use from groundwater source
f_{37}	Domestic water use from groundwater source
f_{38}	Annual groundwater discharge to ZYX
f_{43}	Groundwater recharge from agriculture
f_{78}	Domestic waste water

by 11%. While the overall scale of the network has increased in size, the inner development of the system as reflected in the intensive network properties reveals important changes in the configuration of the network. Specifically, the efficiency (E) variable has increased by 6%, while the redundancy (R) variable has decreased by 6% from 2000 to 2009. These trends warrant a more in-depth examination of the inner dynamics of the water network.

Table 14.3 Trends of ENA variables of the Heihe River Basin network from 2000 to 2009

Year	TST	(*E*)	(*R*)
2000	85.45	1.00	1.53
2001	78.93	0.98	1.60
2002	85.19	1.04	1.50
2003	94.50	1.05	1.45
2004	79.98	1.03	1.51
2005	88.80	1.04	1.52
2006	90.50	1.06	1.46
2007	95.62	1.06	1.43
2008	93.51	1.04	1.49
2009	94.69	1.06	1.43
Average	88.72	1.04	1.49
Lowest	78.93	0.98	1.43
Highest	95.62	1.06	1.60
% change	11%	6%	−6%

The health of the groundwater body is fundamental to the health of the river basin network through time. Excessive extraction or insufficient recharge of the groundwater component damages the health of the network through time. Given the static nature of the annual data snapshots of the Heihe River Basin and the limited understanding of the storage capacities, it is difficult to dynamically examine the dynamics of the groundwater body component and its responses to flow changes in the system. However, some insight can be attained by examining the input and output trends of the groundwater body component. As seen in Table 14.4, it is evident that the output consumption of the groundwater is increasing, while most critically, the groundwater recharge is decreasing. Furthermore, it is evident that the hydrological outputs of the groundwater have also been reduced. These trends indicate a critical negative balance of the groundwater component of the Heihe River Basin.

The results of the trend in the network configuration of the Heihe River Basin reveal two important points relevant to the long-term sustainability of the system. The decreasing trend of firstly the network redundancy and secondly the groundwater recharge flows harms the long-term resilience of the water availability in the middle reaches of the Heihe River Basin. The discussed trends indicate the decreasing capacity of the system to withstand potential shocks and disruptions, for example, hydro-environmental changes or excessive water consumption, and lower the system's flexibility in rerouting water flows in response to such stresses. The weakening of the resilience of the system to potential stresses is perhaps best illustrated in the flow changes directly affecting the groundwater body component. Results indicate excessive extraction and more importantly a significant decrease in recharge, i.e., a reduction of 31%, to this critical component. Considering the input-output flow balance to the groundwater body component, the long-term health of the system has been weakened. This may indicate a strong connection between improvements in the irrigation canals and, consequently, decreasing water seepage and lower groundwater recharge flows.

Table 14.4 Trends of groundwater body input and output flows from 2000 to 2009

			Output							
	Input		Consumption				Hydrological			Balance
Year	I_3	F_{43}	F_{34}	F_{35}	F_{36}	F_{37}	F_{38}	F_{32}	u_1	Inputs – outputs
2000	2.64	12.3	1.92	1.36	0.06	0.04	0.87	7.65	2.9	0.14
2001	2.64	11.64	2.79	1.58	0.1	0.07	0.85	6.01	2.8	0.08
2002	2.64	9.06	2.49	1.17	0.09	0.07	0.94	6.7	0.65	−0.41
2003	2.64	9.76	2.83	0.91	0.12	0.09	1.14	6.81	0.62	−0.12
2004	2.64	9.89	3.25	1.48	0.15	0.11	0.67	5.63	0.37	0.87
2005	2.64	7.67	3.16	1.06	0.15	0.11	1.61	4.98	0.49	−1.25
2006	2.64	9.34	3.25	1.32	0.15	0.11	0.68	5.8	0.43	0.24
2007	2.64	6.46	2.86	0.88	0.09	0.08	0.73	5.5	0.39	−1.43
2008	2.64	8.57	3.46	1.56	0.1	0.08	1.29	4.95	0.85	−1.08
2009	2.64	8.46	3.33	1.76	0.1	0.09	0.49	4.23	1.2	−0.10
Average	2.64	9.32	2.93	1.31	0.11	0.09	0.93	5.83	1.07	−0.30
Lowest	2.64	6.46	1.92	0.88	0.06	0.04	0.49	4.23	0.37	−1.43
Highest	2.64	12.30	3.46	1.76	0.15	0.11	1.61	7.65	2.90	0.87
% Change	**0%**	**−31%**	**73%**	**29%**	**67%**	**125%**	**−44%**	**−45%**	**−59%**	**−167%**
Std. Dev.	0.00	1.73	0.47	0.30	0.03	0.02	0.33	1.01	0.97	0.74

14.4.2 Scenario Analysis

In response to the critical challenges discussed in the previous section, based on the 2009 network, two alternative scenarios are proposed. The first scenario focuses on water recycling, while in the second scenario, in addition to water recycling, water conservation and a groundwater recharge strategy are envisioned. In the first scenario, water recycling is proposed between the agriculture, industry, and household compartments of the system, i.e., between agriculture and industry as f_{46} and f_{64} both for 0.5 10^8 m^3, households and industry as f_{67} and f_{76} both for 0.25 10^8 m^3, and households and agriculture as f_{74} and f_{47} both for 0.25 10^8 m^3. The amount of recycled water reflects annual rates and therefore does not pressure the input and output flows of the system. Furthermore, these small volumes demonstrate how basic changes to the flow structure can affect network indices. In the second scenario, water saving in the agriculture compartment by 0.5 10^8 m^3 and its diversion to the groundwater recharge (f_{43}) is proposed.

The results of the network configuration of these two scenarios can be seen in Table 14.5. The results from the first scenario reveal a significant increase in the redundancy (R) and a small decrease in the efficiency (E) of the system. In the second scenario, these changes were more pronounced. Due to additional water flow from recycled flows, results from both scenarios revealed an increase in the TST value. Results from the second scenario revealed that the increase of groundwater recharge flows (f_{43}) did not significantly change the network configurations of the water system. However, the increasing recharge flows changed the input-output balance of the groundwater body to positive levels and therefore consequently improves the long-term health of the groundwater body compartment.

14.4.3 Policy Considerations for Sustainable Water Resource Management

The results of the trends of the network configurations of the middle reaches of the Heihe River Basin reveal two important water resource management considerations. First, the efficiency of the water network in the middle reaches needs to be maintained at a level which also allows for ample water flow to the lower reaches. The efficiency of the water network can be configured through water resource management strategies promoting conservation, for example, through enhanced

Table 14.5 The system-level and subsystem-level indicators of the two alternative scenarios and their changes relative to the 2009 network configurations

Index	Original (2009)	Scenario 1	Change (%)	Scenario 2	Change (%)
TST	94.6928 10^8 m^3	96.6928	2.11	96.6928 10^8 m^3	2.11
E	1.0575 nats	1.0526	−0.46	1.0454 nats	−1.14
R	1.4339 nats	1.5237	6.27	1.5211 nats	6.08

water canals, less intensive agriculture, and water tariffs and quotas. Second, the redundancy of the water network in the middle reaches needs to be maintained so as not to harm the long-term health and resilience of the water system. As reflected in the results of the case study, these two water resource management considerations may pose trade-offs. In this avenue, the alternative scenarios may provide preferable modifications to the network configurations which better reflect both of the above considerations. However, these scenarios are realistic to a certain extent and do not completely reflect economic or environmental costs.

14.5 Discussion and Conclusion

The ecological network analysis (ENA) approach enables important insights toward the sustainable management of water resource systems. As a holistic approach, the strength of the ENA approach lies in its ability to evaluate system-level trade-offs and better policies in consideration of the resilience of the water system. As illustrated in the case study of the Heihe River Basin, the ENA approach is specifically insightful toward evaluating the health of physically less visible components of water systems such as the groundwater flows and its recharge dynamics. Through the ENA approach, the resilience of a water system is evaluated as a balance between the network efficiency and redundancy of water flows. In the case of the Heihe River Basin, higher network efficiency was achieved through governmental efforts toward the improvement of irrigation canals and water usage quotas. The success of these efforts was confirmed through increases in the value of efficiency (E) of the system and evident through the increase of boundary outflows to the lower reaches of the river basin. This outcome, however, resulted in lower water flow redundancies. A highly efficient water system negates the capacity of the system for resilience and its ability to absorb, for example, climatic and/or socio-economic shocks. The detrimental effect of higher network efficiency resulting from the implementation of the government's water policy was best illustrated through the changes to the input-output flow balance of the groundwater component of the system.

While the ENA approach can raise awareness among policy- and decision-makers of system-level trade-offs critical to resilient water resource management, the limitations and weaknesses of the application of the approach should also be considered. In this avenue, researchers should take caution in utilizing the ENA approach in deriving optimal values for the network configurations and should instead consider optimum values that meet local conditions – see Eq. 16 in Ulanowicz et al. (2009). The discussions surrounding optimal network configurations can be traced to research on natural ecological networks, where these systems were found to maintain configurations within a window of vitality (Ulanowicz 2009). Based on this, some researchers have advanced the idea of biomimicry and the need to change network configurations toward this optimal point for maintaining system resilience. However, this line of reasoning is questionable and its application to systems involving human activities uncertain. Firstly, the literature examining the relationship of natural systems to this optimal point is not systematic, and only 17 ecosystem flow networks have been found to be in close proximity to this optimal point (Ulanowicz 2009). Therefore, to ascertain the existence of such optimality, more natural networks need to be examined. Secondly, while biomimicry in itself is not necessarily without merit, the brutal evolutionary dynamics underlying such optimality may not be possible or desirable in systems involving socioeconomic agents. Therefore, it is questionable whether the evaluation of systems against an optimal point is indeed fruitful – systems may, in fact, maintain heterogeneous optimal points based on their environmental surroundings and the potential shocks and disruptions they may face. Toward this end, it may be insightful to examine whether if networks from different classes of systems, e.g., water, energy, food, and trade, cluster around certain ranges.

The application of the ENA approach to water systems is also weak in considering changes in the quality of the water flows. The issue arises when one considers the various qualities and their suitability for consumption. For example, while water used for agriculture can be reused by industry, water used for the industry may perhaps not pass the quality requirement for household consumption. As the ENA approach assumes the quality of all water flows as equal, researchers need to take caution and also consider the qualitative aspects of water flows in the system under their examination. In the meantime, a number of different models have been developed to investigate material cycles in river basins. They have been developed, respectively, to examine nitrogen cycles (Do-Thu et al. 2014), phosphorus cycles (Strokal et al. 2015), and other chemical compositions. Overcoming existing challenges around these models such as data availability and uncertainty, it becomes possible and worthwhile to integrate ENA with them.

The values of the network configurations resulting from the ENA approach may be affected by how the researcher develops and designs the model reflecting the water system. In this avenue, the discretion of the researcher in considering the scale, boundary, and detail of the components of the water system is important. The inclusion or exclusion of nodes representing environmental, water distribution, and socioeconomic components may, in fact, result in the reduction or increase of pathways within the network and influence the values of the network configuration. Although researchers are limited in their ability to model a system based on data availability and policy relevance, it is necessary to take into account the limitations of taking under consideration all components of a system and the effects that model design may have on network configuration values.

For future research, it is necessary to advance the applicability of the ENA approach for water resource management and make it more communicable to policy- and decision-makers. Toward this end, more scenarios testing the resilience of the water system to various probable and possible socio-environmental shocks are necessary. These scenarios increase the ability to situate the network configurations of a system to relevant policy deliberations. In the same vein, the ENA approach can also be utilized toward the examination of the system-level effects of water policy scenario options. As illustrated by the two alternative scenarios in the case study of the Heihe River Basin, water conservation, recycling, and groundwater recharge policies can alter network configurations and positively affect the resilience of the water system. In the case study of the Heihe River Basin, the alternative scenarios, although not far from reality, were hypothetical in nature and did not consider the social, economic, and environmental feasibilities and costs. Therefore, new research directions are essential in better situating water scenario planning options utilizing the ENA approach. In this avenue, the multi-criteria decision analysis (MCDA) framework is a promising research direction. The MCDA framework allows for the performance ranking of various scenario decisions against multiple criteria which may be even measured through different units (Hajkowicz and Collins 2007). The MCDA framework is also beneficial towards more stakeholder community engagement and decision-making transparency and allows system-level deliberations to be better grounded in social, environmental, and economic realities.

The future application of the ENA approach to water resource management can be strengthened through more abundant and higher-quality data sets. These data sets should include various hydrological and socioeconomic flows reflecting urban, regional, and river basin water systems and enable researchers to better compare findings across different research scales. While the ENA approach has been applied to analyze steady-state system-level properties and trade-offs of water resources, it also has the potential to make more detailed temporal and spatial analytical insights. In this vein, the ENA approach can, for example, be used to integrate material cycle, hydro-ecological, and multi-agent models based on the finite element method. Although the spatial differences among administration, river, and groundwater basin boundaries pose challenges of boundary setting and data collection, it is probable that the integration of these models would be a promising future research area. Furthermore, as water resource management is an issue of optimization of the allocation of water resources and given the increasing amounts of data used for toward their analysis, it is worthy to examine new approaches based in quantum computing as well as quantum annealing to solve relevant combinatorial optimization problems. Finally, new directions leveraging big data, citizen science, low-cost sensors, and hydro-informatics are also promising approaches in this avenue (Chen and Han 2016).

This chapter expanded on the theoretical underpinnings of the ecological network analysis (ENA) and discussed the strengths, weaknesses, and limitation of its application in water resource management research. It is hoped that this article inspires further research in applying the ENA approach toward sustainable water resource management and advancing this holistic approach for practical water policy deliberations.

References

Bodini A (2012) Building a systemic environmental monitoring and indicators for sustainability: what has the ecological network approach to offer? Ecol Indic 15(1):140–148. https://doi.org/10.1016/j.ecolind.2011.09.032

Bodini A, Bondavalli C (2002) Towards a sustainable use of water resources: a whole-ecosystem approach using network analysis. Int J Environ Pollut 18(5):463–485. https://doi.org/10.1504/IJEP.2002.002340

Bodini A, Bondavalli C, Allesina S (2012) Cities as ecosystems: growth, development and implications for sustainability. Ecol Model 245:185–198 Retrieved from http://www.sciencedirect.com/science/article/pii/S0304380012000981

Boltzmann L (1872) Weitere studien über das wärmegleichgewicht unter gasmolekülen. Sitzungsberichte Akademie Der Wissenschaften 66:275–370

Chen Y, Han D (2016) Big data and hydroinformatics. J Hydroinf 18(4):599–614

Chen SQ, Fath BD, Chen B (2011) Information-based network environment analysis: a system perspective for ecological risk assessment. Ecol Indic 11(6):1664–1672

Chen SQ, Chen B, Fath BD (2015) Assessing the cumulative environmental impact of hydropower construction on river systems based on energy network model. Renew Sustain Energy Rev 42:78–92

Cheng GD, Xiao HL, Xu ZM, Li JX, Lu MF (2006) Water issue and its countermeasure in the inland river basins of Northwest China – a case study in Heihe River Basin. J Glaciol Geocryol 28(3):406–413

Chikozho C (2008) Globalizing integrated water resources management: a complicated option in Southern Africa. Water Resour Manag 22(9):1241–1257

Fang D, Chen B (2015) Ecological network analysis for a virtual water network. Environ Sci Technol 49(11):6722–6730. https://doi.org/10.1021/es505388n

Fath BD, Patten BC (1999) Review of the foundations of network environment analysis. Ecosytems 2(2):167–179

Gallego-Ayala J (2013) Trends in integrated water resources management research: a literature review. Water Policy 15:628–647. https://doi.org/10.2166/wp.2013.149

Garcia LE (2008) Integrated water resources management: a 'small' step for conceptualists, a giant step for practitioners. Int J Water Resour Dev 24(1):23–36

Garcia X, Barceló D, Comas J, Corominas L l, Hadjimichael A, Page TJ, Acuña V (2016) Placing ecosystem services at the heart of urban water systems management. Sci Total Environ 563–564:1078–1085. https://doi.org/10.1016/j.scitotenv.2016.05.010

Goderniaux P, Brouyère S, Fowler HJ, Blenkinsop S, Therrien R, Orban P, Dassargues A (2009) Large scale surface-subsurface hydrological model to assess climate change impacts on groundwater reserves. J Hydrol 373(1–2):122–138. https://doi.org/10.1016/j.jhydrol.2009.04.017

Goerner S, Lietaer B, Ulanowicz RE (2009) Quantifying economic sustainability: implications for free-enterprise theory, policy and practice. Ecol Econ 69(1):76–81 Retrieved from http://linkinghub.elsevier.com/retrieve/pii/S0921800909003085

Hai R, Shi H, Zhang B, Zhai Y, Li Y, Wang W (2015) An ecological information analysis-based approach for assessing the sustainability of water use systems: a case study of the Huaihe River Basin, China. Clean Techn Environ Policy 17:2197–2211

Hajkowicz S, Collins K (2007) A review of multiple criteria analysis for water resource planning and management. Water Resour Manag 21(9):1553–1566

Hering JG, Ingold KM (2012) Water resource management: what should be integrated? Science 336:1234–1235

Hering D, Borja A, Carstensen J, Carvalho L, Elliott M, Feld CK et al (2010) The European water framework directive at the age of 10: a critical review of the achievements with recommendations for the future. Sci Total Environ 408(19):4007–4019. https://doi.org/10.1016/j.scitotenv.2010.05.031

Huang J, Ulanowicz RE (2014) Ecological network analysis for economic systems: growth and development and implications for sustainable development. PLoS One 9(6):1–8

Jørgensen SE, Fath BD, Bastianoni S, Marques JC, Müller F, Nielsen SN et al (2007) A new ecology: systems perspective. Elsevier, Amsterdam

Kharrazi A, Rovenskaya E, Fath BD, Yarime M, Kraines S (2013) Quantifying the sustainability of economic resource networks: an ecological information-based approach. Ecol Econ 90:177–186

Kharrazi A, Akiyama T, Yu Y, Li J (2016a) Evaluating the evolution of the Heihe River Basin using the ecological network analysis: efficiency, resilience, and implications for water resource management policy. Sci Total Environ 572:688–696. https://doi.org/10.1016/j.scitotenv.2016.06.210

Kharrazi A, Fath BD, Katzmair H (2016b) Advancing empirical approaches to the concept of resilience: a critical examination of Panarchy, ecological information, and statistical evidence. Sustainability 8(9):935. https://doi.org/10.3390/su8090935

Li Y, Yang Z (2011) Quantifying the sustainability of water use systems: calculating the balance between network efficiency and resilience. Ecol Model 222:1771–1780 Retrieved from http://linkinghub.elsevier.com/retrieve/pii/S0304380011001086

Li Y, Chen B, Yang ZF (2009) Ecological network analysis for water use systems – a case study of the Yellow River Basin. Ecol Model 220(22):3163–3173. https://doi.org/10.1016/j.ecolmodel.2009.08.007

Mukherjee J, Scharler UM, Fath BD, Ray S (2015) Measuring sensitivity of robustness and network indices for an estuarine food web model under perturbations. Ecol Model 306:160–173

Pizzol M, Scotti M, Thomsen M (2013) Network analysis as a tool for assessing environmental sustainability: applying the ecosystem perspective to a Danish water management system. J Environ Manag 118:21–31

Scanlon BR, Keese KE, Flint AL, Flint LE, Gaye CB, Edmunds WM, Simmers I (2006) Global synthesis of groundwater recharge in semiarid and arid regions. Hydrol Process 20 (15):3335–3370

Scott CA, Vicuña S, Blanco-Gutiérrez I, Meza F, Varela-Ortega C (2014) Irrigation efficiency and water-policy implications for river basin resilience. Hydrol Earth Syst Sci 18:1339–1348

Ulanowicz RE (1986) Growth and development: ecosystems phenomenology. Springer, New York

Ulanowicz RE (1997) Ecology, the ascendent perspective. Columbia University Press, New York

Ulanowicz RE (2009) The dual nature of ecosystem dynamics. Ecol Model 220(16):1886–1892

Ulanowicz RE, Norden JS (1990) Symmetrical overhead in flow networks. Int J Syst Sci 21 (2):429–437 Retrieved from http://www.tandfonline.com/doi/abs/10.1080/00207729008910372

Ulanowicz RE, Goerner S, Lietaer B, Gomez R (2009) Quantifying sustainability: resilience, efficiency and the return of information theory. Ecol Complex 6(1):27–36

Xue X, Schoen ME, Ma XC, Hawkings TR, Ashbolt NJ, Cashdollar J, Garland J (2015) Critical insights for a sustainability framework to address integrated community water services: technical metrics and approaches. Water Res 77:155–169

The manufacturer's authorised representative in the EU is Springer Nature Customer Service Centre GmbH, Europaplatz 3, 69115 Heidelberg, Germany. If you have any concerns regarding our products, please contact ProductSafety@springernature.com

Printed and bound by CPI Group (UK) Ltd, Croydon, CR0 4YY
30/09/2026
02225304-0009